Treatment of Water
by Granular Activated Carbon

Treatment of Water
by Granular Activated Carbon

Michael J. McGuire, EDITOR

*The Metropolitan Water District
of Southern California*

I. H. Suffet, EDITOR

Drexel University

Based on a symposium

sponsored by the

Division of Environmental Chemistry

at the 181st Meeting

of the American Chemical Society

Atlanta, Georgia

March 30–April 3, 1981

ADVANCES IN CHEMISTRY SERIES **202**

AMERICAN CHEMICAL SOCIETY
WASHINGTON, D.C.   1983

Library of Congress Cataloging in Publication Data

Treatment of water by granular activated carbon.
  (Advances in chemistry series; 202)

  "Based on a symposium sponsored by the ACS
Division of Environmental Chemistry at the 181st
Meeting of the American Chemical Society, Atlanta,
Georgia, March 30–April 2, 1981."
  Includes index.
  1. Water—Purification—Adsorption—Congresses.
2. Carbon, Activated—Congresses. I. McGuire, Michael
J. II. Suffet, I. H. III. American Chemical Society.
Division of Environmental Chemistry. IV. Series

QD1.A355   no. 202   504s   82–22662
[TD449.5]   [628.1'66]
ISBN 0–8412–0665–1   ADCSAJ

Advances in Chemistry Series

M. Joan Comstock, *Series Editor*

FOREWORD

ADVANCES IN CHEMISTRY SERIES was founded in 1949 by the American Chemical Society as an outlet for symposia and collections of data in special areas of topical interest that could not be accommodated in the Society's journals. It provides a medium for symposia that would otherwise be fragmented, their papers distributed among several journals or not published at all. Papers are reviewed critically according to ACS editorial standards and receive the careful attention and processing characteristic of ACS publications. Volumes in the ADVANCES IN CHEMISTRY SERIES maintain the integrity of the symposia on which they are based; however, verbatim reproductions of previously published papers are not accepted. Papers may include reports of research as well as reviews since symposia may embrace both types of presentation.

ABOUT THE EDITORS

MICHAEL J. MCGUIRE is a Water Quality Engineer with The Metropolitan Water District of Southern California and is responsible for directing the activities of Metropolitan's Water Quality Laboratory. He received his Ph.D. and M.S. in Environmental Engineering from Drexel University, and his B.S. in Civil Engineering from the University of Pennsylvania. Most recently, he and Dr. Suffet coedited "Activated Carbon Adsorption of Organics from the Aqueous Phase," a two-volume set. He is a registered professional engineer in three states, and he has authored more than 40 publications in the environmental engineering field. His research interests are directed toward control of trace contaminants in drinking water.

I. H. SUFFET is Professor of Chemistry and Environmental Science at Drexel University. He received his Ph.D. from Rutgers University, his M.S. in Chemistry from the University of Maryland, and his B.S. in Chemistry from Brooklyn College. In addition to the two-volume set coedited with Dr. McGuire, he has edited a two-volume treatise, "The Fate of Pollutants in the Air and Water Environments," was a journal editor for a special issue of the *Journal of Environmental Science and Health, Part A—Environmental Science and Engineering,* and served on the editorial board of the companion journal, *Part B—Pesticides and Agricultural Products.* He serves on the editorial boards of the journals *Chemosphere* and *Chemtech.* He has coauthored more than 70 research papers and monograph chapters on environmental and analytical chemistry. His expertise in the field of environmental chemistry—analysis, fate, and treatment of pollutants—has involved him in organizing and chairing numerous technical society meetings as well as serving on the Safe Drinking Water Committee of the National Academy of Science, where he chaired the Subcommittee on Adsorption. He is now the treasurer of the ACS Division of Environmental Chemistry.

CONTENTS

To Our Wives and Families
Deborah Marrow and David
and
Eileen Suffet, Alison, and Jeffrey

PREFACE

T$_{\text{HIS}}$ $_{\text{BOOK}}$ $_{\text{IS}}$ $_{\text{A}}$ $_{\text{MODEST}}$ $_{\text{EFFORT}}$ to update the exponentially expanding research output in the field of activated carbon adsorption for water treatment applications. The book presents major sections dealing with theoretical modeling, competitive adsorption, biological/adsorptive interactions, and case histories. Panel discussions at the ends of the sections place the technical information in perspective.

This book is the culmination of an effort to compile the most up-to-date compendium of activated carbon adsorption literature since the publication of Volumes 1 and 2 of "Activated Carbon Adsorption of Organics from the Aqueous Phase" (Ann Arbor Science Publishers, 1980). These first two volumes were produced using papers from a 1978 ACS symposium. Almost all the chapters in the current book were presented at an ACS symposium on activated carbon in 1981.

As was the case with the previous ACS conference, the U.S. Environmental Protection Agency and the American Water Works Association (AWWA) participated as joint sponsors of the program. With these sponsoring organizations, it was possible to attract an audience of both theoreticians and practitioners to the Atlanta symposium. The symposium was organized into half- and full-day sessions that followed the general organization of this book. Each half-day session was concluded by a panel discussion. A significant effort was made to capture the essence of the panel discussions for this book. We feel that these panel discussions are essential to the full understanding of the often complex activated carbon field. Scientists and engineers talked to one another and communicated their ideas and perspectives.

An overview of the regulatory aspects of using activated carbon is presented in the first chapter. While the proposed treatment requirement for contaminated sources of supply has been withdrawn by the U.S. EPA, the agency is still interested in seeing that research work on activated carbon is continued. Also, two large water treatment application sites for activated carbon are under serious consideration at this writing—Denver, Colorado, for direct reuse of reclaimed wastewater and Cincinnati, Ohio, for treatment of Ohio River water.

The first section of this book is devoted to presentations on theoretical approaches to understanding activated carbon adsorption. Three equilibrium adsorption models are presented, as are two discussions of the surface chemistry of carbon.

Modeling and competitive adsorption aspects are covered in the second section. Besides presentations of recent advances in computer modeling, two chapters are devoted to discussions of the minicolumn adsorption method.

The largest growth area in this book, as compared to the two previous volumes, is the third section: biological/adsorptive interactions. U.S. EPA-sponsored projects are represented by three chapters, and European experiences are discussed in two chapters.

The fourth section presents case histories of pilot- and large-scale studies of granular activated carbon (GAC) in the U.S. and the Netherlands. Experiences with full-scale regeneration of GAC are discussed in two chapters.

Each chapter was reviewed technically by the editors and by at least two outside expert referees anonymously selected by the ACS editorial staff to ensure the scientific rigor of a technical journal. The discussion sections were constructed from a transcribed record of the panel discussions at the symposium. The arrangement within the discussion sections varies from the chronology of the symposium panels because it was necessary to draw sections together that were similar in subject matter. We hope that the reader will enjoy the discussion sections, not only for their technical content but also for insight into the personal philosophies of the participants.

For a large project such as this many people have provided significant input. We cannot thank everyone in print because of space limitations, but we know and appreciate how important their contributions were.

The support of David B. Preston, Executive Director, AWWA, and Alan F. Hess, AWWA liaison, is gratefully acknowledged.

The U.S. Environmental Protection Agency provided financial support for this project (Grant No. R–808511–01–0), making it possible to tape the entire symposium and have court stenographers record the panel discussions. We are grateful for the guidance of Joseph A. Cotruvo, Chief of the Criteria and Standards Division, Office of Drinking Water, and Thomas P. Thornton, Project Officer.

The American Chemical Society, Division of Environmental Chemistry, also provided financial assistance for travel funds and guest registrations for foreign scientists. We gratefully acknowledge the support of Leslie B. Laird, Division Chairman, and Roger A. Minear, Division Treasurer.

Students from Georgia Institute of Technology provided the manpower during the symposium for operating the tape recorder and microphone system. Edward S. K. Chian of Georgia Tech is acknowledged for his help in organizing the students to help us.

We are indebted to the many reviewers who spent the hours required to ensure that a high-quality document was produced. We are deeply indebted to the speakers and session chairmen for their contributions and efforts before, during, and after the symposium. We thank the discussion participants for their open and frank opinions; we feel that their contributions and the ideas exchanged will generate better understanding of the subject for the reader.

The task of managing the review process, as well as all production phases of this book, was expertly handled by Janet S. Dodd of ACS Books. We enjoyed working with Ms. Dodd and the entire ACS staff, and we appreciate their patience and professionalism.

At both institutions where we are associated, several people must be acknowledged for their fine editing, typing, and clerical skills. Marguerite R. Kimball, of The Metropolitan Water District of Southern California, was responsible for the excellent editing of the discussion sections. Jane A. Krafka and Emma M. Mallory at Metropolitan and Heather Garrish, Linda Ritter, and Galina Poessl at Drexel University handled the typing chores with professional dedication.

Dr. McGuire expresses his gratitude for the support of this work during his employment at Metropolitan. A. Eugene Bowers was most helpful during the period in which this book was produced.

We both gratefully acknowledge the understanding, support, and encouragement of our wives, Deborah Marrow and Eileen Suffet. They remain a source of perspective as professionals in their own right, and without their concern and help this book would not have been possible.

MICHAEL J. McGUIRE
The Metropolitan Water District of Southern California
La Verne, CA

IRWIN H. (MEL) SUFFET
Drexel University
Philadelphia, PA

December 1982

Overview: Role of Activated Carbon in EPA's Regulatory Program

JOSEPH A. COTRUVO, HUGH F. HANSON, and THOMAS P. THORNTON

Environmental Protection Agency, Washington, D.C. 20460

Past EPA actions affecting the water utility industry and the basis for EPA regulatory actions are reviewed. Ongoing regulatory efforts of the Office of Drinking Water are focusing on implementation of the trihalomethane regulations, development of an Advance Notice of Proposed Rulemaking for volatile organic chemicals, and National Revised Primary Drinking Water Regulations. Current research needs of the EPA, including a call for participation in the EPA Competitive Grants Program, are outlined.

BOTH PAST AND FUTURE EPA ACTIONS may influence the use of granular activated carbon (GAC) as a potable water treatment technology. Past EPA actions affecting the water utility industry include the promulgation of the National Interim Primary Drinking Water Regulations (NIPDWR) (1), the simultaneous proposals to control both trihalomethanes (THMs) and synthetic organic chemicals (SOCs) in drinking water (2), and the subsequent promulgation of the THM regulations (3) and the subsequent withdrawal of the proposal on SOCs. Present and future Office of Drinking Water (ODW) actions impacting the water supply industry include the implementation of the THM regulations and the development of the comprehensive National Revised Primary Drinking Water Regulations (NRPDWR). The first phase of this latter activity may be the publication of regulations for certain synthetic volatile organic chemicals (VOCs) commonly found in groundwater supplies.

Basis for EPA Regulatory Actions

EPA's ODW operates under the authority of the Safe Drinking Water Act (SDWA) (4), which is quite explicit in its intent to protect the public health to the maximum extent possible. Thus, the EPA Administrator is empowered to take action concerning any substance in drinking water which he determines *may* have an adverse effect on human health. Such actions, however, must take into account both technical feasibility and cost.

EPA has initiated five major national monitoring surveys to gain perspective on the frequency and intensity of the organic chemical pollution reaching U.S. citizens through their drinking water. These surveys have focused on different sources of water, different segments of the population, and different contaminants of concern. The results of four surveys [National Organics Reconnaissance Survey (NORS), National Organic Monitoring Survey (NOMS), National Screening Program for Organics in Drinking Water (NSP), and Community Water Supply Survey (CWSS)] form the basis of EPA's current knowledge about the occurrence of organics in the nation's drinking water. Data from the fifth survey, the National Rural Water Survey (NRWS), are now becoming available.

In addition, EPA has initiated another national survey that will further assess the contamination of groundwater sources by VOCs. The Ground Water Supply Survey (GWSS) is a joint venture between EPA and the individual states. Specific segments of this study are geared toward developing a method of predicting the location of potential contamination sites and improving the federal/state response to incidents of contamination. It is hoped that this new study will markedly increase the level of knowledge on the nationwide occurrence and intensity of groundwater contamination. The results of the GWSS should be particularly interesting in light of the changing perceptions of groundwater quality which the previous studies have fostered.

Previous studies detected the presence of a wide range of synthetic organic chemicals in finished drinking waters. Historically, groundwater had been viewed as a relatively uncontaminated resource, unspoiled by the human activities that affect surface waters. Traditional attention of public health officials was focused on those surface waters subject to wastewater discharges, industrial discharges, and nonpoint source run-off. However, data from the described studies are beginning to indicate that, while many surface waters are subject to contamination by a broad spectrum of SOCs at relatively low (microgram per liter) levels, a small percentage of groundwater supplies also are contaminated by one or more discrete compounds, sometimes at much higher concentrations (i.e., 100–1000 μg/L) than have been experienced in surface water. While the data from these surveys may not be representative of the total national picture, significant contamination has been found to occur in numerous locations.

Finding even trace quantities of synthetic organics in drinking water supplies should be a cause of concern since this is a direct indication that a mechanism exists for contamination of that supply by man-made pollutants. To date, the various surveys have detected the presence of more than 1000 individual chemicals in some drinking waters. Concern over the presence of these chemicals arises from the potential human health risks exposure to these chemicals may introduce. Many are suspected carcinogens, while, in some test systems, certain of these chemicals have been shown to be either mutagenic or teratogenic. Vinyl chloride, for example, is a known human carcinogen and has been found in some public water supplies. Also, chloroform has been identified as an animal carcinogen by the National Academy of Sciences (5).

Current Regulatory Efforts

Given these facts and concerns, the ongoing regulatory activities of ODW are focused in the following areas:

- Implementation of the trihalomethane regulations
- Development of an Advance Notice of Proposed Rulemaking for volatile organic chemicals
- National Revised Primary Drinking Water Regulations

The THM regulations were published in the *Federal Register* on November 29, 1979. Those regulations set an enforceable Maximum Contaminant Level (MCL) of 0.10 mg/L for total THMs in drinking water and imposed various monitoring/reporting requirements on affected systems. The effective dates of the individual segments of these regulations vary according to system size, with monitoring requirements for only the largest systems (those serving greater than 75,000 persons) having become effective to date. However, the remaining segments of these regulations (covering systems serving 10,000–75,000 persons) will become effective soon.

To comply with this MCL, it is anticipated that only a small number of affected systems throughout the country will choose to install either granular- or powdered-activated carbon. Such installations will either remove THM precursors from source waters or remove the actual THMs after their formation.

The other half of the original joint proposal, the use of GAC as a required treatment for SOC contaminated systems, was not finalized. The GAC proposal was withdrawn by *Federal Register* Notice in March 1981. The proposal envisioned GAC as a comprehensive treatment technique for removing contaminants from those surface water supplies considered to be vulnerable to SOC pollution. Comments received on that proposal, as well as new data made available from EPA-sponsored research projects, caused ODW to reconsider the GAC proposal and eventually to withdraw it.

ODW is presently re-evaluating that original proposal based upon concerns over the definition of a vulnerable water system, the performance levels, regeneration frequencies and costs of GAC, and the specific criteria for alternate treatment techniques. Future ODW actions will emphasize MCLs for individual chemicals or groups of chemicals rather than a general treatment technique requirement.

ODW's more immediate focus is on the development of the National Revised Primary Drinking Water Regulations (NRPDWR) that are intended by law to "contain a comprehensive program of control of drinking water contamination" (6). These regulations are intended to provide a set of enforceable standards that will assure the healthfulness of the drinking water provided at the taps of American consumers. These regulations are intended to be an evolution of the already promulgated NIPDWR.

The first phase of the NRPDWR is anticipated to be the determination of whether MCLs are needed for VOCs such as trichloroethylene and carbon tetrachloride. An Advance Notice of Proposed Rulemaking (ANPRM) on this subject is being developed for *Federal Register* publication. The alleviation and/or prevention of groundwater contamination problems is currently capturing much of EPA's attention, as evidenced by the Underground Injection Control (UIC) regulations, the Resource Conservation and Recovery Act (RCRA) regulations, the Ground Water Strategy Initiative, and the ongoing Superfund activities. Since the chief drinking water related contamination problems appear to involve VOCs in groundwater-supplied systems, ODW's concentration on possible regulatory action for those VOCs is largely consistent with EPA's current outlook. Current data suggest that aeration and/or GAC will provide effective controls in many cases of VOC-contaminated groundwater. Adsorption of organics by synthetic macroreticular resins also is a promising treatment technique. Given the nature of the VOC contamination problem, package plants or even home treatment devices in very small communities also might be an appropriate removal method.

Current Research Needs

In addition to many ongoing in-house research projects, EPA is currently supporting several pilot and field scale projects to evaluate the feasibility of and obtain operating and cost data for GAC. These field projects are located in: New Orleans, Shreveport, and Jefferson Parish, La.; Passaic Valley, N.J.; Philadelphia, Pa.; Kansas City, Mo.; Miami, Fla., Cincinnati, Ohio; Glen Cove, N.Y.; Manchester, N.H.; and Evansville, Ind.

ODW's Technical Support Division (TSD) is currently conducting a survey of existing GAC water treatment plant units (used mostly for control of tastes and odors) to determine their applicability for removing a broad spectrum of organics from source waters. This study will specifically attempt to determine the reasonable service life to be expected from such a filter. Limited data indicate that GAC, when utilized in the sand-replacement mode at an existing conventional water treatment plant, can be useful in removing certain SOCs. Also, ODW recently funded a project at the University of North Carolina which will attempt to develop an analytical technique that can identify and quantify the individual chemicals adsorbed by GAC filters during long periods of usage.

These are two areas where ODW's current knowledge concerning GAC is not extensive. ODW needs advice on exactly what role the activated-carbon technology can play in EPA's comprehensive approach to implementing the SDWA. Ideally, those aspects of activated-carbon treatment that require additional explanation or further research will be identified. Relevant technical information should be provided to assist ODW in evaluating the performance and practicality of GAC in potable water treatment.

Finally, ODW encourages participation in EPA's Competitive Grants Program in those areas for which they have interest. This program is a relatively new mechanism for funding long-range research activities which support ongoing EPA regulatory efforts. Through this competitive process, EPA hopes to increase the number of proposals available for consideration, stimulate greater competition among proposal writers, and improve the quality of any research grants eventually funded. Many significant aspects of GAC treatment technology would seemingly be eligible for such funding. Qualified individuals or organizations that desire to conduct research into areas of concern to EPA are encouraged to submit formal proposals to the appropriate EPA research office. All grant applications will be reviewed by the appropriate Science Peer Review Panel, composed primarily of non-EPA scientists. That review will evaluate the scientific merit of each proposal, rank each according to its merit, and result in a recommendation to EPA.

Conclusion

ODW currently perceives that GAC will become an increasingly important weapon in the fight to protect the quality of drinking water. GAC, when employed with other unit treatment operations, has demonstrated a capability for groundwater contaminant removal. However, questions remain regarding this technology's drinking water applications, e.g., frequency of required regeneration and chromatographic effects.

Literature Cited

1. *National Interim Primary Drinking Water Regulations*, U.S. EPA-570/9-76-003, Office of Water Supply, December 24, 1975.
2. *Federal Register*, February 9, 1978, pp. 5756–5780.
3. *Federal Register*, November 29, 1979, pp. 68,624–68,707.
4. *The Safe Drinking Water Act*, as amended (42 U.S.C., §300 f *et. seq.*).
5. "Drinking Water and Health," National Academy of Sciences: Washington, D.C., June 1977; p 6–256.
6. *The Safe Drinking Water Act* (42 U.S.C., §1401).

RECEIVED for review August 3, 1981. ACCEPTED for publication March 4, 1981.

THEORETICAL APPROACHES

Adsorption of Multicomponent Liquids from Water onto Activated Carbon: Convenient Estimation Methods

MILTON MANES and MICK GREENBANK

Kent State University, Chemistry Department, Kent, OH 44242

The conceptual simplicity of the Polanyi model is demonstrated by using a macroscopic gravitational analogy. Its application to the adsorption of multicomponent mixtures of organic liquids partially miscible in water and completely miscible in each other is considered. The use of the Polanyi model is compared to various aspects of the ideal adsorbate solution (IAS) model; whereas in many systems both models give similar predictions, the Polanyi model works for a number of systems that are outside the scope of the IAS model. It also may be used to supplement the IAS model and to improve convenience where it applies.

Previously (*1*), we considered the Polanyi adsorption potential theory and its application to the adsorption, from water solution onto activated carbon, of single organic liquids and single and multiple organic solids. We now consider its application to the adsorption of multi-component mixtures of organic liquids partially miscible in water and completely miscible in each other. An extensive treatment of the theory and a considerable body of data on binary and ternary solutes is presented elsewhere (*2*); a representative sampling of the experimental results will be discussed. The conceptual simplicity of the Polanyi model is demonstrated by showing how it is directly analogous to a macroscopic gravitational model.

The Polanyi model is of interest as a means toward understanding a wide diversity of adsorption phenomena on activated carbon (*1*). The practical applications of interest may be illustrated by the fact that we have a computer program to incorporate, as a data base, appropriate data (molar volume, refractive index, and water solubility) for a number of adsorbates, together with the characteristic curve for the carbon. This program practically immediately yields estimates of the adsorbate mole numbers of a multicomponent mixture (up to 25 components) from the individual equilibrium concentrations. Except for some rather exotic systems (which can also be handled), the calculations are simple enough to be carried out for simple systems on a hand calculator.

Whereas we continue to maintain that the Polanyi-based model is deserving of more widespread use, we have earlier noted some of its limitations (1). The model can be used both to predict individual isotherms (3) and to predict multicomponent adsorption (2, 4) from either estimated or experimental isotherms; therefore, any errors in the estimation of individual isotherms [as might come about, for example, from steric effects (5)] may be corrected by incorporating more accurate experimental isotherms into the model. For a wide variety of systems, however, the isotherms estimated by the Wohleber–Manes (3) method should suffice. For others, one can use the model to extrapolate isotherms.

Of the alternative approaches to adsorption from water solution, the most popular approach in recent years has been the ideal adsorbate solution (IAS) model of Radke and Prausnitz (6), which has recently been applied, e.g., by DiGiano et al. (7) and Jossens et al. (8) and which is considered here only in its application to liquid adsorbates. For many systems in which one has the individual adsorption isotherms over the required range, this model leads to practically the same calculations as the Polanyi-based model and the choice between the two turns out to be one of computational convenience; given the improved power of modern computers, the superiority of the Polanyi-based model in this respect may not be important. We found some systems for which the two models give quite different results; some of these systems are handled quite routinely by the Polanyi model but are completely outside the scope of the IAS model. Although these systems are important for comparing the physical validity of alternative models, the distinction may not necessarily be significant for many practical systems. We have found (2) that the Polanyi approach may be used to supplement the IAS model to estimate some awkward integrals that appear in the IAS model and therefore to improve its convenience and applicability for those who prefer to use it.

Two criticisms have been made of at least some applications of the Polanyi model (in addition to the limitations already noted): excessive mathematical complexity in multicomponent systems (7) and thermodynamic consistency (9). For multiple liquids, the first was met some 10 years ago in Wohleber's dissertation (4), which gives a straightforward Polanyi-based model for multicomponent liquid solutes. The second applies, not to the full Polanyi-based model, but to a simplifying assumption that nevertheless works well in many systems (2). The latter criticism is considered in the *Discussion*.

The following components have been studied in binary and ternary mixtures at 25°C (2): diethyl ether (EE), ethyl acetate (EA), propionitrile (PN), dichloromethane (DCM), 1,2-dichloroethane (DCE), and 1-pentanol (PEN). With the exception of PEN, all had been studied as single components (on the same carbon) by Wohléber and Manes (3). The PEN was incorporated into this study because it was well suited for liquid-solid studies (to be reported later). The systems studied were: EE–EA, PN–EA, EE–DCE, DCM–DCE, DCE–PEN, EE–EA–PN, EE–EA–PEN, and DCM–DCE–PEN.

The study also included coumarin (COU), phthalide (PHL), and *p*-nitrophenol (PNP) as liquid components above their (underwater) melting points. Each of these molten adsorbates was studied in binary mixtures with EA and PEN. Each molten liquid adsorbate had been studied previously by Chiou and Manes (10). They were chosen for study in the expectation that they would exhibit unequivocal adsorbate nonuniformity because of their high refractive indices; an additional factor was their convenient solubilities in water. They provided an interesting challenge to adsorption models.

Theoretical

The details of the theoretical approach are given elsewhere (1, 2). A brief outline is presented together with a gravitational analogy that illustrates the essential simplicity of the model. Following the basic model of Polanyi (11, 12), a carbon surface is postulated in which practically all of the volume active in adsorption is in the form of crevices or pores of varying size and unspecified shape. Polanyi showed the model schematically as one pore of varied cross-section (1); it may be represented equally well as a distribution of pores of varying cross-section and shape. Rather than characterizing an element of the adsorption space by an assumed pore shape or size (as in the various approaches to pore size distribution that use the Kelvin equation), the Polanyi model characterizes it by the negative of the adsorption energy (which Polanyi calls the adsorption potential, ε) of some specified adsorbate. The relative adsorption energies of any single adsorbate at different locations depend on relative proximity to carbon and may be expected to be higher in finer pores because of proximity to more carbon. A plot of the cumulative volume of pores (or in Polanyi language, "adsorption space") with adsorption energy to equal or exceed some given value, against that value as abscissa, is referred to as a "characteristic curve." If one assumes pores of specified shape, the characteristic curve of Polanyi becomes a cumulative pore size distribution with a somewhat remapped abscissa.

In vapor phase adsorption, the Polanyi model postulates that the (location-dependent) adsorption potential results in a corresponding location-dependent concentration of the vapor, to an extent described by the Boltzmann equation; condensation to a liquid results wherever the adsorption potential (or energy loss due to the adsorptive forces) suffices to concentrate the vapor to its saturation concentration. In effect, the attractive forces of the carbon for the adsorbate molecules reinforce their normal attractive forces for each other. The model is therefore very specific about the interactions of adsorbate molecules, setting them as equal to normal bulk interactions. With some exceptions (e.g., for very dilute fixed gases at elevated temperatures), the condensation accounts for practically all of the observed adsorption.

When a carbon sample, initially in a vacuum, is exposed to increasing pressures of a single vapor, condensation begins in the finer pores and progresses to the coarser ones until the adsorption volume is filled at saturation pressure. A plot of adsorbate volume against the equilibrium adsorption potential reproduces the characteristic curve for the carbon with that adsorbate. (Characteristic curves for other adsorbates differ only by an abscissa scale factor.) To relate this characteristic curve to the variables in an adsorption isotherm (mass adsorbed versus pressure or relative pressure), one uses the adsorbate liquid density [usually approximated by the bulk liquid density, with some modifications (13)] to relate the adsorbate mass to adsorbate volume. In the abscissa scale, the equilibrium adsorption potential is related to the inverse relative pressure, p_s/p, by the Polanyi (or Boltzmann) equation:

$$\varepsilon = RT \ln (p_s/p) \tag{1}$$

For adsorption of partially miscible organic liquids from water, the model is essentially the same (3, 12), except that the effective or net adsorption potential of the adsorbate is its (gas phase) adsorption potential corrected for the adsorption potential of an equal volume of the water it must displace. The correction is quite analogous to Archimedes' principle.

A gravitational analogy is now considered. Consider (Figure 1) a vessel in a very powerful gravitational field, which for simplicity is assumed to be uniform. The vessel may be of any shape or volume; for this illustration, we arbitrarily pick a vessel of the approximate size and shape of a centrifuge tube of 10 cm in length. If the top of the tube is designated as the level of zero potential energy and distances are measured downward from this level, then the "gravitational potential," ε', of a molecule in the tube ("adsorption space") may be defined as:

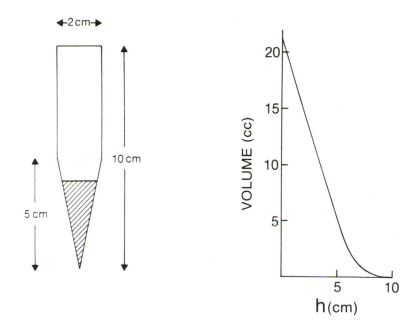

Figure 1. Schematic of gravitational analogy to Polanyi model. The plotted curve of cumulative volume versus ε/V (=ρg'h) is analogous to the characteristic curve in the Polanyi model and obviously related to the dimensions of the vessel or to the "structure of the adsorption space."

$$\varepsilon' = Mg'h \qquad (2)$$

where M is the molecular weight and g' is the assumed acceleration of gravity. (To attain values of ε' at the bottom of a 10-cm tube that would be comparable to those in the fine pores of a commercial activated carbon would require that g' be some 500,000 times the normal acceleration of gravity. For a system with normal gravity, the difference in height would have to be of the order of 50 km.)

If a vapor is maintained at partial pressure, p, at the top of the tube, its pressure varies within the tube according to the Boltzmann equation, and it condenses wherever:

$$Mg'h = \varepsilon' \geq RT \ln p_s/p \qquad (3)$$

If the pressure at the top is increased from zero to saturation, vapor first concentrates nonuniformly in the tube and then starts condensing from the bottom up until the entire tube is filled when $p = p_s$. A plot of the "adsorbate" volume versus ε' is the "characteristic curve" for that adsorbate. It clearly depends on the shape of the centrifuge tube and for a given tube will be the same for all "adsorbates" except for an abscissa scale factor that, in this case, depends on the molecular weight of the adsorbate. In this gravitational analogy, the molecular weight is analogous to the molar polarizability in adsorption if the adsorptive forces are all London forces.

In this gravitational analogy, the adsorption from water of a liquid solute of higher density is considered next. One would expect the concentration of this solute to increase in the tube with increasing downward distance from the top. The liquid solute would be expected to come out as a separate phase wherever its "net" gravitational potential, ε_s, suffices to concentrate it to saturation, i.e., wherever:

$$\varepsilon_s = \varepsilon'_s - \frac{\varepsilon'_l V_s}{V_l} \geq RT \ln c_s/c^0 \tag{4}$$

where ε'_s is the molar gravitational potential of the solute as a pure substance (given by Equation 2), ε'_l is the molar gravitational potential of the (water) solvent at the same location, V_s and V_l are the respective molar volumes, and c_s and c^0 are the saturation and zero level concentrations. The relation to Archimedes' principle may be seen by noting that the left side of Equation 4 may be rewritten as:

$$\varepsilon'_s - \varepsilon'_l V_s/V_l = M_s g'h(1 - \rho_l/\rho_s) \tag{5}$$

where ρ_l and ρ_s are the respective densities of solute and solvent. Again one could plot a "characteristic curve" for this system as the volume of organic condensate against the net gravitational potential of the liquid at that volume. Except for an abscissa scale factor that depends on the nature of the adsorbate, the "characteristic curve" in this analogy is nothing more than a plot of volume versus distance as one fills the vessel, which obviously is determined by the "structure" of the "adsorption space."

The analogy can be extended to mixed liquids of different densities, each higher than water. Again imagine the separation of an organic phase wherever the local concentrations suffice. If the net potential for each liquid solute is written as ε_i, the conditions at the equilibrium interface are:

$$x_i = (c^0/\gamma c_s)_i \tag{6}$$

$$\Sigma x_i = 1 \tag{7}$$

$$c_i^\circ = c_i^0 \exp(\varepsilon_i/RT) \tag{8}$$

If one assumes that the organic phase is an ideal solution, then the (Raoult's law convention) activity coefficients, γ_i, become unity and Equations 6–8 may be solved readily by iterative procedures to give both the total (organic) volume and the mole fractions, x_i, at the organic side of the organic–water interface. In the special case where the organic liquids have equal densities (the more general case is considered later), the organic phase is of uniform composition; the number of moles of each component in the organic phase can be calculated readily if one makes the assumption of negligible deviation from volume additivity. The mole numbers of each "adsorbate" component have now been calculated, given only the "bulk" concentrations, the molecular weights and densities, and a "characteristic curve" for the vessel, or "adsorbent."

The calculations just outlined become identical with the Polanyi-based calculations of Wohleber (*4*), Greenbank (*14*), and Greenbank and Manes (*2*) for the adsorption of multiple organic liquids from water (on the assumption of adsorbate uniformity) when one substitutes relative adsorption potentials (Polanyi model) for molecular weights (gravitational model) and adsorption potentials per unit volume for densities. The same gravitational analogy applies to the earlier calculations by Grant and Manes (*15*) for the adsorption of multiple gases or vapors (with no provision for adsorbate nonuniformity).

We now have to deal in our gravitational model with the possibility that the individual organic "adsorbates" are of significantly unequal densities, in which case the components of higher density will tend to concentrate toward the bottom of the tube. This now poses the additional problem of calculating the composition of the separated phase as a function of distance from the interface. Given this functional dependence and the cross-sectional area as a function of distance (or the "characteristic curve"), we can calculate the differential mole numbers, $dn_i(h)$, at each value of h, and then the total mole number of each component by integration over the entire adsorbate volume. The gravitational calculation for local composition is quite straightforward. One method that is a bit shorter than that of Hansen and Fackler (*16*) is to set as a condition of equilibrium that:

$$\Delta G_i = -\Delta \varepsilon_i' + V_i \Delta P + RT \ln \frac{x_i}{x_i^{\circ}} = 0 \qquad (9)$$

where $\Delta \varepsilon_i'$ is the difference in gravitational potential for the ith component between any location and the interface location, V_i is the molar volume, ΔP is the corresponding difference in hydrostatic pressure ($= 0$ at the interface), and x_i and x_i° are the location-dependent and interfacial mole fractions. The condition that ΔP be the same for all components at any given location gives:

$$\frac{-\Delta\varepsilon_i}{V_i} + \frac{RT}{V_i} \ln \frac{x_i}{x_i^{\circ}} = \frac{-\Delta\varepsilon_j}{V_j} + \frac{RT}{V_i} \ln \frac{x_j}{x_j^{\circ}} \qquad (10)$$

(and one may correctly use either $\Delta\varepsilon$ or $\Delta\varepsilon'$). Although $\Delta\varepsilon_i'/V_i = \rho_i g\Delta h$, Equation 10 is left in its present form because it is precisely the same as the modification of the Hansen–Fackler (16) equation for nonuniform adsorption applied by Greenbank and Manes (2) to the general case of adsorption of liquids of different adsorption potentials per unit volume. The equations in the Polanyi-based models may therefore be considered as analogous to the more familiar equations of atmospheric concentration under gravity and of sedimentation equilibrium. One would hope that this should take much of the mystery out of the Polanyi model.

Consider now a simple extension of the IAS model. It has been shown ($2, 15$) that the IAS postulate that each component be evaluated at equal "spreading pressure" is the same as the condition of equal area under a plot of V versus ε/V, as shown in Figure 2. To evaluate these areas one needs an adsorption isotherm for each component that extends to low capacities; such data frequently cannot be determined for strongly adsorbed components because the concentrations may become unmeasurably small at low capacities. If, however, one has a characteristic curve for the carbon (which may be determined from gas phase data or from liquid phase data on a weakly adsorbed component), together with an abscissa scale factor [which can be estimated from adsorption data at relatively high capacities or from refractive indices by the method of Wohleber and

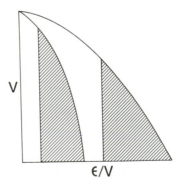

Figure 2. Schematic illustrating the calculation of "spreading pressure" in the IAS model, showing its relation to a correlation curve in the Polanyi model. Note that the ordinate is equal to the adsorbate volume in a linear scale.

Manes (*3*)], then one has a quite convenient means of estimating the required isotherm data and therefore the indicated integrals; this application of the Polanyi model conveniently facilitates the application of the IAS model.

As Figure 2 indicates, the conditions of equal total adsorbate volume and of equal "spreading pressure" (in the IAS model) become equivalent when the components have equal abscissa scale factors, in which case the two models give identical results. They differ when the scale factors are not identical. Finally, at high adsorbate volumes of the component with the higher scale factor, it attains a "spreading pressure" that cannot be matched by other components, in which case the calculation goes out of bounds. Examples of all three kinds of behavior are given in the experimental results.

Experimental

Carbon.   The activated carbon (Calgon CAL 2131) was the same carbon that was used in earlier investigations (*1*); the use of a single carbon batch facilitates the testing of alternative models on a large and consistent body of data.

Apparatus and Experimental Method.   All experiments were carried out in a minicolumn apparatus (*2, 14*) that was essentially as described by Rosene (*17*) and Rosene and Manes (*11*). Similar equipment has been described by Rosene et al. (*18*). The column technique makes it possible to fix the equilibrium (influent) concentrations quite accurately and to vary them systematically without any difficulty; for multicomponent systems, this is much more convenient than with shaker-bath techniques. By using column tubing of different sizes [down to 0.16-cm (1.16-in.) stainless steel tubing], one could vary the sample size [down to 5 mg; the wide range of sample size was important in studying adsorption over a wide range of experimental conditions (as much as seven orders of magnitude of relative mole fractions). Equilibrium was achieved in runs taking several hours by fine grinding of the samples (which were at 200–320 mesh before grinding); in all cases, the ground sample was completely transferred to the column to avoid any possible nonuniformity problems.

In most cases, the monitored component was fed to the carbon at concentrations such that its adsorbate mole fraction would be too low to affect substantially the total adsorbate volume, and it was fed to a carbon sample that had been previously equilibrated with a fixed concentration(s) of the other component(s). This procedure limits somewhat the concentrations that can be measured, but allows one to vary the mole fractions by orders of magnitude in the low ranges.

The amount of adsorption was determined from the influent concentrations and the effluent volumes; effluent fractions were collected separately so that by far the largest fraction of the effluent contained zero concentration of the monitored adsorbate. This procedure increased considerably the precision of measurement. Achievement of equilibrium was verified routinely by letting the column sit overnight and then checking the concentration of the liquid content, which was usually within a percent or two of the equilibrium (influent) concentration.

Results

Figures 3–7 illustrate some of the approximately 120 runs carried out with binary and ternary systems. Figure 3 shows some results with the EA–PN and EE–DCE systems, and Figure 4 shows results with DCM–DCE and PN-DCE. Figure 5 illustrates results with the ternary systems: DCM-DCE–PEN and PEN–EE–EA. Figure 6 shows the EA–PHL system, with EA adsorbing from a higher concentration of PHL, and Figure 7 shows results of the adsorption of COU and PNP from EA and PEN.

In all cases, the solid theoretical curves were calculated from a hydrocarbon correlation curve determined for the carbon by butane

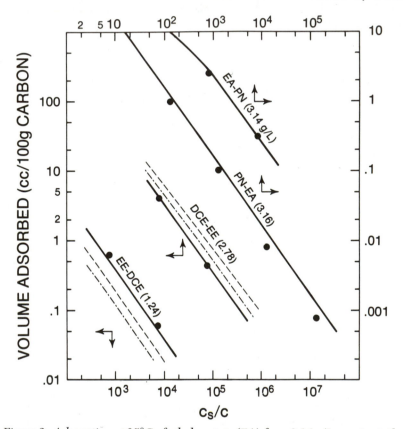

Figure 3. Adsorption at 25°C of ethyl acetate (EA) from 3.14 g/L propionitrile (PN); PN from 3.6 g/L EA; 1,2-dichloroethane (DCE) from 2.78 g/L diethyl ether (EE); and EE from 1.24 g/L DCE. The ordinate in this and subsequent figures is the volume of the minor component. Numbers in parentheses on figures indicate fixed concentrations in grams per liter. Key: —, uniform adsorbate model; -----, nonuniform adsorbate model; and – · –, IAS model. (In the upper two plots, all models give essentially similar results.) (Reproduced from Ref. 2. Copyright 1981, American Chemical Society.)

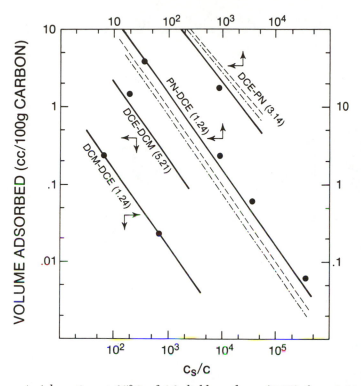

Figure 4. Adsorption at 25°C of 1,2-dichloroethane (DCE) from 3.14 g/L propionitrile (PN); PN from 1.24 g/L DCE; DCE from dichloromethane (DCM); and DCM from DCE. Legend as in Figure 3. (In the lower two plots, the theoretical lines for the three models were too close for convenient presentation.) (Reproduced from Ref. 2. Copyright 1981, American Chemical Society.)

desorption data obtained by the method of Semonian and Manes (*19*) at the Calgon Corp. (*20*). For the room temperature systems, the solid curves were calculated from adsorption isotherms estimated from the hydrocarbon curve by the method of Wohleber and Manes (*3*), which uses refractive indices to determine the abscissa scale factors. Although the fit could be improved somewhat by using experimental rather than estimated adsorption isotherms, the improvement was not considered enough to justify complicating a quite simple calculation method. For the high temperature systems, in all cases, the adsorption isotherms were derived from the hydrocarbon line by a best fit of single-component isotherm data, in which the parameters were the abscissa scale factor and the adsorbate density. (For these systems, the calculation already was complicated by the necessity of integration of the adsorption space.) In all cases, the solid theoretical curves are for the assumption of adsorbate uniformity or the "uniform adsorbate model."

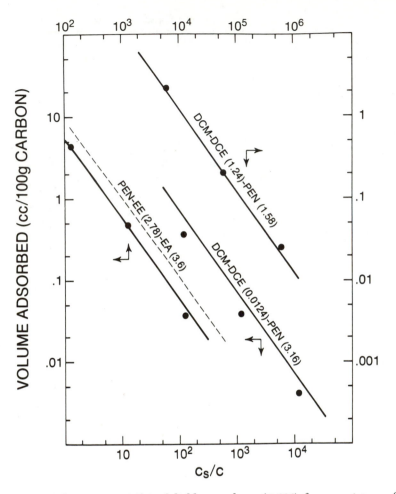

Figure 5. Adsorption at 25°C of dichloromethane (DCM) from a mixture of 1.24 g/L dichloroethane (DCE) and 0.158 g/L pentanol (PEN); DCM from 0.0124 g/L DCE and 3.16 g/L PEN; and PEN from 2.78 g/L diethyl ether (EE) and 3.6 g/L ethyl acetate (EA). Legend as in Figure 3. (Plots for uniform and nonuniform models nearly coincided in upper two cases. IAS model was not applied to ternary systems.) (Reproduced from Ref. 2. Copyright 1981, American Chemical Society.)

The dashed lines were all calculated for the "nonuniform adsorbate" model, i.e., for the Polanyi-based model that uses the Hansen–Fackler treatment for adsorbate nonuniformity. The alternately dotted and dashed lines were calculated for the IAS model, where the Polanyi treatment was used to extrapolate the individual adsorption isotherms.

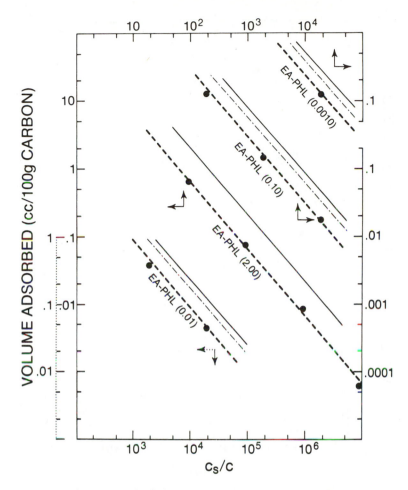

Figure 6. Adsorption of ethyl acetate (EA) at 63°C from phthalide (PHL) solutions ranging from 0.0010 to 2.00 g/L. Legend as in Figure 3. The IAS model is out of bounds for the PHL solution at 2.00 g/L (second from bottom). The multiplicity of scales reflects the wide range of concentrations and capacities. (Reproduced from Ref. 2. Copyright 1981, American Chemical Society.)

Details of the calculations are given by Greenbank and Manes (2) and by Greenbank (14).

Theoretical curves are omitted in the figures where they practically coincide with the curve for the uniform adsorbate model. Curves for the IAS model were not calculated for the ternary adsorbate systems. Their omission on occasion from a system with a molten solid (Figure 6) is an indication that the system is out of bounds for the model.

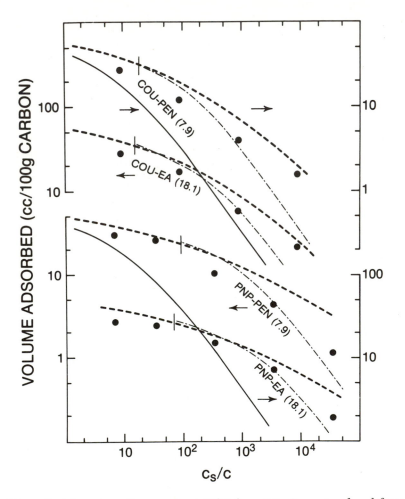

Figure 7. Adsorption of coumarin at 62°C from 7.9 g/L pentanol and from 18.1 g/L ethyl acetate; and p-nitrophenol from 6.9 g/L pentanol and 18.1 g/L ethyl acetate. Legend as in Figure 3. Note multiplicity of ordinate scales (capacity ranges for given c_s/c are all of the same order of magnitude). Vertical lines indicate upper limit of validity of the IAS model. Only two curves are shown for the uniform adsorbate model; all are similar in shape. (Reproduced from Ref. 2. Copyright 1981, American Chemical Society.)

Discussion

We shall first consider the experimental results and their likely applicability, and then consider some features of the Polanyi-based approach and its relation to others.

In both the binary and ternary systems at 25°C (Figures 3–5), we first note that the adsorption of a trace of one component in the presence of larger amounts of another is usually described by a linear dependence (Henry's law) of capacity on concentration. The model has to predict not only the linear relationship but also the magnitude of the coefficient. Thus, in the figures, the theoretical lines all depict a linear relationship except where they show some curvature, usually at the higher capacities. For these systems, the difference between the various models may be judged by the extent of vertical displacement of the various lines. For systems of approximately equal refractive index (or abscissa scale factor) such as EA–PN (Figure 3) and DCM–DCE (Figure 4), the Polanyi-based models (with either assumption of uniformity) and the IAS model give essentially the same results and may be used interchangeably. Indeed, although the IAS model was originally limited by its authors (6) to dilute solution, it should extend quite readily even to saturated solution for liquid solutes of comparable refractive index. The largest deviations between the models and between the alternative assumptions in the Polanyi-based model are for systems with DCE. Here a considerable part of the discrepancy was due to the refractive index approximation for the DCE isotherm and the fact that DCE turned out (for as yet unknown reasons) to give an isotherm with a poorer fit to the hydrocarbon correlation line than most other adsorbates.

The maximum error of any prediction method is a factor of about 2, and for most systems the prediction is much better. In some cases, the uniform adsorbate model gives better results than the more rigorous nonuniform adsorbate model; the differences, however, are not considered sufficient to make a judgment on the relative validity of either assumption. For these systems, which are probably representative of many systems to be met with in practice, all models should be of about equal utility and the choice between the Polanyi and the IAS formalism can be made on the basis of computational convenience (which favors the Polanyi model for many components) or personal convenience. However, since all models give similar results, we cannot use systems of this sort to judge the relative merits of different models.

We now consider the systems with molten COU, PHL, and PNP (Figures 6 and 7) and we first consider the adsorption of EA and PEN in relatively larger amounts of the molten solids (Figure 6). Again, a factor of about 2 separates all of the various models over orders of magnitude of concentration, and the IAS model could be considered satisfactory where it applies. However, for relatively concentrated solutions of the more strongly adsorbed adsorbate, it goes out of bounds; the Polanyi-based model handles all of these systems quite well.

The difference between the uniform and nonuniform assumptions becomes most striking for traces of the molten solids in larger concen-

trations of pentanol and ethyl acetate (Figure 7). Here there is ample
opportunity for concentration of the more strongly adsorbed component
within the adsorbate, and the difference in capacity between the nonuniform adsorbate treatment and the uniform adsorbate approximation
reaches some two orders of magnitude. The assumption of adsorbate
uniformity is clearly inadequate for these systems. The (Polanyi-
supplemented) IAS model predicts a Henry law dependence of capacity
on concentration, which is not in accord with the Polanyi-based model or
with experiment. However, the predictions of the nonuniform adsorbate
model at low capacities are sensitive to the values of the abscissa scale
factors and the agreement of experiment with theory is approximate for
both models. We are, however, dealing with quite extreme systems over
more than three orders of magnitude of concentration.

The systems with the molten solids, although they may be considered
as somewhat exotic for many practical applications (they may not be so
exotic for PCBs, for example), have been of interest because they
illustrate the extent to which one must go to find systems that are poorly
represented by the uniform adsorbate approximation, and because they
show the kind of systems that can be handled by the full Polanyi model.
We have yet to find a liquid adsorbate system for which the full model
breaks down badly or goes out of bounds. Whereas the IAS model does
well for some systems and reasonably well for others, there are systems, as
we have seen, for which it breaks down badly. Although these systems are
outside of the originally stated scope of the model, they are not outside the
scope of the Polanyi-based treatment.

A likely reason for the occasional breakdown of the IAS model is
the essential artificiality of the postulate that individual adsorbates be
compared at equal "spreading pressure." Although one may determine
"spreading pressure" on a film balance, the concept is not operational on a
highly heterogeneous surface such as activated carbon. Although the
thermodynamic derivation on which the IAS model is based is supposed to
be "model-free" (and therefore supposedly rigorous), the "spreading-
pressure" criterion, however useful it may be in providing a calculation
method, must be regarded as a purely speculative postulate with no more
thermodynamic justification than any other.

The systems with molten solids also are of interest to the question of
the thermodynamic consistency of the Polanyi model, which has been
questioned. The proper point raised by Sircar and Myers (9) is in effect
that the adsorption from a saturated aqueous solution of two liquids should
be the same as from the bulk liquid mixture in equilibrium with the
saturated water (one assumes for simplicity that the solubility of water in
this organic phase is not significant). The uniform adsorbate approximation of the Polanyi treatment [which was in effect used by Grant and
Manes (15) in their treatment of multicomponent vapor adsorption, and

which was criticized by Sircar and Meyers (9)] leads to the expectation that the adsorbate at saturation would have the same composition as the organic mixture at saturation, and therefore, that there would be no adsorption selectivity from the liquid phase. The uniform adsorbate assumption is therefore in error to the extent that such selectively takes place. The full Polanyi treatment [incorporating the Hansen-Fackler (16) modification], as used by Greenbank and Manes (2) and earlier by Schenz and Manes (21) in their treatment of the adsorption of mixed liquids, does lead to the expectation of selective adsorption from the liquid phase. The incorporation of the Hansen-Fackler treatment (as illustrated in our gravitational analogy) therefore resolves the question of thermodynamic consistency. For many practical applications, the errors associated with the thermodynamic inconsistency of the approximate model frequently are not experimentally significant. What this means in the model is that differences in the solubilities and molar volumes of the various adsorbates are the dominant factors in adsorbate composition from dilute water solution and that these effects dominate the relatively small selectivity that would be observed from anhydrous liquid mixtures. Thus far we have seen no evidence that the Polanyi model is thermodynamically inferior.

We now consider the questions of group functionality and steric effects on adsorption. Considering first the steric effects, as one would expect in highly branched adsorbates, we have been for some time aware of their possible significance, as has been illustrated by the work of Chiou and Manes (5) on highly sterically hindered adsorbates. We would expect effects of branching to be exhibited in the reduction of the magnitude of the abscissa scale factors and therefore as a complication in the prediction of adsorption isotherms from refractive indices by the Wohleber–Manes (3) method. Given the experimental adsorption isotherms, we could readily incorporate them into our Polanyi-based model, either as experimental isotherms or as hydrocarbon correlation curves with appropriate abscissa scale factors. There may very well be approximately constant group contributions to branching, and these could be very well accommodated by the Polanyi model. Indeed, the effect of branching on the abscissa scale factors (in the absence of molecular sieving effects) might best be expressed in terms of their effects on the abscissa scale factors.

Group functionality effects do not appear explicitly in either the Polanyi-based or the IAS treatments of multicomponent adsorption, although the latter model does not deal with the estimation of individual isotherms. Although one may use a Polanyi-based treatment that starts with experimental adsorption isotherms, estimated isotherms also can be used. Thus far, it appears that the solubility, molar volume, and refractive indices can account for most of the variability in adsorption isotherms for a wide variety of systems.

We note, finally, that although the results presented here have been all determined on one carbon sample, they should be readily applicable to most commercial carbons, once each such carbon is characterized. The most meaningful characterization would be either a correlation curve (such as the hydrocarbon correlation curve) for the carbon, preferably expressed in polynomial form as a function of log (adsorbate volume) versus $(T/V)\log(p_s/p)$ or else some other isotherm, either vapor or liquid, that covers the capacity range of interest (e.g., saturation to 0.1 g/100 g carbon). Laboratories that do not have gas phase adsorption apparatus but do have a gas chromatograph can readily determine the isotherm of propionitrile, for example, over the capacity range of interest. It would add greatly to our ability to make comparisons of different adsorbates on different carbons in different laboratories if a single wide range adsorption isotherm were reported for each carbon. The isotherm should in time replace the surface area, which has been useful for many years but is now becoming obsolete as a means of characterizing activated carbons. As more laboratories have their own computer programs for estimating adsorption isotherms, they may well expect manufacturers to supply a more informative characterization of their carbons than has been customary.

Acknowledgments

We thank the Calgon Corp. and the National Science Foundation (Grant CME-7909247) for supporting this work. We also thank the Calgon Corp. for the butane desorption data on our carbon.

Note Added in Proof. A general solution of the problem of sedimentation equilibrium for multicomponent liquids in a gravitational field appears in the 1923 edition of G. N. Lewis and M. C. Randall's "Thermodynamics," which cites Gouy and Chaperon, *Ann. Chim. Phys.* **1887**, *12*(6), 384.

Literature Cited

1. Manes, M. In "Activated Carbon Adsorption of Organics from the Aqueous Phase," Suffet, I. H.; McGuire, M. J., Eds.; Ann Arbor Science: Ann Arbor, Mich. 1980; Vol. 1, Chapter 2 and references cited therein.
2. Greenbank, M.; Manes, M. *J. Phys. Chem.* **1981**, *85*, 3050.
3. Wohleber, D. A.; Manes, M. *J. Phys. Chem.* **1971**, *75*, 61.
4. Wohleber, D. A. Ph.D. Dissertation, Kent State University, Kent, Oh., 1970.
5. Chiou, C. C. T.; Manes, M. *J. Phys. Chem.* **1972**, *77*, 809.
6. Radke, C. J.; Prausnitz, J. M. *AIChE J.* **1971**, *17*, 186.
7. DiGiano, F. A.; Baldauf, G.; Frick, B.; Sontheimer, H. *Chem. Eng. Sci.* **1978**, *33*, 1667.
8. Jossens, L.; Prausnitz, J. M.; Fritz, W.; Schlünder, E. U.; Myers, A. L. *Chem. Eng. Sci.* **1978**, *33*, 1097.

9. Sircar, S. R.; Myers, A. L. *Chem. Eng. Science* **1973**, *28*, 489.
10. Chiou, C. C. T.; Manes, M. *J. Phys. Chem.* **1974**, *78*, 622.
11. Rosene, M. R.; Manes, M. *J. Phys. Chem.* **1975**, *79*, 604; **1976**, *80*, 953.
12. Polanyi, M. *Verh Dtsch. Phys. Ges.* **1914**, *16*, 1012; *Z. Elektrochem.* **1920**, *26*, 370; *Z. Phys.* **1920**, *2*, 111.
13. Lewis, W. K.; Gilliland, E. R.; Chertow, B.; Cadogan, W. P. *Ind. Eng. Chem.* **1950**, *42*, 1326.
14. Greenbank, M. Ph.D. Dissertation, Kent State University, Kent, Ohio, 1981.
15. Grant, R. J.; Manes, M. *Ind. Eng. Chem. Fundam.* **1966**, *5*, 490.
16. Hansen, R. S.; Fackler, W. V. *J. Phys. Chem.* **1953**, *57*, 634.
17. Rosene, M. R. Ph.D. Dissertation, Kent State University, Kent, Oh. 1977.
18. Rosene, M. R.; Deithorn, R. T.; Lutchko, J. R.; Wagner, N. J. In "Activated Carbon Adsorption of Organics from the Aqueous Phase,"; Suffet, I. H.; McGuire, M. J., Eds.; Ann Arbor Science: Ann Arbor, Mich., 1980; Vol. 1, p 309.
19. Semonian, B. P.; Manes, M. *Anal. Chem.* **1977**, *49*, 991.
20. Urbanic, J. E. Calgon Corp. (personal communication).
21. Schenz, T. W.; Manes, M. *J. Phys. Chem.* **1975**, *79*, 604.

RECEIVED for review August 3, 1981. ACCEPTED for publication March 23, 1982.

3

Selective Adsorption of Organic Homologues onto Activated Carbon from Dilute Aqueous Solutions

Solvophobic Interaction Approach— Branching and Predictions

GORDON ALTSHULER and GEORGES BELFORT

Rensselaer Polytechnic Institute, Department of Chemical and Environmental Engineering, Troy, NY 12181

The testing of the solvophobic (cφ) theory was extended to include adsorption of branched alkyl alcohols and slightly ionized linear carboxylic acids. Correlations with adsorption capacity for both the comprehensive and the simplified cφ theories were compared with similar correlations for other independent variables including molecular weight, density, index of refraction, molar volume, and group contribution parameters such as molar refraction, octanol–water partition coefficient, and parachor. The best correlative parameters were the octanol–water partition coefficient, three solvophobic terms, and the total cavity surface area. A simplified expression requiring only the total cavity surface area was used to predict the unknown adsorbability of two branched alcohols.

THE ABILITY TO PREDICT THE EFFECTS of even simple structural modifications on the adsorption of organic molecules from dilute aqueous solutions onto activated carbon (or other adsorbents) could be valuable in the design and operation of large-scale commercial water and wastewater treatment plants. This capability would assist in gaining a better understanding of competitive adsorption between and chromatographic elution by different organic solutes during adsorption.

This is Part III in a series.

0065-2393/83/0202-0029$09.25/0

To accomplish this goal, a general comprehensive solution interaction approach, originally developed by Sinanoğlu (1), was recently adapted to the adsorption of organic homologues from dilute aqueous solutions by Belfort (2–4). By including solvent effects in the equilibrium adsorption (association) process, this solvophobic theory (cφ) differs from most other equilibrium theories in that the latter were originally derived from gas and vapor phase adsorption and thereby a priori ignored the presence of the solvent during adsorption (5).

Previously (3) the solvophobic theory was adapted to adsorption and the free energies of solvent-mediated adsorption for a series of linear aliphatic alcohols, ketones, aldehydes, and acids. Both the comprehensive ($\Delta G^{net}/RT$) and simplified (total surface area, TSA, $\mathring{A}^2$) theories were tested with adsorption of linear aliphatic and aromatic alkyl compounds for single and solute solutions.

In this chapter, the testing of the solvophobic theory is extended to include the adsorption of branched aliphatic alcohols and slightly ionized linear carboxylic acids. Correlations from the comprehensive and simplified cφ theories are compared with correlations of other independent variables such as the molecular weight, density, index of refraction, molar volume, molar refraction, octanol–water partition coefficient, parachor, and polarizability. Total solute surface cavity areas (TSA) are calculated with a newly developed molecular-build program coupled to Hermann's method of intersecting atomic spheres with appropriate crowding factors (6). The theory also can be used to predict the adsorption capacity of organic compounds whose experimental adsorption capacity has not been measured but whose homologues have.

With respect to solid–liquid adsorption, a major limitation of the various equilibrium theories of adsorption (4,5) is that they were originally derived from gas and vapor phase adsorption, and thereby a priori ignored the presence of the solvent during solute adsorption. Also, single solute adsorption data are needed to predict multisolute competitive adsorption, while other empirical mathematical equations represent the data without attempting to establish a physical model (5).

Freundlich, in his classic monograph (7), describes the first attempts 90 years ago by Traub and others to include the solvent–solute interaction effect through interfacial tension at the solid–liquid interface. Although Defay et al. (8) updated this approach with a major emphasis on the simplified Gibbs adsorption model especially under nonequilibrium conditions, a comprehensive interaction theory between solute, solvent, and sorbent was still lacking. Recent attempts to include the solvent effect in aqueous phase adsorption include a semi-empirical quasi-theoretical approach based on partial solubility parameters called the net adsorption energy approach (9–11). In spite of the limitations of this approach (see

Reference *4* for a detailed discussion) it does provide a useful semi-quantitative screening method for estimating the relative adsorption capacity (rank order) of different organics.

Although these approaches provide useful insight with respect to solute–solvent adsorption onto solids, a comprehensive formalism of aqueous phase adsorption including fundamental formulations of all known interactions among solute, solvent, and sorbent is needed. By invoking and adapting the $c\phi$ theory of Sinanoğlu (*1*), Belfort (*2–5*) and more recently Melander and Horváth (*12*) attempted to provide such a formalism for describing the solvent effect on the solute–solid adsorption association reaction. The results presented in this chapter represent a continuing effort to use, test, and adapt this formalism to predict a priori a ranking order of adsorption capacity especially for homologues without previously measured single solute adsorption data. Here, specifically the testing of the $c\phi$ theory is extended to branched and slightly ionized organics in dilute aqueous solution.

Theory

General Solvophobic Approach. The solvophobic ($c\phi$) theory describes the tendency of a surrounding solvent medium to influence aggregation or dissociation of molecules with considerable microsurface areas exposed to the solvent medium. Examples of these molecules include amino acids, nucleotide bases in biopolymers, various drug molecules, antigens and substrates, and relatively low molecular weight organic homologues and an adsorption surface. The $c\phi$ theory also has introduced a new measurable quantity called "the thermodynamic microsurface area change of a reaction" (*13*) which is now finding applications in protein structure (*14*), high-performance liquid chromatography (HPLC) (*15*), protein salting-in and salting-out (*16*), the conformational structure of organic compounds (*17*), isomerization reactions (*18*), stacking of bases and solvent denaturation of DNA (*19*), and association adsorption reactions (*3*).

The approach used here follows closely that used by Belfort (*3*) in which he adapted the $c\phi$ theory, originally presented by Sinanoğlu (*1*) and applied to HPLC by Horváth et al. (*15*), to the association adsorption reaction. For a comprehensive review of the adaptation of this approach to the adsorption of organic substances from dilute aqueous solutions by nonpolar adsorbents, the reader is referred to the literature (*3, 4, 12*). Since this is an equilibrium theory, kinetic effects such as convective mass transfer and solute diffusion processes are not considered.

In the solvophobic treatment, the effect of the solvent on the reversible association reaction of the adsorbate molecules S_i, with

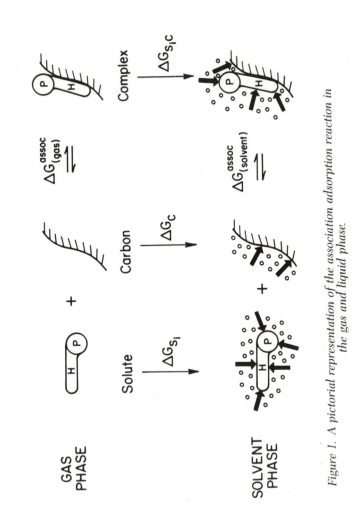

Figure 1. A pictorial representation of the association adsorption reaction in the gas and liquid phase.

activated carbon adsorbent, C, at the surface to produce adsorbed complex S_iC is obtained by subtracting the standard free energy change for the reaction in the gas phase from that in the solution phase under unitary standard state ($\chi°_k \equiv 1$, $p°_k \equiv 1$ atm ideal gas). This process is diagrammed in Figure 1 and gives rise to a net free energy, $\Delta G^{net}_{(solvent\ effect)}$, expressing the effect of the solvent in the association adsorption process. Conceptually, Sinanoğlu proposed a two-step dissolution process (1). First, a hole or cavity needs to be prepared in the solvent to accommodate the solute, carbon, or adsorbed complex "molecule"; second, after the "molecule" is placed into the cavity, it interacts with the solvent. Quantitatively this process is expressed as follows:

$$\Delta G^{net}_{(solvent\ effect)} = \Delta G^{assoc}_{(solvent)} - \Delta G^{assoc}_{(gas)} = RT \ln [k^H_{S_iC}/k^H_{S_iC}k_C^{H}] \qquad (1)$$

or

$$\Delta G^{net}_{(solvent\ effect)} = \Delta G^{net}_{j,S_iC} - \Delta G^{net}_{j,S_i} » \Delta G^{net}_{j,C} \qquad (2)$$

where $k^H_k = p_k/\chi_k$, are the Henry's constants for the kth species and j represents each type of interaction. After specifying each interaction (such as the cavity, van der Waals, and electrostatic terms plus two corrections terms) for polymer mixing and reduced electrostatic effects due to the presence of the solvent, the following expression is obtained from Equations 1 and 2 for the overall standard free energy change:

$$\Delta G^{assoc}_{(solvent)} = \Delta G^{assoc}_{(gas)} + [\Delta G_{cav} + \Delta G_{vdw} + \Delta G_{es} + \Delta G_{mix} + \Delta G_{red}]^{net}_{S_iC-S_iC}$$
$$- RT \ln (RT/P_0 V) \qquad (3)$$

where the last term is called the cratic term and results from an entropy or free volume reduction. The term $\Delta G^{assoc}_{(solvent)}$ is related to the experimental equilibrium constant $K_{solvent,i} = \chi_{S_iC}/\chi_{S_i}\chi_C$ which itself will be related to the experimental adsorption capacity $\bar{p}_i$ for solute S_i later in this analysis. Each term in the bracket in Equation 3 can be calculated explicitly from known physico-chemical parameters obtained from the literature (20). Explicit formulas to do this and a discussion on the relevance of each term to the adsorption association reaction were presented previously (3). The dominant interactive forces and important parameters for each interactive process are, however, summarized in Table I.

Since the expressions for all terms but one are identical to those presented previously (3), only the changed electrostatic term will be discussed here.

The electrostatic term differs in this analysis to that presented previously (3) because instead of considering only simple dipoles, ionized

Table I. Interactive Free Energy Changes

Interaction Process	Free Energy Designation	Interaction Forces	Important Parameters
Cavity formation	$\Delta G_{cav,i}^{net}$	surface forces	solute molecular surface area, TSA corrected macroscopic surface tension, γ_i
Solvent–solute interaction van der Waals	$\Delta G_{vdw,i}^{net}$	dispersive (attractive) forces	ionization potential and molecular volume v_i
Electrostatic	$\Delta G_{es,i}^{net}$	electronic interaction	(a) Simple dipole. Solute dipole moment μ_i and dielectric constant of solvent ε (b) Charged species. Ionic charge ze and ionic strength, I.
Mixing effects (polymers)	ΔG_{mix}^{net}	free energy of mixing	
Dispersive reduction due to solvent	ΔG_{red}^{net}	masking of the dispersive forces	
Free volume reduction (cratic)	$RT \ln (RT/P_o V)$	entropic reduction	solvent molar volume, V, and pressure, P

Table II. Unitary Free Energy Change for Electrostatic Interactions

Interactive Process	Limitations	$\Delta G_{es,i}$	Equation Number	Reference
Dipole–dipole[a]	Simple dipoles with no formal charges	$-\dfrac{N}{2}\dfrac{\mu_i}{v_i}DP$	4a	20
Charged species[b]	Very low ionic strength $I \leq 0.1$	$\dfrac{Z^2e^2N}{2\varepsilon}\left[\dfrac{1}{b_i}-\dfrac{\kappa}{1+\kappa a_i}\right]$	4b	21,22

[a] μ_i = static dipole moment of i, $D = [2(\varepsilon - 1)/2\varepsilon + 1]$, ε = static dielectric constant of the solvent, $P = [4\pi\varepsilon_0(1 - D(\alpha_i/v_i))]^{-1} \approx [4\pi\varepsilon_0(1 - (\alpha_i/v_i)]D \approx 1$. α_i = polarizability of i, ε_0 = permittivity constant.

[b] Ze = electronic charge of the ion, b_i = ionic radius, a_i = distance of closest approach. κ = Debye screening parameter and is obtained from $\kappa^2 = 8\pi e^2 I/\varepsilon kT$, where the ionic strength $I = 1/2 \ \Sigma n_i Z_i^2$, where n_i is the number of ions of species i per unit volume.

(charged) species also will be studied. The expressions are summarized in Table II. Equation 4b in Table II derives from the simple Debye–Hückel analysis and is most appropriate for low ionic strength, I, which is relevant for organic removal from dilute aqueous solutions by activated-carbon adsorption. Expressions for higher I can be found (12, 21, 22). Values of b_i are estimated from molecular volumes V_i assuming a sphere $b_i = [(3/4) (v_i/\pi)]^{1/3}$, $a_i \equiv b_i + 2$ Å where v_i is given in cubic Angstrom units. Sometimes 3 Å is used in place of 2 Å in the definition of a_i.

For the general case of an incompletely ionized solute with a degree of dissociation of $\alpha°$, the unitary free energy change for electrostatic interactions can be expressed as follows:

$$\Delta G_{es,i} = (1 - \alpha°)\, \Delta G_{es,i}(\text{dipole–dipole}) + \alpha°\, \Delta G_{es,i}(\text{charged species}) \quad (5)$$

where $\Delta G_{es,i}(\text{dipole–dipole})$ and $\Delta G_{es,i}(\text{charged species})$ are given by Equations 4a and 4b (Table II), respectively.

In summary, the cavity formation and van der Waals dispersive interactions are roughly proportional to molecular surface area, while the electrostatic interaction is inversely proportional to molecular volume (for dipole–dipole interactions).

Assumptions. For nonpolymer molecules where Henry's law is expected to hold, the correction ΔG_{mix}^{net} can be neglected in Equation 3. The dispersion correction for the presence of water ΔG_{red}^{net} is only relevant to the adsorbed complex S_iC, and although it can vary in value by about 15–30% of the $\Delta G_{vdw,\,S_iC}^{net}$ value, it is probably fairly constant for a homologous series and will thus be neglected in the present calculations. Also, the solute–sorbent gas phase interaction, $\Delta G_{(gas)}^{assoc}$, can, in general, include van der Waals electrostatic, hydrogen-bonding, or covalent interactions. For a homologous series, this term is thought to vary negligibly and is thus considered constant in the present analysis.

For activated-carbon adsorption, the solvent molecules (water) and their cavity surface area are usually significantly smaller than the solute molecules, their cavity surface area, and the exposed nonpolar surface area of the activated carbon. Then the following assumptions appear to be reasonable (15):

$$\mu_C = 0 \text{ (nonpolar carbon)} \quad (6)$$

$$\mu_{S_iC} = \mu_{S_i} \text{ (solute could be polar)} \quad (7)$$

$$W_{S_iC} = W_{S_i} = W_C = 0 \text{ (if } K_i^s = K_i^e \text{ or if } d_i/d > 1) \quad (8)$$

$$\Delta G_{vdw,S_iC} = \Delta G_{vdw,C} \text{ (dilute solute concentration)} \quad (9)$$

To simplify the algebra, it is convenient to express the molecular volume of the carbon–solute complex, S_iC, as a multiple of the solute molecular volume:

$$v_{S_iC} = \lambda v_{S_i} \tag{10}$$

where λ is the proportionality factor. For HPLC, Horvath et al. (*15*) estimated λ to be about 3.7. In the equations below, the ratio $(\lambda-1)/2\lambda$ appears and varies in value from 0.25 to 0.5 as λ varies from 2 to infinity. Previously, it was shown that changes in the magnitude of the electrostatic term as a function of λ or $(\lambda-1)/2\lambda$ were very small in comparison to the total change in free energy of association $\Delta G^{assoc}_{(solvent)}$(*3*). Because of this, in in all the calculations that follow, λ was arbitrarily chosen as equal to 100 ($v_{S_iC} \gg v_{S_i}$, *see* Equation 10).

Because of the association adsorption reaction of solute S_i onto sorbent C, the total surface area in contact with the free solvent is reduced by the contact surface area:

$$\Delta A = A_{S_i} + A_C - A_{S_iC} \tag{11}$$

As mentioned earlier, Sinanoğlu calls this "the thermodynamic micro-surface area change of the reaction" (*13*).

With respect to the activated-carbon surface, essentially two types of surface interactions are thought to predominate. The first type of surface interaction, which occurs on most of the available surface area (on basal planes), can be characterized by van der Waals physical interactions and is apolar. The second type of surface interaction, which occurs at the more reactive edges of the microcrystallites, can be characterized by positive physical (and maybe even chemical) attractive interactions due to hydrogen bonding and electrostatic forces. Although the second type of surface interaction probably occurs at a small fraction of the total surface area, the specific adsorptions result from the surface heterogeneity and to the presence of oxides, hydroxyls, and other groups on the surface. Adsorption differences among activated carbons made from different raw materials (wood, lignite, coal, bone petroleum residues, and nut shells) are probably the result of their different energy potential and the extent of these heterogeneous sites. Since hydrophobic or apolar interactions play a dominant role in activated-carbon adsorption of nonpolar (or nonpolar moieties of) solutes, the reduction in area, ΔA, given in Equation 9 is most likely the result of the first type of interaction discussed.

Comprehensive Model.   After substituting for each relevant free energy term in Equation 3 and making some reasonable assumptions concerning the nonpolar nature and large surface area of the sorbent, the standard unitary free energy change for the overall process is given by (3):

$$\Delta G^{\text{assoc}}_{(\text{solvent})} = \Delta G^{\text{assoc}}_{(\text{gas})} - \Delta G_{\text{vdw},S_i} + \Delta G_{es,S_i} + N(K^e_{S_iC}A_{S_iC}\gamma_{w,S_iC}$$
$$- K^e_{S_i}A_{S_i}\gamma_{w,S_i} - K^e_C A_C\gamma_{w,C}) - RT \ln RT/(P_0V) \qquad (12)$$

where $\Delta G_{es,i}$ is obtained from Equation 5.

To take into account the highly curved rather than flat surface of the solute molecule, a curvature correction is included in the general solvophobic theory (1). This correction is shown in Equation 9 of Reference 3 and is given by $K^e_i (1 - W)$. Sinanoğlu uses $\gamma_{w,i} = \gamma$ (neat solvent) with this correction, thereby obviating the need to distinguish between surface tensions, $\gamma_{w,i}$. The major complication in this approach is to obtain a reasonable estimate of K^e_i and W (see footnote a in Table II of Reference 3). Thus, K^e_i values for species i can be estimated from:

$$K^e_i = 1 + (V/V_i)^{2/3} (K^e - 1) \qquad (13)$$

where K^e is the microscopic cavity factor for the pure solvent (in itself). For large solute molecules relative to the solvent molecules ($V/V_i < 1$), K^s_i, $K^e_i \to 1$, and $W \to 0$. Thus, as per Horváth et al. (15):

$$K^e_{S_iC} = K^e_C = 1 \qquad (14)$$

for a large adsorbent volume (S_iC and C) relative to the solvent molecular volume. With these assumptions, the fourth term on the right-hand side of Equation 12 can be replaced by: $N\gamma (A_{S_iC} - K^e_{S_i}A_{S_i} - A_C)$, and with the assumptions described, the following is obtained:

$$\Delta G^{\text{assoc}}_{(\text{solvent})} = \Delta G^{\text{assoc}}_{(\text{gas})} - \Delta G_{\text{vdw},S_i} + \Delta G_{es,S_i} - N\gamma\Delta A - N\gamma V^{2/3}(K^e - 1) \frac{A_{S_i}}{V^{2/3}_i}$$

$$- RT \ln RT/(P_0V) \qquad (15)$$

where $\Delta G_{es,i}$ is obtained from Equation 5. A difficulty with this analysis lies in the fact that, although Equation 13 is most applicable for nonpolar liquids, it also has been used for polar liquids (18,23). It can be shown, however, that for relatively small symmetrical molecules $A_{S_i}/v_{S_i}^{2/3} \approx K_a \approx 4.8364$ for spheres, $K_a = 5.8588$ for right circular cylinders ($r = h$), and $K_a = 7.355$ for right circular cones ($r = h$). Thus, the fifth term on the right-hand side of Equation 15 is probably fairly constant, and the correction due to curvature appears to fall away. Because this conclusion is somewhat unsettling, Altshuler and Belfort (24) derived an alternate phenomenological approach in which the curvature correction does not fall away.

In applying the solvophobic theory to *cis–trans* isomerization of azobenzene in several solvents, Halicioğlu and Sinanoğlu (18) calculated each term in Equation 15 without the use of an adjustable parameter. In most cases, they found that the solvation of inverse volume forces (third term on the right-hand side of Equation 15) and the surface forces (fourth term on the right-hand side of Equation 15) both strongly affect the molecular association equilibrium or rate constants. The treatment indicates that these two solvent effects usually oppose each other. For small solute molecules relative to the size of the solvent molecules, the solvation or inverse volume forces may dominate; however, as the relative size of the solute molecules increases, surface forces may predominate (1, 18, 23).

Before simplifying the comprehensive model expressed through Equation 15, it is appropriate to note that:

$$\Delta G^{assoc}_{\substack{(solvent) \\ (gas)}} = RT \ln K_{\substack{(solvent)i \\ (gas)i}} \text{ (unitary)} \tag{16}$$

where:

$$K_{(solvent)i} = \chi_{S_iC}/(\chi_{S_i}\chi_C) \text{ or } K_{(gas)i} = p_{S_iC}/(p_{S_i}p_C) \tag{17}$$

for dilute solutions assuming $a_i \approx \chi_i$ and $\gamma_i \equiv 1$.

Simplified Model (3). In considering Equation 15 at least two special cases of interest are relevant for aqueous phase activated-carbon adsorption studies. The first case concerns the adsorption on different activated-carbon surfaces (i.e., different types of activated carbon) of the same organic compounds at constant temperature, composition, and flow conditions. Although the theoretical expressions for this case are available (3), they have not been tested with data yet. The second case and the one of interest here concerns the adsorption of different solutes, such as those belonging to various homologous series, at constant temperature and flow rate and with the same solvent and carbon. For this case, the standard unitary free energy change in the solvent is given by the following expression for:

$$\Delta G^{\text{assoc}}_{\text{(solvent)}} = K'_1 + \Delta G_{es,i} - K'_2 \, \Delta A \tag{18}$$

where the constants K'_j are given by:

$$K'_1 = \Delta G^{\text{assoc}}_{\text{(gas)}} - \Delta G_{\text{vdw},i} - N\gamma V^{2/3}(K^e - 1)\frac{A_i}{V_i^{2/3}} - RT \ln RT/(P_0 V) \tag{19}$$

$$K'_2 = N\gamma \tag{20}$$

and where $\Delta G_{es,i}$ is given by the expressions from Equation 4a (Table II) for dipole–dipole interactions and from Equation 4b (Table II) for charged ionized species interactions. In Equation 18, ΔA is later replaced by g (HSA) as in Reference 3 and by g' (TSA) for correlations of the simplified model.

Equilibrium Adsorption Isotherms

Many well-known and not so well-known expressions have been formulated to describe single solute adsorption from gas and liquid phases onto liquid and solid surfaces. Some models are based on a proposed physical description of the adsorption process (e.g., Polanyi, Langmuir, and BET), while other models are empirical. For an assessment of equilibrium theories, vis a vis single and multisolute liquid phase adsorption onto activated carbon the reader is referred to Reference 5.

A general empirical equation for calculating the adsorption equilibria of organic solutes in aqueous solutions has been proposed (25):

$$Y_i = \frac{a_{i0}X_i^{b_{i0}}}{c_i + \sum_{j=1}^{n} a_{ij} X_j^{b_{ij}}} \tag{21}$$

where Y_i is the amount of solute, i, adsorbed per unit weight of carbon at equilibrium concentration, X_i, in a solution containing n solutes; X_i and X_j are the equilibrium concentrations for solute i and solutes j–n, respectively; and n is total number of solutes present. Equation 21 reduces to several well-known relationships as special cases. These include (1) the Langmuir equation for bisolute adsorption (25,26) ($b_{i0} = b_{ij} = c_i = 1$); (2) the Jäger and Erdös equation for a solute in a bisolute system (27) ($a_{i0} = Y_i^\circ$, $b_{i0} = a_{11} = b_{12} = 1$, $c_i = 0$); (3) the Radke and Prausnitz equation for single solutes (28) ($b_{11} = b_{10} - 1$, $c_1 = 1$); (4) the Freundlich equation for single solutes ($c_i = 0$); and (5) the Langmuir equation for single solutes ($b_{10} = b_{11} = c_1 = 1$).

Both the Langmuir equations (*29–31*) and Freundlich equations (*30*) have been used widely for organic adsorption from water. Several objections have arisen in using both the Langmuir and Freundlich equations including:

1. The Langmuir mixture model violates the Gibbs adsorption equation and consequently is thermodynamically inconsistent (*28*).

2. When both the Langmuir and Freundlich models fit the data, the underlying physical pictures derived from each model are mutually exclusive (*33*).

3. At vanishing concentration, the Freundlich equation does not approach Henry's law (*34*). Dobbs et al. (*35, 36*) applied the Freundlich equation to the activated-carbon adsorption of many toxic and nontoxic organic compounds. Several compounds showed poor adsorptivity by activated carbon. Jossens et al. (*34*) also discussed four equations and all of them reduce to Henry's law at very low single-solute concentrations.

From Equations 2, 16, and 17, the relationship between the equilibrium constant for unitary adsorption from the liquid phase, $K_{(\text{solvent})i}$, and that from the gas phase, $K_{(\text{gas})i}$, is obtained as a function of the net free energy for the solvent effect:

$$K_{(\text{solvent})i} = K_{(\text{gas})i} \exp\left[-\Delta G^{\text{net}}_{(\text{solvent effect})} / RT\right] \tag{22}$$

The intention now is to relate the equilibrium coefficient, $K_{(\text{solvent})i}$, to a measurable adsorption capacity, $\bar{p}_i$, characteristic of species i. Unfortunately, at present, no single parameter is considered reliable in characterizing the entire adsorption isotherm. However, Myers (*37*) recently suggested that the free energy of immersion on placing a sorbent into the pure liquid phase solute, ΔG_{im}, could be used with the saturated adsorption capacity, $Q°$, to characterize this association reaction and to transform the Polanyi potential into a dimensionless function that should be the same for all adsorbates in a given adsorbent. Although ΔG_{im} should be obtainable from the $c\phi$ theory directly, further work in this area is necessary.

By using the general empirical Equation 21 to characterize and calculate the adsorption equilibrium of organic solutes in dilute aqueous solutions, the adsorption capacity coefficient can be written as:

$$\bar{p}_i = [S_iC]/[S_i] = Y_i/X_i = f(X_i,X_j)_{a_{ig}b_{ij}c_i} = [C]\,K_{(\text{solvent})i} \tag{23}$$

where $f(X_i,X_j)_{a_{ij}b_{ij}c_i}$ results from a parametric fit to the adsorption equilibria for multisolute adsorption. Explicit expressions for f for single solute Langmuir and Freundlich expressions are given in Reference 3. At infinitely low bulk solute concentration, X_1, $f_1 = \bar{p}_1$ should be equal to Henry's constant, k_1^H for liquid phase adsorption.

Ideally, to characterize the extent of adsorption, $\bar{p}_i$ should be constant for a particular solute i. Failing this, many workers have arbitrarily chosen some convenient X_i (and X_{j0} initial values) to obtain a corresponding Y_i from Equation 21. For single solute adsorption $Y_{i=1}$ was chosen to represent $\bar{p}_{i=1}$ the capacity of a carbon and evaluated at $X_1 = 10^{-3}M$ by McGuire et al. (10); at $X_1 = 1$ mg/L by Dobbs et al. (35,36) to obtain a_{10} (or K in their nomenclature for the Freundlich coefficient); at fixed initial $X_1 = 1000$ mg/L with 5 g carbon by Giusti et al. (32); and when $b_{10} = b_{11} = 1$ and $a_{11} X_1 > c_1$ for the Langmuir equation, at $Y_1 = a_{10}/a_{11}$ by Martin and Al-Bahrani (30).

Thus, in general from Equations 22 and 23:

$$- RT \ln \bar{p}_i = \Delta G^{assoc}_{(solvent)} + \theta \tag{24}$$

where $\bar{p}_i$ is an adsorption capacity coefficient for solute i and $\theta = RT \ln [C]$ is a constant for dilute solute adsorption and a given activated-carbon column. With the requirement for dilute solute adsorption, the complexity of solute–solute interaction in the adsorbed phase is bypassed.

Equation 18 can now be expressed in terms of the measurable adsorption capacity coefficient, $\bar{p}_i$, for comparison of different solutes as follows:

$$- RT \ln \bar{p}_i = K_1'' + \Delta G_{es,i} - K_2' \Delta A \tag{25}$$

where $K_1'' = K_1' + \theta$, $K_2' = N\gamma$, as before, and $\Delta G_{es,i}$ is given by Equation 5. The subscript S_i is replaced by i.

Adsorption Data

Adsorption data of typical aliphatic organics on activated carbon were obtained from two sources (32, 38). Single-solute equilibrium adsorption data (32) were obtained by contacting 100-mL aliquots of 1000 mg/L stock solution with 5 g/L of pulverized Westvaco Nuchar GAC, Grade WV-G, for 2 h. The molar amount of aliphatic solute adsorbed during this period is designated Q_m'.

The single solute Freundlich (K_f,n) and Langmuir ($Q°,b$) parameters for the isothermal adsorption of aliphatic organics onto pulverized Filtrasorb 400 (Calgon Corp., Pittsburgh, PA) were obtained by contacting 200 mL of a 500 mg/L phosphate-buffered solute in distilled water for 2 h with 0.05–16.0 g carbon (*38*). Solute concentrations were measured for total organic carbon calibrated with standard solute solutions.

Solvent–Solute Data

Various parameters that characterize the solvent (water), solute, and adsorbent (activated carbon) are needed to calculate the free energy terms in Equation 15. In addition, a series of solute and solute–solvent parameters, such as molecular weight (MW), density (ρ), index of refraction (n_D) molar volume (V), molar refraction (MR), octanol–water partition coefficient log *P*, parachor [*P*], and polarizability (α), are correlated with adsorption capacity (ln $\bar{p}_i$). A summary of the values used for the parameters of each solute is presented in Table III. The major reference sources are included in the table, while the definitions of several parameters are given in the footnotes.

Group Contribution Parameters. Group contribution methods have been successfully used to predict among other properties, the critical temperature (*47*), critical pressure (*47, 48*), critical volume (*48*), and normal boiling point (*47*). They have also been used to predict heat capacity (*50*), enthalpy of formation (*51*), heat of vaporization (*52*), and activity coefficients (*53*). Another application of group contribution methods is quantitative structure and activity relationship (QSAR) where structure of a solute is correlated with its biological or pharmacological effect. The appearance of MR, log *P*, [*P*], and α in some or many of these group contribution methods suggests that these parameters might be considered with respect to the adsorption of homologous series.

The Clausius–Mosotti equation for atomic and electronic polarization (P_a and P_e) may be written as follows:

$$P = P_a + P_e = \frac{\varepsilon - 1}{\varepsilon + 2} \frac{MW}{\rho} = \frac{4}{3} \pi N \alpha \tag{26}$$

By using Maxwell's equation which equates the dielectric constant, ε, with the square of the index of refraction, n_D^2, for nonpolar molecules without a permanent dipole, Equation 26 is transformed into the Lorenz–Lorentz equation to give molar refraction:

$$MR = \frac{n_D^2 - 1}{n_D^2 + 2} \frac{MW}{\rho} = \frac{4}{3} \pi N \alpha \tag{27}$$

where *P* is total polarization, ρ is density, *N* is Avogadro's number, α is polarizability, and MW is molecular weight. MR has been widely used in QSAR (*54, 55*).

Table III. Solute

No. Compounds	Molecular Weight, MW	Ionization Potential, I (39,60) eV	Refract. Index, n_D (40,61)	Density ρ (40,61) g/cm^3	Molecular[a] Volume v Å^3
Aliphatic Alcohols					
1. Methanol	32.0	10.97	1.32652	0.7866	67.55
2. Ethanol	46.1	10.65	1.35941	0.7851	97.51
3. 2-Propanol	60.1	10.49	1.3752	0.7813	127.74
4. 1-Propanol	60.1	10.52	1.3837	0.7998	124.79
5. 2-Methyl-2-propanol	74.1	10.5	1.3851	0.7812	157.51
6. 2-Butanol	74.1	10.35	1.3950	0.8026	153.31
7. 2-Methyl-1-propanol	74.1	10.47	1.3939	0.7978	154.24
8. 1-Butanol	74.1	10.44	1.3973	0.8060	152.67
9. 2-Methyl-2-butanol	88.2	10.16	1.4024	0.8044	181.97
10. 1-Pentanol	88.2	10.42	1.4079	0.8112	180.33
11. 1-Hexanol	102.2	—	1.4161	0.8159	208.00
12. 3-Methyl-1-pentanol	102.2	—	1.4175	0.8200	206.96
Carboxylic Acids[f]					
13. Formic acid	46.0	11.05	1.3714	1.220	62.61
14. Acetic acid	60.1	10.35	1.3716	1.0492	95.12
15. Propanoic acid	74.1	10.24	1.3869	0.9930	123.92
16. Butanoic acid	88.1	10.16	1.3980	0.9577	152.76
17. Pentanoic acid	102.1	10.12	1.4085	0.9391	180.54
Solvent					
18. Water	18.0	15.59	1.33250	0.9971	29.99

[a]$v = (MW/N\rho)\,10^{24}$ in Å^3.
[b]For nonpolar hydrocarbon of same geometry.
[c]$MR = (MW/\rho)(n_D^2 - 1)/(n_D^2 + 2) = (4/3)\pi N\alpha$.
[d]$[P] = \gamma^{1/4}MW/(\rho_l - \rho_v)$, where γ = surface tension, ρ_l = liquid density, and ρ_v = vapor density. For $\rho_l + \rho_v$, $[P] \approx \gamma^{1/4}V$ and for $\gamma = 1$, $[P] = V$.
[e]$\log P = \log$ (concentration of solute in octanol phase/concentration of solute in water phase), i.e., P is octanol–water distribution coefficient.
[f]The acid dissociation constants $10^5 K_{HA}(H_2O)$ and 25°C for Compounds 13–17 are 17.12, 1.76, 1.34, 1.50, and 1.38, respectively.
[g]Estimated.

The partition coefficient, log P, of a solute distributed between a nonpolar phase such as octanol and a polar phase such as water is an obvious choice for correlating the adsorption of organics onto the nonpolar activated-carbon sorbent from the polar aqueous solution.

The parachor $[P]$ was developed in the 1920's as a means for calculating molecular volumes (56). It is not dimensionally correct to consider $[P]$ as a volume. More accurately, the ratio of the parachors of

and Solvent Properties

Accentric[b] Factor, w (39)	Dipole Moment, μ (40) Debye	Molar[c] Refraction, MR cm³/mol	Parachor[d] [P] calc (42)	Partition[e] Coeff. log P (43,44)	Polarization, α
0.098	1.78	8.2194	85.3	−0.75	3.26
0.152	1.68	12.9402	125.3	−0.21	5.13
0.176	1.72	17.6182	161.6	0.11	6.98
0.193	1.63	17.5579	165.3	0.33	6.96
0.197	1.67	22.2339	197.9	0.43	8.81
0.227	1.66	22.1341	201.6	0.65	8.77
0.227	1.67[g]	22.2123	201.6	0.65	8.81
0.251	1.60	22.1543	205.3	0.87	8.78
0.231	1.59[g]	26.7219	237.9	0.97	10.59
0.296	1.59	26.8171	245.3	1.41	10.63
—	1.55	31.4288	285.3	1.95	12.46
—	—	31.3755	281.6	1.82	12.44
0.152	1.46	8.5572	94.0	−1.09	3.39
0.176	1.05	13.0064	131.2	−0.20	5.16
0.227	0.89	17.5642	169.0	0.34	6.96
0.279	0.94	22.2023	208.6	0.88	8.80
0.330	0.94[g]	26.8502	247.0	1.42	10.64
0.023	1.84	—			

two compounds is equal to the ratio of the molecular volumes under conditions such that the surface tension is equal or unity. Hence, the parachor may be thought of as a molecular volume at a corresponding state with respect to surface tension, rather than temperature or pressure (*44*).

It should also be noted that the MR, α, and molecular volume, ν, parameters are included in the comprehensive $c\phi$ theory. (*See* Equations 4 and 15).

Molecular Cavity Surface Area. As described during the development of the $c\phi$ theory, the thermodynamic microsurface area change, ΔA, resulting from the association of solute and sorbent is an important newly introduced measurable quantity. This contact area between solute and sorbent is not directly known. It depends not only on the interactive forces between solute and sorbent but also between solute and solvent. It also depends on the position of a polar moiety on the solute, on the solute's morphological characteristics (flexibility and length to diameter ratio), and the sorbents' surface morphology and cavity volume. Thus, the system will try to maximize both the contact area and the solvent interaction (12).

To calculate $\Delta G_{cav,i}^{net}$ for species i an estimate of ΔA is necessary. See, for example, the fourth term on the right-hand side of Equation 15 ($N\gamma\Delta A$). Horváth et al. (15) plotted log $\bar{p}_i$ (for HPLC) versus the hydrocarbonaceous surface area (HSA) of the solutes and obtained linear parallel plots for closely related aromatic solutes, carboxylic acids, amino acids, and amines. The identity of the slopes for constant solvent surface tension $\gamma(= 32.6$ dynes/cm) indicated that ΔA was proportional to HSA [i.e., $\Delta A = g(\text{HSA})$] for homologous series. By using the same approach, very low g values are obtained for adsorption of alkyl aliphatic compounds. This is probably due to two factors: the unlikely assumption that the g is constant, and the fact that the curvature correction to the surface tension falls away. Using a phenomenological approach, Altshuler and Belfort (24) derived an equation of the form

$$\log \bar{p}_i = mA_i + b' \tag{28}$$

where the slope is given by:

$$m = - \frac{N\gamma_{wh}}{RT} [(1 - q) - g] \tag{29}$$

and the curvature correction is given by:

$$q = \left[1 + \frac{2}{R} \right]^{-1} = \left[1 + \frac{2}{\left(\dfrac{A_i}{4\pi} \right)^{1/2}} \right]^{-1} \tag{30}$$

where γ_{wh} is the interfacial tension between water and a hydrophobic surface (51.1 ergs/cm^2), k is the Boltzmann's constant of 1.38×10^{-16} ergs/°K, T is the absolute temperature in °K, R is the radius of the molecules in Å and equals $(A_i/4\pi)^{1/2}$ for a sphere, and A_i is TSA or HSA in Å^2.

Thus, q, the curvature correction, can be estimated for a spherical compound from its A_i in Equation 30. From a plot of log $\bar{p}_i$ versus A_i, the slope m is obtained allowing g to be obtained directly. This approach was taken here in calculating explicit values of q and g and thus $\Delta G_{cav, i}^{net}$ ($\equiv$ CAV).

The molecular cavity surface areas of each solute [i.e., total surface areas (TSA) or hydrocarbonaceous surface areas (HSA)] were calculated using the computer program developed and supplied by Hermann (57, 58). The method, based on intersecting atomic spheres with appropriate crowding factors, requires as input atomic radii, bond lengths, and bond angles, and calculates the detail coordinates for each atom in the molecule and the surface areas. The preparation of input data is time-consuming and becomes more so with complicated molecules. A molecular-build program was therefore developed to provide these data automatically.

Ionization of Carboxylic Acids. The initial concentrations used by Giusti et al. (32) for their adsorption measurements for the carboxylic acids (No. 13–17) listed in Table III were 0.0217, 0.0166, 0.0135, 0.0114, and 0.0098 g-mol/L, respectively. With these values and the acid dissociation constants $K_{HA}^{25°C}$ (H_2O) given in footnote f of Table III, their respective degrees of dissociation $\alpha°$ are 8.5, 3.2, 3.1, 3.6, and 3.7%. Since these weak acids are hardly dissociated for a nonbuffered system, they only represent a tentative test of the theory for the charged species contribution in Equations 4b and 5.

Equation 15 is used to calculate the contribution of each free energy change term to the overall standard unitary free energy change for the adsorption process as a function of carbon number for normal aliphatic alcohols and organic acids. Combining Equations 15 and 24 gives:

$$\ln \bar{p}_i = \text{NET} - (\Delta G_{(\text{gas})}^{\text{assoc}} + \theta)/RT \qquad (31)$$

where:

$$\text{NET} = \text{VDW} + \text{ES} + \text{CAV} + \text{CRAT}$$
$$\text{NET} = + \Delta G_{\text{vdw},i}/RT - \Delta G_{es,i}/RT$$
$$+ [4.8364 N^{1/3} V^{2/3} (K^e - 1)\gamma + N\gamma\Delta A]/RT + \ln RT/(P_0 V) \qquad (32)$$

where:

$$A_{S_i} = 4.8364 v_{S_i}^{2/3} \text{ (for a sphere)} \qquad (33)$$

and $\Delta G_{es,i}$ is given by Equation 5.

The adsorption data for a homologous series of linear and branched aliphatic alcohols and of carboxylic acids, together with the solvophobic calculations, are presented in Table IV.

Table IV. Adsorption Data of Aliphatic Alcohols and Carboxylic Acids with Solvophobic Calculations

| No. Adsorbate | Adsorbability[a] | | | Cavity Surface Area[b] | | Solvophobic Free Energy Change, $\Delta G/RT$[c] | | | | Curvature[d] Correction, q | Fractional[e] Contact Area, g |
	$\ln 10^4 Q_m$	$\ln Q_0 b$	$\ln K_f$	TSA, Å²	HSA, Å²	CAV	VDW	ES	NET		
Aliphatic Alcohols											
1. Methanol	0.7839	—	—	171.74	105.47	8.03	−6.27	−0.58	8.39	0.6489	0.3023
2. Ethanol	1.4679	—	—	209.19	149.96	8.86	−10.07	−0.36	5.64	0.6711	0.2802
3. 2-Propanol	1.4255	−2.3313	−2.3860	236.20	186.53	9.41	−13.02	−0.29	3.32	0.6843	0.2689
4. 1-Propanol	1.8437	−1.9285	−2.0636	241.05	181.82	9.50	−14.04	−0.27	2.41	0.6865	0.2647
5. 2-Methyl-2-propanol	2.0744	—	—	260.25	216.86	9.87	−15.79	−0.22	1.07	0.6947	0.2565
6. 2-Butanol	2.4283	−0.6906	−1.1712	263.67	217.92	9.94	−17.94	−0.23	−1.02	0.6961	0.2552
7. 2-Methyl-1-propanol	—	—	—	265.33	206.10	9.97	−18.07	−0.23	−1.12	—	—
8. 1-Butanol	2.670	0.0112	−0.7508	272.89	213.66	10.10	−19.69	−0.21	−2.59	0.6997	0.2515
9. 2-Methyl-2-butanol	—	—	—	284.96	241.57	10.32	−19.93	−0.17	−2.57	—	—

10. 1-Pentanol	2.8662	0.8942	0.1655	304.75	245.52	10.67	−24.46	−0.19	−6.77	0.7112	0.2401
11. 1-Hexanol	2.9280	1.4664	0.5653	336.53	273.30	—	—	—	—	0.7212	0.2300
12. 3-Methyl-1-pentanol	2.8112	—	—	317.73	258.50	—	—	—	—	0.7154	0.2358
Carboxylic Acids											
13. Formic acid	2.3243	—	—	163.07	36.91	8.28	−8.44	−0.61	6.44	0.6430	0.3309
14. Acetic acid	2.0782	—	—	205.14	91.31	9.35	−11.08	−0.21	5.27	0.6589	0.3050
15. Propanoic acid	2.1713	—	—	236.98	101.63	10.09	−15.72	−0.14	1.44	0.6877	0.2893
16. Butanoic acid	2.6034	—	—	278.96	204.44	11.00	−21.36	−0.14	−3.29	0.7020	0.2719
17. Pentanoic acid	2.7453	—	—	300.68	198.36	11.44	−28.02	−0.13	−9.49	0.7008	0.2641

[a] Adsorbability data, Q'_m, obtained from Giusti et al. (32), and Q_0, b, K_f, and n values obtained from Arbuckle and Romagnoli (38).
[b] TSA = total surface area of molecule, HSA = total surface area (TSA) − functional surface area (FSA) = hydrocarbonaceous surface area.
[c] Terms are defined in Equations 31 and 32. CRAT = $\ln(RT/P_oV)$ = 7.21.
[d] See Equation 30.
[e] See Equation 29.

Results and Discussion

In our previous paper (3), the components or terms comprising the net standard unitary free energy shown in Equation 32 were calculated for linear members of the following aliphatic homologous series: alcohols (six compounds), organic acids (four), aldehydes (four), and methyl ketones (three). Correlations with the logarithm of adsorption capacity for each term (NET, VDW, CAV, and ES) and for molecular weight were compared. Although the number of compounds compared in each group was limited (by the availability of reliable adsorption data), the correlation coefficient, r^2, for the cavity term (CAV), on an average ranked the highest. Because of this and because the cavity term is proportional to the cavity surface area, the simplified analytical expression for different solutes on the same carbon as expressed by Equation 25 was invoked and tested. In an overwhelming number of cases, the simplified approach successfully correlated (with $r^2 > 0.9$) the adsorption data within homologous aliphatic and aromatic groups.

Here we extend the calculations, and hence the test of the theory, to include linear and branched members of a relatively larger set of aliphatic alcohols (eight compounds). In addition, since the theory has been modified to include ionized and partially ionized species, calculations similar to the alcohols are presented for carboxylic acids (five compounds).

In Figure 2, the normalized free energy change for each term in Equation 32 is plotted as a function of carbon number for alkyl alcohols and carboxylic acids. In both systems, the van der Waals dispersion term, VDW, is dominating and always opposite in sign to the cavity term. The effect of branching on the free energy changes (VDW and NET) also is shown in Figure 2a for the alcohols. In all cases, branching (Compounds 3, 5, 6, 7, and 9) results in an increase in free energy change from the normal linear alcohols (Compounds 1, 2, 4, 8, and 10). Thus, dispersion forces dominate during activated-carbon adsorption, and the free energy change is increased as a result of branching. This increase should increase solute solubility and decrease adsorption, and the surface and inverse volume forces counterbalance each other.

By using the results in Table IV and Figure 2, the comprehensive solvophobic model and each term therein are correlated with adsorption capacity ($\ln \bar{p}_i$). Correlations with two sets of data (32, 38) giving three adsorbability parameters ($Q'_m, \overset{\circ}{Q}b$, and K_f) are summarized in Table V. For comparison purposes, several group contribution and other parameters also are correlated with adsorption capacity ($\ln \bar{p}_i$). These results also are included in Table V.

As an example, the extent of adsorption ($\ln \bar{p}_i$) is plotted for two different sets of adsorption data versus molecular weight, total surface

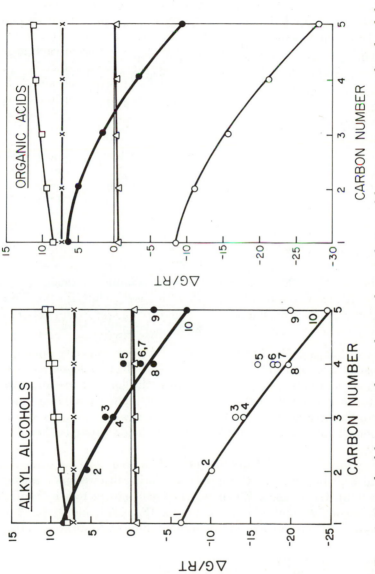

Figure 2. Normalized free energy changes showing the contributions of the interaction-terms to the solvophobic adsorption model (NET = VDW + ES + CAV + CRAT see Equation 32). Key: □, CAV; ×, CRAT; ●, NET; △, ES; and ○, VDW.

area, and $\Delta G/RT$ (NET and VDW) in Figure 3. The effects of branching are clear, since compounds of the same molecular weight show different adsorption capacities. TSA and the solvophobic terms compensate for branching and consequently result in higher linear correlation coefficients, r^2. The plotted point for 2-propanol (Compound 3) deviates from the linear plot for the Giusti et al. (32) data (Q'_m) but less so for the Arbuckle and Romagnoli data (Q^ob) (38). The deviation is thus probably due to experimental error.

Branching. On careful analysis of the data presented in Table V, the experimentally determined log P values followed by NET, VDW, and CAV gave, on an average for all alcohol correlations (Groups 1a–1d)[1], the highest linear correlations efficients, r^2. TSAs always correlated better than HSAs and other than the solvophobic terms (NET, VDW, and CAV) and log P, TSA correlates better than all other parameters.

The only solvophobic term that gives poor correlation with adsorption is ES. For the Giusti et al. (32) data (unbuffered), r^2 for ES is ranked 12th twice (Groups 1a and 1b); for the Arbuckle and Romagnoli (38) data (buffered to pH 7), r^2 for ES is ranked much higher at 1st and 5th place (Group 1c and 1d).

The results for Group 1b represent the most stringent test for correlating branched molecules for the data set in Table IV. From the r^2 and their ranking in Table V, it is clear that both the comprehensive solvophobic model (NET and VDW) and the simplified model (TSA) exhibit the highest correlations together with log P^2. Thus, for the group contribution and other parameters, log P and to a lesser extent [P] show significantly higher correlations than molecular weight.

Ionized Solutes. In Figure 3, the adsorption capacity (ln $\bar{p}_i$) is plotted against TSA and VDW and NET for the solvophobic model. In all cases, formic acid (Compound 1) is not near the linear correlation and appears to have a much higher adsorbability than expected. Unusual behavior of the first member of a homologous series has been noted (59). For the correlations shown in Table V, the result for formic acid was not included.

Since Giusti et al. (32) failed to buffer their solutions, the carboxylic acids were probably only slightly ionized (*see* earlier discussion in solvent–solute section). Once again, ES shows very poor correlation in Table V (Group 2a), while the highest r^2 values are for TSA, CAV, and V_{molar}. These results should only be considered preliminary since not enough data (only

---

[1]Average rankings for the four alcohol correlations are obtained by summing the individual rankings and dividing by four and then reranking.

[2]The good correlation with log P may be biased for the aliphatic alcohol homologous series because log P is a measure of the distribution of a solute between water and octanol which is itself a member of the series.

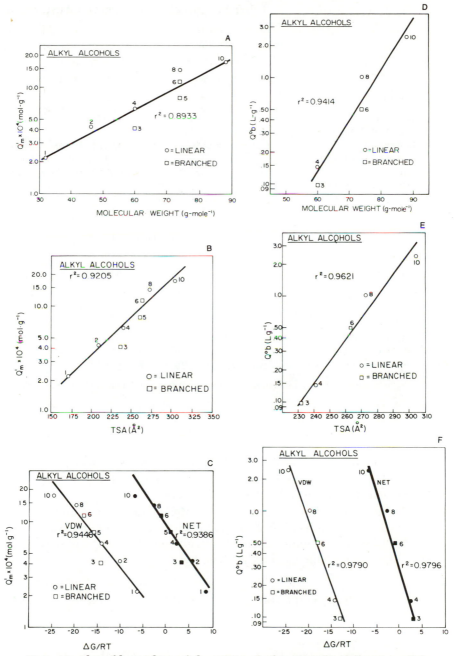

Figure 3. Plot of logarithim of the extent of adsorption for different sets of data vs. molecular weight, total surface area (TSA) and solvophobic free energy changes (NET and VDW) for linear and branched aliphatic alcohols (Q'_m and $Q°b$ adsorption data from Refs. 32 and 38, respectively).

Table V. Correlations Between Adsorption

Group Compounds[a]	No. of Compounds	Adsorp. Capac.[b] $\ln p_i$	r^2	MW	ρ	n_D
Aliphatic Alcohols						
1a. Linear	5	$\ln 10^4 Q'_m$	r^2	0.9733	0.8686	0.9507
(1,2,4,8.10)			rank[d]	8	13	11
1b. Linear and	8	$\ln 10^4 Q'_m$	r^2	0.8933	0.6024	0.8950
branched (1–6,8,10)			rank	8	13	7
1c. Linear and	5	$\ln Q° b$	r^2	0.9414	0.7609	0.9683
branched (3,4,6,8,10)			rank	9	13	4
1d. Linear and	5	$\ln K_f$	r^2	0.9652	0.7534	0.9726
branched (3,4,6,8,10)			rank	9	13	8
			1/4 Σ			
			rank	8.5	13	7.5
			Ave. Ranking[e]	10	13	7
Carboxylic Acids						
2a. Linear (14–17)	4	$\ln 10^4 Q'_m$	r^2	0.9356	0.8650	0.9062
			rank	8	12	11

[a]Compounds listed by number in Table IV.
[b]Data for adsorption isotherm models. Freundlich (K_f, n) and Langmuir $(Q°, b)$ values and adsorption extent (Q'_m) are found in Table IV.
[c]Data for correlation parameters MW, ρ, n_D, V_{molar}, MR, [P], and log P are given in Table III, and the solvophobic components and cavity surface areas are given in Table IV.
[d]Ranked according to r^2 value.
[e]Average rankings obtained by summing individual rankings and dividing by four, and then reranking.

four compounds) were available and the adsorption experiments were not buffered. The theory (*see* Equation 5) accounting for polar–polar or ionized species must still be thoroughly tested.

Predictions. The correlations presented in Table V can now be used to estimate the adsorbability of compounds whose experimental values are not available. For example, the TSA correlation for Group 1b in Table V for branched aliphatic alcohols is given by:

$$\ln 10^4 Q'_m = 0.0166 \text{ TSA} - 2.1137 \qquad r^2 = 0.9205 \qquad (34)$$

Equation 34 can be used to estimate the adsorbability $10^4 Q'_m$ directly from chemical structure. For example, using the molecular-build program coupled to Hermann's cavity area program surface areas (TSA), 2-methyl-1-propanol, 2-methyl-2-butanol, and 3-methyl-1-pentanol are calculated

Capacity and Solute Characteristic Properties

				Correlation Parameters[c]					
V_{molar}	MR	[P]	log P	CAV	ES	VDW	NET	TSA	HSA
0.9752	0.9735	0.9735	0.9735	0.9786	0.8912	0.9680	0.9583	0.9766	0.9769
4	5	5	5	1	12	9	10	3	2
0.8673	0.8900	0.9138	0.9458	0.9133	0.7786	0.9446	0.9386	0.9205	0.8625
10	9	5	1	6	12	2	3	4	11
0.9246	0.9396	0.9576	0.9683	0.9681	0.9871	0.9790	0.9796	0.9621	0.9048
11	10	8	4	6	1	3	2	7	12
0.9497	0.9641	0.9807	0.9892	0.9900	0.9871	0.9956	0.9959	0.9865	0.9289
11	10	7	4	3	5	2	1	6	12
7.5	8.5	6.25	3.50	4.00	7.5	4.00	4.00	5.00	9.25
7	10	6	1	2	7	2	2	5	12
0.9537	0.9367	0.9395	0.9356	0.9553	0.5615	0.9361	0.9280	0.9622	0.9499
3	6	5	8	2	13	7	10	1	4

to be 265.33, 284.96, and 317.73 Å^2, respectively. Substitution into Equation 34 for TSA gives the estimated adsorbabilities of 2.2908, 2.6166, and 3.1606 g-mol/g (as $10^4 Q'_m$), respectively. Comparison of $10^4 Q'_m = 3.1606$ g-mol/g with the experimental value of 2.8112 g-mol/g given in Table IV for 3-methyl-1-pentanol results in an error of about 12%.

Statistics. Because of the small number of observed data pairs (i.e., number of compounds studied in a group), the probability that the hypothesis $r_1 = r_2 = , \ldots , = r_i$ cannot be rejected increases significantly with a decrease in the value of r. Assuming a bivariate normal distribution, the hypothesis $r = 0$ is however rejected in all cases studied (62). As a result of this, less weight should be given to differences in r^2 values than to the fact that the $c\phi$ theory parameters for different data banks and adsorption isotherm models consistently show the highest correlation among the correlation parameters shown in Table V. Clearly, more reliable data are needed to test the $c\phi$ theory with greater confidence. Such a data bank is currently being assembled from equilibrium adsorption experiments conducted in the authors' laboratory.

Conclusions

The work reported here has to do with the development of a comprehensive theoretical basis for predicting a priori the preferential

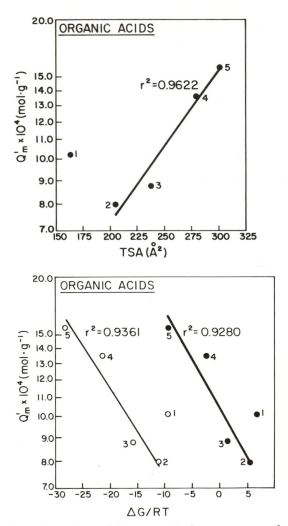

Figure 4. Plot of logarithim of the extent of adsorption vs. total surface area (TSA) and solvophobic free energy changes (NET and VDW) for linear carboxylic acids (Q'_m adsorption data from Ref. 32). Key: ○, VDW; and ●, NET.

adsorption of organic compounds onto activated carbon from dilute aqueous solutions. A detailed analysis of the effect of the solvent on this process yields an expression for the extent of adsorption with essentially no adjustable constants. Branching, cyclization, and position isomerism are automatically accounted for without introducing additional terms or adjustable constants. In principal, the theory incorporates the variation of a wide span of experimental conditions including a comparison of the adsorption of similar organic compounds on different activated carbons and a comparison of the adsorption of different organic solutes on the same activated carbon. Under special conditions and assumptions, simplified analytical expressions result, allowing the prediction of the extent of adsorption as a function of the simple geometric characteristics of the hydrocarbonaceous moiety of the solute.

In this paper, the testing of the solvophobic ($c\phi$) theory is extended to include the adsorption of branched alkyl alcohols and slightly ionized linear carboxylic acids. Correlations with adsorption capacity for both the comprehensive and the simplified $c\phi$ theories are compared with similar correlations for other independent variables including molecular weight, density, index of refraction (or polarizability), octanol–water partition coefficient, and parachor. On the average for the alcohol homologous series, the best correlative parameters were octanol–water partition coefficient ($\log P$), three solvophobic terms (NET, VDW, and CAV), and the total cavity surface area (TSA). A simplified expression requiring only the total cavity surface area is used to predict the unknown adsorbability of two branched alcohols and the known adsorbability within 12% of another branched alcohol.

Although not tested convincingly for ionized and partially ionized solutes because of the lack of adequate adsorption data, the $c\phi$ theory has been extended to include ionization.

A molecular-build program has been developed and used to automatically provide the input data (molecular structure, atomic radii, bond lengths, and bond angles) to the Hermann surface area program. This program greatly simplifies the complexities in obtaining the total cavity surface areas needed for estimating solute adsorbabilities.

With regard to the curvature correction needed to account for microscopic surface tension effects, a new phenomenological method was introduced and used. This circumvents a potential problem discovered in the original theoretical development.

The major limitation in testing and developing the $c\phi$ theory is the lack of adequate reliable equilibrium adsorption data. In this regard, the authors are at present involved in developing a comprehensive experimental data base for specific homologous groups. Competition, branching, aliphatic–aromatic mix of a solute, and position of an organic group relative to the functional group on an aromatic ring are some variables being tested. Additional work is needed to study the adsorption of similar organic homologues on different adsorbents.

Nomenclature

a_i distance of closest ionic approach
A_i surface area of species i

Nomenclature—Continued

b_i ionic radius

b Langmuir isotherm constant

D $[2(\varepsilon - 1)/2\varepsilon + 1]$

e electronic charge

g ratio of microsurface area change of reaction to solute hydrophobic surface area

g' ratio of microsurface area change of reaction to solute total area

HSA   hydrophobic surface area of molecule

I ionization potential

k Boltzmann's constant

k_i^H Henry's constant for the ith species

K_i^e correction factor for energy of expanding highly curved solute cavity surface

K^e same as above, for solvent

K_a geometric ratio of surface area to volume$^{2/3}$

$K_{(solvent)i}$ equilibrium constant for the adsorption association reaction

K_f Freundlich isotherm constant

K_{HA} acid dissociation constant

$\log P$ logarithm of octanol–water partition coefficient

MR   molar refraction

MW   molecular weight

n Freundlich isotherm constant

n_D index of refraction

N Avogadro's number

p partial pressure

$\bar{p}_i$ adsorption capacity

P_0 pressure

$[P]$ parachor

q alternative correction factor for surface energy at curved surface

Q° Langmuir isotherm constant

R universal gas constant

TSA   total surface area of solute cavity

V molar volume

W curvature correction factor

X mole fraction

z ionic valency

Greek

α atomic polarizability

Nomenclature—Continued

$\alpha°$ ionic degree of dissociation
γ interfacial surface energy
ΔA microsurface area change of reaction
ΔG Gibbs free energy change of reaction
ε_i static dielectric constant of solute i
ε_0 permittivity of vacuum
θ experimental adsorption capacity constant
κ Debye screening length
λ ratio of volume of carbon–solute complex to that of solute
μ_i static dipole moment
ν_i molecular volume
ρ density

Subscripts

cav   cavity formation
es   electrostatic interactions
gas   gas phase process
j j-th interaction
mix   mixing effects
red   dispersive reduction due to solvent
solvent   solvent phase
solvent effect   net effect of solvent on process
vdw   van der Waals interactions

Superscripts

assoc   for adsorption association process
net   sum of all interactions

Acknowledgment

The authors thank James Nicoletti and Bob Tonti for computer programming assistance and Jesse M. Cohen, James J. Westrick, and Richard A. Dobbs of the U. S. Environmental Protection Agency,

Literature Cited

1. Sinanoğlu, O. In "Molecular Associations in Biology," Pullman, B. Ed., Academic, New York, N.Y., 1968, pp 427–45.

2. Belfort, G., *Environ. Sci. Technol.* **1979**, *13*(8), 939–46.
3. Belfort, G., "Selective Adsorption of Homologues onto Activated Carbon from Dilute Aqueous Solutions. Solvophobic Interaction Approach—II, Development and Test of Theory," presented before the Env. Chem. Symp. ACS 179th National Meeting, Houston, Tex. (March 23–28, 1980) to be published in "Chemistry in Water Reuse", Cooper, W. J., Ed., Ann Arbor Science, Ann Arbor, Mich., 1981, Volume 2.
4. Miller, S., "Adsorption on Carbon: Solvent Effects on Adsorption (An Interview with Dr. Georges Belfort)," *Environ. Sci. Technol.* **1980**, *14*(9), 1037–49.
5. Ibid, **1980**, *14*(8), 910–14.
6. Hermann, R. B. *J. Phys. Chem.* *76*(19) 2754–2759, **1972**, *76*(19), 2754–9.
7. Freundlich, H., "Colloid and Capillary Chemistry," Methuen and Co., Ltd., London, England, 1926.
8. Defay, R.; Prigogine, I.; Bellemans, A.; D. H. Everett (Translator), "Surface Tension and Adsorption," Wiley, New York, N.Y., 1966.
9. McGuire, M. J. Ph.D. Thesis Drexel University, Philadelphia, Pa., 1977.
10. McGuire, J. J.; Suffet, I. H.; Radziul, J. V. *J. Am. Water Works Assoc.* **1978**, *70*, 565.
11. McGuire, M. J.; Suffet, I. H. In "Activated Carbon Adsorption of Organics from the Aqueous Phase," Suffet, I. H., McGuire, M. J., Eds., Ann Arbor Science, Ann Arbor, Mich., 1980, Volume 1, Chapter 4, pp 91–115.
12. Melander, W.; Horváth, C. In "Activated Carbon Adsorption of Organics from the Aqueous Phase," Suffet, I. H.; McGuire, M. J., Eds., Ann Arbor Science, Ann Arbor, Mich., 1980, Volume 1, Chapter 3, pp 65–89.
13. Sinanoğlu, O. *Int. J. Quantum Chem.* **1980**, *XVIII*, 381–92.
14. Richards, F. M. *Ann. Rev. Biophys. Bioeng.* **1977**, *6*, 1151–76.
15. Horváth, C.; Melander, W.; Molnar I. *J. Chromatog.* **1976**, *125*, 129.
16. Melander, W.; Horváth, C. *Arch. Biochem. Biophys.* **1977**, *183*, 200–15.
17. Birnstock, F.; Hofmann, H. J.; Köhler, H. J. *Theoret. Chim. Act (Berlin)*, **1976**, *42*, 311–23.
18. Halicioğlu, T.; Sinanoğlu, O. *Ann. N.Y. Acad. Sci.* **1969**, *158*, 308–17.
19. Sinanoğlu, O.; Abdulnur, S. *Photochem. Photobiol.* **1964**, *3*, 333–42.
20. Sinanoğlu, O.; Abdulnur, S.; Kestner, N. R. In "Electronic Aspects of Biochemistry," Pullman, B., Ed., Academic, New York, N.Y., pp 301–11.
21. Edsall, J. T.; Wyman, J. "Biophysical Chemistry," Academic, New York, N.Y., 1958, pp. 282–96.
22. Gordon, J. E. "The Organic Chemistry of Electrolyte Solutions," Wiley, New York, N.Y., 1975, pp 38–42.
23. Halicioğlu, T. Ph.D. Thesis, Yale University, New Haven, Conn., 1969.
24. Altshuler, G.; Belfort, G. "Surface Tension and Aqueous Activated Carbon Adsorption" presented at the Fifteenth Biennial Conference on Carbon, The University of Pennsylvania, Pa., June 22–26, 1981.
25. Fritz, W.; Schluender, E. U. *Chem. Eng. Sci.* **1974**, *29*, 1279.
26. Morris, J. C.; Weber, W. J., Jr., "Adsorption of Biochemically Resistant Materials from Solution, Part 1," U.S. Public Health Service, AWTR-9, Dept. 999-WP-11, 1964.
27. Jäger, I.; Erdös, E. *Coll. Czech. Chem. Comm.* **1959**, *24*, 3019.
28. Radke, C. M.; Prausnitz, J. M. *Am. Inst., Chem. Eng. J.* **1972**, *18*(4): 761–8.
29. Al-Bahrani, K. S.; Martin, R. J. *Water Res.* **1976**, *10* 731–6.
30. Martin, R. J.; Al-Bahrani, K. S. *Water Res.* **1977**, *11*, 991–9.
31. Martin, R. J.; Al-Bahrani, K. S. *Water Res.* **1978**, *12*, 879.
32. Giusti, D. M.; Conway R. A.; Lawson C. T. *J. Water Poll. Control Fed.* **1974**, *46*(5), 947–65.
33. Weber, W. J., Jr.; Van Vliet B. M. presented at the symposium on Activated Carbon Adsorption of Organics from the Aqueous Phase, American Chemical Society 176th National Meeting, Miami Beach, Fla., Sept 10–15, 1978.

34. Jossens, L.; Prausnitz, J. M.; Fritz, W.; Schluender E. U.; Meyers, A. L. *Chem. Eng. Sci.* **1978**, *33*, 1097.
35. Dobbs, R. A.; Middendorf R. J.; Cohen J. M. "Carbon Adsorption Isotherms for Toxic Organics," U.S. EPA, MERL, Cincinnati, Ohio, 1978.
36. Dobbs, R. A.; Cohen, J. M. "Carbon Adsorption Isotherms for Toxic Organics," U.S. EPA, MERL, Cincinnati, Ohio, EPA-600/8-80-023, 1980.
37. Myers, A. private communication, 1981.
38. Arbuckle, W. B.; Romagnoli, R. J. "Predication of the Preferentially Adsorbed Compounds in Bisolute Column Studies," presented at 87th American Institute of Chemical Engineers National Meeting, Boston, Mass. August 19–22, 1979.
39. Vedeneyev, A. L.; Gurvich, L. V.; Kondrat'yev, V. N.; Medvedev, V. A., Frankevich, Y. L. "Bond Energies, Ionization Potentials, and Electron Affinities," St. Martin's Press, New York, N.Y., 1966.
40. "Handbook of Chemistry and Physics," 57th ed., CRC Press, Inc., Cleveland, Ohio, 1976.
41. Reid, R. C.; Sherwood, T. K. "The Properties of Gases and Liquids," 2nd ed., McGraw-Hill, New York, N.Y., 1966.
42. McClellan, A. L. "Tables of Experimental Dipole Moments" Freeman, San Francisco, Calif., 1963.
43. Hill, N. E. "Dielectric Properties and Molecular Behavior," Van Nostrand Reinhold, London, England, 1969, Chapter 3.
44. Quayle, O. R. *Chem. Revs.* **1953**, *53*, 439.
45. Leo, A.; Hansch, C.; Elkins, D. *Chem. Revs.* **1971**, *71*(6), 525.
46. Leo, A.; Jow P.Y.C.; Silipo, C.; Hansch, C. *J. Med Chem.* **1975**, *18*(9), 865.
47. Meissner, H. P. *Chem. Eng. Prog.* **1949**, *45*(2), 149.
48. Gold, P. I.; Ogle, G. J. *Chem. Eng.* 185 (Nov. 4, 1968).
49. Ibid, 119 (Jan. 13, 1969).
50. Ibid, 130 (April 7, 1969).
51. Reid, R. C.; Prausnitz, J. M.; Sherwood, T. K. "The Properties of Gases and Liquids," 3d Ed., McGraw-Hill, New York, N.Y., 1977.
52. Fedors, R. *Polym. Sci. Eng.* **1974**, *14*(2), 147.
53. Abrams, D. S.; Prausnitz, J. M. *AIChE J.*, **1975**, *21*(1), 116.
54. Yoshimoto, M.; Hansch, C. *J. Org. Chem.* **1976**, *41*(13), 2269.
55. Leo, A.; Hansch, C.; Church, C. *J. Med. Chem.* **1969**, *12*, 766.
56. Sugden, S. *J. Chem. Soc.* **1924**, *125*, 1177.
57. Hermann, R. B. *J. Phys. Chem.* **1971**, *75*, 363.
58. Hermann, R. B. *J. Phys. Chem.* **1972**, *76*, 2754–9.
59. Benjamin, L. *J. Phys. Chem.* **1964**, *68*, 3575–81.
60. Peel, J. B.; Willett, G. D. *Aust. J. Chem.* **1975**, *28*, 2357.
61. "Selected Values of Properties of Chemical Compounds," Manufacturing Chemists Association Research Project, Chemical Thermodynamics Properties Center, Texas A & M University, College Station, Tex., 1966.
62. "Documenta Geigy, Scientific Tables," 6th ed., Diem, K., Ed., J.R. Geigy, S. A., Basle, Switzerland, 1962.

RECEIVED for review August 3, 1981. ACCEPTED for publication March 5, 1982.

Theory of Correspondence for Adsorption from Dilute Solutions on Heterogeneous Adsorbents

A. L. MYERS

University of Pennsylvania, Chemical Engineering Department, Philadelphia, PA 19104

S. SIRCAR

Air Products and Chemicals, Inc., Allentown, PA 18105

The thermodynamics of adsorption from dilute solutions can be formulated analogous to that for adsorption from pure vapors. A principle of correspondence for adsorption of dilute solutes is derived. It is found that K functions [K = (mRT/ΔG) ln a] of different solutes coincide. Adsorption isotherms can be predicted from the K function of a reference solute. The constants of the theory are the saturation capacity of the adsorbent for the solute and the free energy of immersion of the adsorbent in the pure solute.

E QUILIBRIUM PROPERTIES OF ADSORBENTS play an important role in the performance of adsorbers and chromatographs. Experimental determination of equilibrium data is tedious and costly. Design engineers are interested in the development of theories that can be used to predict equilibrium adsorptive data from a minimum of information. Several theories have been proposed for the calculation of multicomponent vapor–solid (1) and liquid–solid (2–4) adsorptive behavior. The starting point for all of these methods is either pure vapor adsorption isotherms or adsorption from binary liquid mixtures. Thus, the minimum information referred to is the adsorption of each component of a vapor mixture or, in the case of liquid adsorbates, the selectivity of the solid surface for each solute–solvent pair.

0065-2393/83/0202-0063$06.00/0

In a search for a simpler approach, several workers (5–9) tried to generate "characteristic adsorption isotherms" for the vapor–solid interface, so that isotherms of different vapors can be coalesced into a single curve for a particular adsorbent. The idea is to search for certain constants called affinity coefficients that can be used to define new, reduced adsorptive properties. If such a characteristic curve could be defined and if the affinity coefficients could be correlated with the properties of the adsorbate, then it would be possible to predict adsorption isotherms using a single measured isotherm of a reference substance. This process would provide an enormous simplification of the described design problem.

The procedure of finding affinity coefficients that will produce a characteristic isotherm for different adsorbates is empirical. The underlying Polanyi potential theory (10) is not actually a theory because its derivation is based upon so-called equipotential surfaces whose existence can be neither proved nor disproved. In this work, the objective of correlating experimental data is the same but the theory is based upon specific assumptions about the interactions of adsorbate molecules with the surface. Instead of adjustable parameters like affinity coefficients, the reducing variable is a thermodynamic quantity: the free energy of immersion. If our result is called a characteristic adsorption isotherm, it will be classified as another in a long line of extensions of the Polanyi theory (11). The Polanyi theory has indeed been useful, but after nearly 70 years of study its applications probably have been exhausted. To avoid the empirical connotation associated with Polanyi plots and characteristic isotherms, the approach discussed here is called a principle of correspondence because of its similarity with the principle of corresponding states for pure fluids. In both cases, the intention is to cast the volumetric properties of a fluid (adsorbate) in terms of dimensionless functions that are the same for all fluids (adsorbates).

Recently (8), we proposed a principle of correspondence for adsorption of pure vapors. Extension to the case of adsorption from liquids is complicated because of competition of the solvent for available surface area. Neglect of this competition is justifiable when the solvent is sterically excluded from the surface of the adsorbent, e.g., adsorption of water on 3A molecular sieve from a dilute solution of water in cyclohexane (12). In general, however, solvent molecules compete effectively for the surface even when the solute is much more strongly adsorbed than the solvent, because the solvent is present in relatively high concentration. Consider, for example, the mole fractions of solute and solvent in the adsorbed and bulk liquid phases in Table I. Here, the solute is much more strongly adsorbed than the solvent as indicated by a selectivity of 4000:

$$s_{1,2} = \frac{x_1/x_{10}}{x_2/x_{20}} = \frac{0.8/0.001}{0.2/0.999} = 4000$$

Table I. Typical Concentrations in Bulk Liquid and Adsorbed Phases for Preferential Adsorption of a Dilute Solute from a Liquid Solvent

	Adsorbed Phase, x_i	Liquid Phase, x_{io}
Solute 1	0.8	0.001
Solvent 2	0.2	0.999

but comparable amounts of solute and solvent are present in the adsorbed phase.

Three assumptions will be utilized in the theoretical development of the conditions necessary for the existence of a principle of correspondence or a "universal" adsorption isotherm for different vapors adsorbed on a particular adsorbent:

1. The solute is only sparingly soluble in the solvent, e.g., no more than a few mole percent.
2. The solute is strongly adsorbed relative to the solvent so that $s_{1,2} \gg 1$.
3. The surface of the adsorbent is heterogeneous.

These conditions are usually satisfied in practical separation problems for the separation of trace impurities such as dehydration of hydrocarbons and purification of water containing organic substances. It will be shown that these systems can be treated as the adsorption of a single component (the solute) by subtracting the properties of the solvent from the adsorption isotherm. This procedure permits the adsorption of single solutes to be treated in a manner analogous to the adsorption of single vapors. Therefore, procedures developed for the vapor–solid interface (8) may be applied to adsorption from liquids. First, it is necessary to consider the equations of thermodynamic equilibrium for adsorption from dilute solutions.

Thermodynamics of Adsorption: Excess Properties

The Gibbsian model of the liquid–solid interface is adopted (13). The surface excess for adsorption of Solute 1 or Solvent 2 is:

$$n_i^e = n(x_i - x_{i0}) \tag{1}$$

where n is the total amount adsorbed per unit mass of adsorbent, x_i is the mole fraction of solute or solvent in the adsorbed phase, and x_{i0} is the corresponding mole fraction in the bulk liquid phase at equilibrium. It follows from the definition that:

$$n_1^e + n_2^e = 0 \tag{2}$$

so that a positive surface excess of solute (n_1^e) implies an equal in magnitude but negative surface excess of solvent (n_2^e). Excess Gibbs free energy for adsorption from solution is defined as:

$$\Delta G^e = n(g - g_0) - \Sigma n_i^e \mu_i^\circ \tag{3}$$

where g is the molar Gibbs free energy of the adsorbed phase and g_0 is the value for the equilibrium liquid solution. Excess properties defined this way are consistent with the definition of surface excess in Equation 1. The terms g and g_0 refer to different compositions, and the purpose of the summation term is to cancel the standard-state chemical potentials (μ_i°) which arise due to this difference in composition. ΔG^e is always negative for the spontaneous process of adsorption. Similar excess functions can be defined for enthalpy and entropy. The Gibbs free energy of the adsorbed phase is defined by (10):

$$G = ng = \gamma A + n_1 \mu_1 + n_2 \mu_2 \tag{4}$$

Thus, the free energy of the adsorbed phase can be considered as the sum of contributions from the solid surface (γA), the solute $(n_1 \mu_1)$ and the solvent $(n_2 \mu_2)$. In Equation 4, γ is the surface tension of the liquid–solid interface, and A is the surface area per unit mass of adsorbent. In the case of microporous adsorbents, the meaning of A loses its significance but the product (γA) is always a measurable quantity (14).

The molar Gibbs free energy of the bulk liquid is:

$$g_0 = x_{10} \mu_{10} + x_{20} \mu_{20} \tag{5}$$

Because of equilibrium:

$$\mu_1 = \mu_{10} \text{ and } \mu_2 = \mu_{20} \tag{6}$$

Substitution of Equations 4–6 into Equation 3 yields:

$$\Delta G^e = \gamma A + n_1^e(\mu_1 - \mu_1^\circ) + n_2^e(\mu_2 - \mu_2^\circ) \tag{7}$$

For adsorption of pure solvent, $n_1^e = n_2^e = 0$ and:

$$\Delta G = \gamma_0 A \tag{8}$$

where γ_0 refers to the surface tension of the solvent–solid interface. To subtract the properties of the solvent from Equation 7, ΔG^e is redefined relative to the adsorption of pure solvent:

$$\Delta G^e = (\gamma - \gamma_0)A + n_1^e(\mu_1 - \mu_1^\circ) + n_2^e(\mu_2 - \mu_2^\circ) \tag{9}$$

so that $\Delta G^e = 0$ for adsorption of solvent alone. According to the assumption that the solute is present at low concentration, the mole fraction of the solvent is close to unity and $\mu_2 \simeq \mu_2^\circ$. Equation 9 reduces to:

$$\Delta G^e = (\gamma - \gamma_0)A + n_1^e(\mu_1 - \mu_1^\circ) \tag{10}$$

Since the solute concentration is assumed to be less than a few mole percent, its activity is given by Henry's law:

$$\mu_1 = \mu_1^\circ + RT\ln a_1 \tag{11}$$

where:

$$a_1 = \gamma_1^\infty x_{10} = \frac{x_{10}}{x_{10}^{\text{sat}}} = \frac{c_1}{c_1^{\text{sat}}} \tag{12}$$

and γ_1^∞, the activity coefficient at the limit of infinite dilution, is assumed to be constant up to the mole fraction at saturation, x_1^{sat}. Equation 10 becomes:

$$\Delta G^e = (\gamma - \gamma_0)A + n_1^e RT\ln a_1 \tag{13}$$

By using the fact that $x_{20} \simeq 1$, it follows from Equations 1 and 2 that:

$$n_1^e = -n_2^e = -n(x_2 - x_{20}) = nx_1 = n_1 \tag{14}$$

Thus, the surface excess of solute, n_1^e, is the actual amount of solute in the adsorbed phase, n_1, so Equation 13 may be written as:

$$\Delta G^e = (\gamma - \gamma_0)A + n_1 RT\ln a_1 \tag{15}$$

The surface tension of the interface may be calculated from the Gibbs adsorption isotherm (*10*):

$$-Ad\gamma = n_1 d\mu_1 + n_2 d\mu_2 \tag{16}$$

Since the selectivity of the adsorbent for the solute relative to the solvent, $s_{1,2}$, is large, the magnitude of the ratio:

$$\frac{n_2 d\mu_2}{n_1 d\mu_1} = \frac{x_2 d\ln a_2}{x_1 d\ln a_1} = \frac{1}{s}\frac{x_{20} d\ln x_{20}}{x_{10} d\ln x_{10}} =: \frac{1}{s}\frac{dx_{20}}{dx_{10}} = -\frac{1}{s}$$

is small and $n_2 d\mu_2$ may be neglected in comparison with $n_1 d\mu_1$. Therefore, Equation 16 reduces to:

$$-A d\gamma = n_1 d\mu_1 = n_1 RT d\ln a_1 \tag{17}$$

Integration of Equation 17 from immersion in pure solvent to a particular loading of solute gives:

$$-A(\gamma - \gamma_0) = RT \int_{n_1 = 0}^{n_1} n_1 d\ln a_1 \tag{18}$$

Substituting Equation 18 into Equation 15 gives:

$$\Delta G^e/RT = n_1 \ln a_1 - \int_{n_1 = 0}^{n_1} n_1 d\ln a_1 \tag{19}$$

Integration by parts of the integral in Equation 19 gives the simple result:

$$\Delta G^e/RT = \int_{n_1 = 0}^{n_1} \ln a_1 dn_1 \tag{20}$$

In spite of the discontinuity of the integrand at the lower limit, this improper integral exists and is a well-defined thermodynamic quantity. For example, at very low solute loading, Henry's law for adsorption (15) is obeyed:

$$n_1 = K a_1$$

and in this range the integral is:

$$\Delta G^e/RT = K \int_{x=0}^{a_1} \ln x dx = K(a_1 \ln a_1 - a_1) = n_1(\ln a_1 - 1)$$

From Equation 15, it is found that at saturation of the solute ($a_1 = 1$):

$$\lim_{a_1 \to 1} \Delta G^e = (\gamma_{sat} - \gamma_0)A \equiv \Delta G \tag{21}$$

so that ΔG^e at saturation is the free energy of immersion of the adsorbent in pure liquid solute, relative to the free energy of immersion in pure solvent. Both free energies are negative and since the solute is more strongly adsorbed, ΔG is also a negative quantity. We propose that, at saturation of the solute, the last of the solvent is displaced from the adsorbent surface due to the immiscibility of the solute–solvent pair and the preference of the surface for the solute. Therefore, a fractional filling of the micropores of the adsorbent may be defined:

$$\theta = n_1/m_1 \tag{22}$$

where m_1 is the saturation capacity of the adsorbent for the solute. The fractional filling of the micropores by solute varies from zero to unity, while the portion of the surface not filled by solute contains solvent molecules. The solvent molecules are displaced by solute during the loading process.

Surface Heterogeneity

The final and most important of our assumptions is that the adsorbent is heterogeneous (*16*). If the surface has a distribution $f(E)$ of energies such that adsorption of a solute molecule on a portion of the surface with energy E is:

$$\theta = \theta(a_1, T, E) \tag{23}$$

then E is a positive quantity corresponding to the decrease in energy for the transfer of one mole of solute from the liquid phase to the adsorbed phase. Thus, $\theta(a_1, T, E)$, the local adsorption isotherm, is distinguished from $\theta(a_1, T)$, the overall adsorption isotherm for the entire surface. During adsorption of solute, solvent molecules must be displaced so E refers to the energy exchange process of replacing solvent molecules by solute molecules. The total adsorption isotherm is given by an integration of Equation 23:

$$\theta(a_1, T) = \int_{E_{min}}^{E_{max}} \theta(a_1, T, E) f(E) \, dE \tag{24}$$

where $f(E)$ is the energy density function and is normalized:

$$\int_{E_{min}}^{E_{max}} f(E) \, dE = 1 \tag{25}$$

It has been shown (*8*) that when the dimensionless dispersion of the energy distribution:

$$\sigma^\circ = \frac{\sqrt{(\overline{E^2}) - (\bar{E})^2}}{RT}$$

is large, as is the case for adsorbents such as active carbon and silica gel, the adsorption isotherm has the form:

$$-RT\ln a_1 = F^{-1}(1 - \theta) \tag{26}$$

where F^{-1} is the inverse function of the cumulative energy distribution function:

$$F(z) = \int_0^z f(z)\,dz \tag{27}$$

and:

$$z = E - E_{\min} \tag{28}$$

Equation 26 may be written in the form:

$$\ln a_1 = J(T) \cdot \psi(\theta) \tag{29}$$

The theoretical significance of $\psi(\theta)$ is that the cumulative energy distribution F for the energy z is given by:

$$F(z) = 1 - \psi^{-1}(z)$$

In the case of physical adsorption, the function F should have the same shape for different solutes on a particular adsorbent. Specifically, the distribution function for the dimensionless energy $(z/\bar{z})$ is the same for all solutes. Different solutes need not have equal energies of adsorption on a particular site, but the ratio of energies on any pair of sites must be the same for all solutes. This is reasonable for physical interactions such as dispersion energies between the solid and solute molecules, but it does not apply to solutes that exhibit specific chemical interactions with the surface.

Several important adsorption isotherms have the form of Equation 29: the equation of Dubinin–Radushkevich (6) and that of Frenkel–Halsey–Hill (15). The first of these applies to microporous solids having type I isotherms with a well-defined saturation capacity. The second applies to multilayer adsorption on nonporous adsorbents. However, several well-known isotherms do not satisfy Equation 29: the Langmuir equation (17), the Hill–deBoer equation (15), the BET equation (18), and the Toth equation (19). The first three of these equations were derived for nonheterogeneous surfaces, so it is not expected that they should obey Equation 29. Even though the Toth equation was specifically derived for heterogeneous surfaces, it does not satisfy Equation 29 because it is a generalization of the Langmuir equation.

It is desired to calculate the function ψ in Equation 29 and, therefore, the energy distribution on the surface, without making any assumptions about the value of the temperature function J. This can be accomplished by means of thermodynamics using Equation 20. If Equations 22 and 29 are substituted into Equation 20:

$$\Delta G^e = m_1 RTJ \int_0^\theta \psi(\theta)\,d\theta \qquad (30)$$

The free energy of immersion in pure solute, relative to the free energy of immersion in pure solvent, is:

$$\Delta G = m_1 RTJ \int_0^1 \psi(\theta)\,d\theta \qquad (31)$$

The values of m_1 and J are different for each solute, but both quantities vanish from the quotient:

$$\frac{\Delta G^e}{\Delta G} = \frac{\displaystyle\int_0^\theta \psi(\theta)\,d\theta}{\displaystyle\int_0^1 \psi(\theta)\,d\theta} \qquad (32)$$

The derivative of this quotient is proportional to the ψ function sought:

$$\frac{d}{d\theta}\left(\frac{\Delta G^e}{\Delta G}\right) = \frac{\psi(\theta)}{\displaystyle\int_0^1 \psi(\theta)\,d\theta} \equiv K(\theta) \qquad (33)$$

Substitution of Equations 12 and 29 into Equation 33 gives the result:

$$K(\theta) = \frac{m_1 RT\ln(c_1/c_1^{\mathrm{sat}})}{\Delta G} \qquad (34)$$

The dimensionless K function depends on θ only. Adsorption isotherms of different solutes, after transformation according to Equation 34, should coincide. This coalescence into a single universal curve is called the principle of correspondence, which should be obeyed for adsorption of dilute solutes on heterogeneous adsorbents. Different adsorbents have similarly shaped but different K functions according to their energy distributions. The K function is normalized according to Equation 33.

Before testing Equation 34, we discuss its use for predicting adsorption isotherms of single solutes using the K function of one solute as a reference. The solubility (c_1^{sat}) of the solute in the solvent, saturation capacity (m_1), and free energy of immersion (ΔG) are needed for each solute. Adsorption isotherms at temperatures other than the reference temperature can be calculated from the heat of immersion of the solute relative to the solvent (ΔH) using the Gibbs–Helmholtz relation:

$$\Delta H = -T^2 \frac{d}{dT} \left(\frac{\Delta G}{T} \right)$$

Therefore, the problem reduces to the determination of ΔG and, when necessary, ΔH.

It may be possible to predict ΔG for similar or homologous series of adsorbents using the group contribution methods that have been successful in correlating thermodynamic properties of liquid mixtures. Experimentally, there are two methods of obtaining ΔG. The first and most direct approach is to measure one point on the adsorption isotherm. Then ΔG can be calculated using the estimated value of m_1 (see below) and Equation 34. The value of the K function is known from the adsorption of the reference substance. Of course, the accuracy of the value of ΔG increases with the number of experimental points for the adsorption isotherm. The second experimental method (20) is to measure the differences in free energy of immersion for pure solute vapor 1 and pure solvent vapor 2:

$$-\Delta G = RT \int_{P=0}^{P_1^{sat}} n_1 d\ln P - RT \int_{P=0}^{P_2^{sat}} n_2 d\ln P$$

This requires accurate experimental data on the adsorption isotherms for the pure vapors. Since the solute is more strongly adsorbed, the first integral is larger than the second one.

The saturation capacity of the adsorbent for each solute (m_1) may be estimated using Gurvitch's rule (21). According to this rule, which is remarkably accurate for microporous adsorbents, the saturation capacity of a microporous adsorbent, V, expressed as volume of liquefied adsorbate is a constant:

$$V = mv_0 = \text{constant} \tag{35}$$

where v_0 is the molar volume of the adsorbate in the state of saturated liquid at the same temperature as the adsorption isotherm.

Since m_1 can be estimated for any solute using Equation 35 and the value of V for the reference substance, Equation 34 contains only one constant for each solute, ΔG. This constant, the free energy of immersion, is not a correlating parameter. It is a well-defined physical constant for the solid–solute pair and can be measured experimentally from liquid or gas phase adsorption.

Test of Principle of Correspondence

Wohleber and Manes (22) measured the adsorption of several organic solutes from aqueous solution at 25°C on Pittsburgh CAL activated carbon (BET surface area, 1140 m^2/g). Their data satisfied the assumptions of the theory of correspondence: the adsorbent is heterogeneous, the solutes are strongly adsorbed relative to the solvent, and the solute concentration is less than 1 mol %. Moreover, loading was determined over 5–6 decades of concentration so that accurate values of thermodynamic properties can be derived.

Free energies of immersion were calculated using Equation 20 and are given in Table II. The uncertainty in ΔG is estimated to be less than 5%. The saturation capacities of the solutes in Table II were obtained by extrapolation of the adsorption isotherm to saturation and the estimated error in m is about 10%.

The dimensionless free energy of adsorption $(-\Delta G/mRT)$ is a measure of the relative strength of adsorption. According to Table II, the order of increasing strength of adsorption for these solutes is propionitrile < methylene chloride < diethyl ether < ethyl acetate. However, this principle is ambiguous for these solutes because the adsorption isotherms, plotted as loading versus mole fraction, cross over indicating that the relative strength of adsorption changes with coverage. This effect is called adsorption azeotropy. At low concentration in the parts per million range, the order of increasing strength of adsorption is propionitrile < methylene chloride < ethyl acetate < diethyl ether.

The theory of correspondence was tested for these data by comparing K functions calculated from Equation 34 using the constants in Table II. As shown on Figure 1, the average discrepancy between the experimental loading and the K function (solid line) is 10%, which is good considering that the loading varies over three orders of magnitude.

In the future, we shall publish additional tests of the theory of correspondence for adsorption of dilute solutes on heterogeneous adsorbents. Preliminary calculations indicate that the K functions of different solutes coincide within experimental error.

In addition to its usefulness as a method of correlating and predicting adsorption isotherms, the theory of correspondence can be applied to the prediction of mixed solute adsorption. The useful simplification of treating the adsorption of single solutes similar to adsorption of pure gases should not obscure the fact that the adsorbed phase is a binary mixture of solute and solvent molecules. This simplification deserves further study.

Summary

For adsorption of dilute solutes on a heterogeneous adsorbent, the K function defined by Equation 34 is independent of the solute. The key variable required for the transformation from individual solute isotherms to a universal curve is the dimensionless group $(\Delta G/mRT)$. This group,

Table II. Properties of Solutes Adsorbed from Aqueous Solution at 25°C on Pittsburgh CAL Activated Carbon

Property	Diethyl Ether $CH_3CH_2OCH_2CH_3$	Ethyl Acetate $CH_3COOCH_2CH_3$	Methylene Chloride CH_2Cl_2	Propionitrile CH_3CH_2CN
M.W.	74.12	88.11	84.93	55.08
Solubility, mole fraction	0.00604	0.01783	0.00338	0.03320
$-\Delta G$, J/g	35.4	42.6	53.7	48.8
m, mmol/g	4.03	4.50	7.38	7.54
$-\Delta G/mRT$	3.55	3.81	2.94	2.61

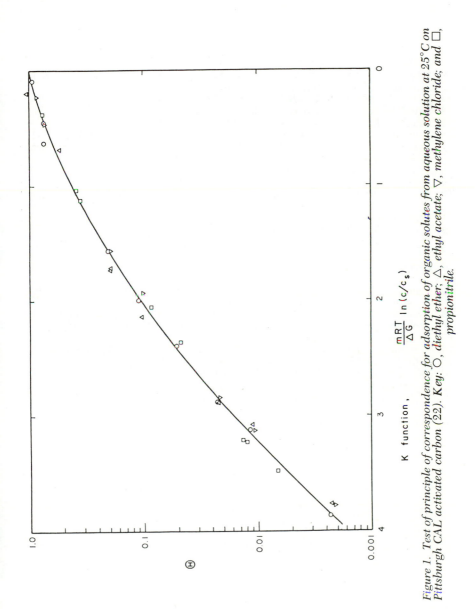

Figure 1. Test of principle of correspondence for adsorption of organic solutes from aqueous solution at 25°C on Pittsburgh CAL activated carbon (22). Key: ○, diethyl ether; △, ethyl acetate; ▽, methylene chloride; and □, propionitrile.

which is the ratio of the average molar free energy of adsorption to the thermal energy, is proportional to the strength of adsorption of the solute.

Literature Cited

1. Sircar, S.; Myers, A. L. *Chem. Eng. Sci.* **1973**, *28*, 489.
2. Jossens, L.; Prausnitz, J. M.; Fritz, W.; Schlünder, E. U.; Myers, A. L. *Chem. Eng. Sci.* **1978**, *33*, 1097.
3. Minka, C.; Myers, A. L. *AIChE J.* **1973**, *19*, 453.
4. Sircar, S.; Myers, A. L. *AIChE J.* **1973**, *19*, 159.
5. Grant, R. J.; Manes, M.; Smith, S. B. *AIChE J.*, **1962**, *8*, 403.
6. Dubinin, M. M.; Radushkevich, L. V. *Dokl. Akad. Nauk SSSR* **1947**, *55*, 331.
7. Lewis, W. K.; Gilliland, E. R.; Chertow, B.; Cadogan, W. P. *Ind. Eng. Chem.* **1950**, *42*, 1362.
8. Myers, A. L.; Sircar, S. *AIChE J.* **1981**, in press.
9. Potter, C.; Sussman, M. V. *Ind. Eng. Chem.* **1957**, *49*, 1763.
10. Aveyard, R.; Haydon, D. A. "An Introduction to the Principles of Surface Chemistry"; Cambridge University Press: London, 1973; pp 12, 15, and 166.
11. Polanyi, M. *Verh. Dtsch. Phys. Ges.*, **1914**, *16*, 1012.
12. Sircar, S.; Myers, A. L.; Molstad, M. C. *Trans. Faraday Soc.*, **1970**, *66*, 2354.
13. Sircar, S.; Novosad, J.; Myers, A. L. *Ind. Eng. Chem. Fundamentals* **1972**, *11*, 249.
14. Bering, B. P.; Myers, A. L.; Serpinsky, V. V. *Dokl. Akad. Nauk SSSR*, **1970**, *193*, 119.
15. Young, D. M.; Crowell, A. D. "Physical Adsorption of Gases"; Butterworths: London, 1962; pp. 64–70, 137, and 167.
16. Zolandz, R. R.; Myers, A. L. In "Progress in Filtration and Separation" Wakeman, R. J., Ed.; Elsevier: New York, 1979; pp. 1–29.
17. Langmuir, I. *J. Amer. Chem. Soc.*, **1918**, *40*, 1361.
18. Brunauer, S.; Emmett, P. H.; Teller, E. *J. Amer. Chem. Soc.*, **1938**, *60*, 309.
19. Toth, J. *Acta Chim. Acad. Sci. Hung.*, **1971**, *69*, 311.
20. Myers, A. L.; Sircar, S. *J. Phys. Chem.*, **1972**, *76*, 3415.
21. Gurvitsch, L. *J. Phys. Chem. Soc. Russ.* **1915**, *47*, 805.
22. Wohleber, D. A.; Manes, M. *J. Phys. Chem.*, **1971**, *75*, 61.

RECEIVED for review August 3, 1981. ACCEPTED for publication June 29, 1982.

Physicochemical Aspects of Carbon Affecting Adsorption from the Aqueous Phase

BALWANT RAI PURI

India Carbon Limited, Gauhati, Assam (India)

Treatments such as deashing with a hot HF–HCl mixture, burn-off in oxygen under low pressures of 10–20 torr at 600°C, and exposure to ozonized oxygen at ambient temperature increase surface areas of activated carbons as well as sugar and coconut shell charcoals which were also examined. These treatments enhanced the adsorption of phenol from the aqueous phase. Adsorption isotherms (35°C) of carbon tetrachloride vapor on activated carbons of known nitrogen surface areas can be helpful in characterizing carbons for relative proportions of transitional pores and micropores, an important parameter influencing carbon adsorption from the aqueous phase. Activated carbons when oxidized in concentrated nitric acid, acidified potassium persulfate, aqueous hydrogen peroxide, or moist air at 285°C neutralize ammonia and aliphatic amines in aqueous solutions due to development of acidic surface oxides and may be used as effective adsorbents for alkaline pollutants in wastewaters. The used carbons can be regenerated on flushing with dilute acid solutions. Adsorption of weakly basic aromatic amines, however, is affected negatively by acidic surface oxides but positively by quinonic and carbonyl surface oxides. This phenomenon is similar to that observed in the case of adsorption of phenols and substituted phenols from the aqueous phase and is attributed to a specific interaction between electrons of the benzene nucleus and the partial positive charge on the quinonic or carbonyl oxygen. This view receives support from adsorption isotherms of benzene vapor on variously modified carbons. Carbons associated with large concentrations of unsaturated sites generated on outgassing at temperatures around 700°C and extended on surface

0065-2393/83/0202-0077$06.00/0

*oxidation followed by outgassing are highly effective in
chemisorbing chlorine and hydrogen sulfide from the aqueous
phase and may find increasing applications in wastewater
and drinking water treatment processes.*

A CTIVATED-CARBON ADSORPTION is a recognized technology in wastewater
and drinking water treatments. There are, however, no general rules
or uniform standards for the effective removal of the broad spectrum of
pollutants and carcinogens known to be present in such waters. In view of
the growing importance of the subject, there is need to have clearer
understanding of the adsorbate–adsorbent interactions involved. Carbon
is a versatile material. The magnitude of its surface and its surface
characteristics can be altered substantially by giving it suitable treatments.
There is thus ample scope for continued improvement in carbon
preparation and carbon selection and for development of materials of
desired physicochemical features capable of promoting enhanced adsorption from the aqueous phase. The present work was undertaken with these
objectives in view.

Experimental

Materials. Eight commercial samples of activated carbons were used in the
various experiments. A few typical samples of carbon blacks and two chars
obtained by carbonization of cane sugar and coconut shells also were used in some
cases for comparison. Surface areas (nitrogen) of the various carbons before and
after some treatments were determined by the conventional BET technique.

Treatments. The following treatments were used to change the surface
area characteristics.

DEASHING. The activated carbons, as received, contained about 2–5% ash
(*see* Table I). To see the effect on available surface (N_2 and BET), these carbons
were given the usual "deashing" treatment in hot HF–HCl. This treatment
lowered their ash content to values between 1.0 and 1.4%.

BURN-OFFS. Some carbons were given different burn-offs in oxygen under
low pressures of 10–20 torr at 600°C (*1*). The amount of burn-off was obtained
from loss in weight. The products were outgassed at 1000°C to eliminate
chemisorbed oxygen, if any, and then stored in bottles flushed with nitrogen.

TREATMENT IN OZONIZED OXYGEN. Carbon (5 g) was taken in a porcelain
tube (12.5 mm diameter) placed in a thermostat, maintained at 30°C and
connected on one side to an ozonizer (Gallenkamp GE-150) and on the other side
to a series of conical flasks containing known volumes of standard (0.1 *N*) barium
hydroxide (VI). Ozonized oxygen (2–3% ozone) was passed at the rate of 2 L/h
over the carbon bed which was rotated at the rate of 18–20 rpm. The treated
product was withdrawn after a given interval of time, outgassed at 1000°C to
eliminate chemisorbed oxygen, if any, and stored in bottles, as already described.

Treatment in Oxidizing Solutions. Some activated carbons were given
oxidation treatments with concentrated nitric acid (*2*), aqueous hydrogen
peroxide (*3*), and acidified potassium persulfate (*3*), and by passing moist air at the

Table I. Effect of Deashing Treatment on Surface Areas of Carbons

	Ash (%)		Surface Area (m^2/g)	
Carbon	Before[a]	After[b]	Before[a]	After[b]
A	4.01	1.31	1122	1242
B	3.20	1.38	1046	1208
C	2.78	1.09	832	917
D	3.06	1.26	922	1035
E	1.94	1.05	549	619
G	5.24	1.14	912	1027
H	4.88	1.25	901	1046
J	3.80	1.10	1011	1131
Coconut shell Charcoal	3.56	0.58	372	436

[a]Before any treatment.
[b]After deashing treatment.

rate of 3 L/h over a 10-g carbon bed held in a rotating porcelain tube maintained at 285°C.

For treatment with nitric acid, carbon (10 g) was mixed with 150 mL of pure concentrated nitric acid in a 250-mL beaker and heated on a low Bunsen flame until the volume was reduced to about 15 mL. The contents were transferred to a filter paper over a funnel, and the residual carbon was washed repeatedly in hot distilled water, then dried in an electric oven (120°C), and stored under nitrogen.

For treatments with aqueous hydrogen peroxide and potassium persulfate, carbon (5 g) was mixed with 250 mL of 3 N hydrogen peroxide or 500 mL of nearly saturated potassium persulfate solution in 2 N sulfuric acid. The suspensions were shaken for 24 h and then filtered. The residue was washed repeatedly, then dried, and stored in the usual way.

Adsorption Isotherms of Phenol from Aqueous Phase. Adsorption isotherms (35 ± 0.05°C) of phenol from the aqueous phase were determined by mixing carbon (0.5 g) with a known weight (5 g) of phenol solution in water of a given concentration and allowing the suspension to stand in a thermostat with occasional shaking for 24 h. The fall in concentration was determined interferometrically (Carl Zeiss laboratory interferometer).

Adsorption Isotherms of Carbon Tetrachloride and Benzene. Adsorption–desorption isotherms (35°C) of carbon tetrachloride and benzene vapor were determined by using McBain's adsorption balance technique. The sensitivity of the quartz spring used was 20 cm/g.

Neutralization of Aqueous Ammonia and Aliphatic Amines. Neutralization of aqueous ammonia and aliphatic amines was studied by mixing carbon (1 g) with 100 mL of about 0.075 M of a given solution and shaking the suspension for about 4 h. The fall in concentration was determined by titrating an aliquot of the clear supernatant liquid against standard HCl in the usual way.

Neutralization of ammonia under flow conditions was examined by allowing the solution to flow through a 10-cm carbon column, placed between two glass wool plugs in a 3.8-cm diameter Pyrex tube. The solution was kept at a constant level, 5 cm above the upper glass wool plug. The liquor was allowed to pass through the column at constant flow rates of 80 and 150 mL/min. The concentration of the elute was checked at every 1-min interval until it approached that of the effluent.

Adsorption of Aniline and Methylaniline from Aqueous Phase. Carbon (0.5 g) was mixed with a known weight (5 g) of a given solution contained in a small glass tube drawn out at one end and subsequently sealed. The contents were shaken mechanically in a thermostat maintained at $25 \pm 0.1°$ C for 24 h. The tube was then allowed to stand in the same thermostat to allow the particles to settle. An aliquot of the clear supernatant liquid was then examined interfero-metrically to estimate the fall in concentration.

Interaction with Aqueous Chlorine. Carbon (2.5 g) was mixed with 250 mL of about $0.05 M$ aqueous chlorine, and the suspension shaken for 4 h. The residual product was washed, dried, and examined for combined oxygen and chlorine. The product was treated in a current of hydrogen at 600°C, and the amount of water vapor and hydrogen chloride evolved was estimated by the usual methods.

Interaction with Aqueous Hydrogen Sulfide. Carbon (2 g) was mixed with 250 mL of $0.05 M$ aqueous hydrogen sulfide, and the suspension was shaken for 4 h. The residual product, after washing and drying, was examined for its sulfur content by treating it in a current of hydrogen at 800°C and estimating iodometrically the amount of H_2S evolved.

Results and Discussion

The effect of lowering the ash content of activated carbons to values within 1.5% on the surface area is shown in Table I. It is seen that deashing leads to an appreciable increase in surface area, amounting to about 10–16% of the initial value, in every case. It appears that treatment with the hot HF–HCl mixture causes adequate cleansing of the micro-pores making additional space available for adsorption.

The effects of using different burn-offs for three activated carbons and coconut charcoal which had been outgassed at 1000°C are shown in Table II. Surface areas increase appreciably with an increase in burn-off in every case, although the values tend to level off ultimately. At about 10% burn-off, the increase in surface area amounts roughly to about 30% of the initial value in the case of activated carbons. The increase in the case of coconut shell charcoal is phenomenal as the value rises almost by an order of 3.5 and becomes almost the same as those of activated carbons. However, this value represents nearly a 38% loss of carbon.

The effect of treating in ozonized oxygen at ambient temperature is shown in Table III. In this case, it is seen that, even for a small loss of carbon, there is an appreciable rise in surface area. The increase in the case of the charcoal is noteworthy. The value becomes almost double for just about 4% loss in weight of carbon due to gassification.

There is, however, an anomaly in the case of activated carbons. The

Table II. Effect of Different Burn-offs on Carbon
for Their Surface Areas

Carbon	Burn-off (%)	Surface Area (m²/g)
A	Nil	1122
	2.5	1258
	5.9	1380
	10.4	1472
	12.4	1488
B	Nil	1046
	2.5	1109
	4.8	1198
	9.6	1285
	13.2	1296
D	Nil	922
	6.5	1057
	10.4	1172
Coconut shell	Nil	320
charcoal outgassed	10.8	750
at 1000° C	20.2	845
	37.6	1162

value, after the initial rise, progressively falls if the treatment is continued beyond 4–6 h. Similar trends were observed in the case of two other activated carbons (A and D) which were also examined. It appears that by continuing the oxidation treatment in ozone beyond a certain stage the extremely fine particles, contributing largely toward the surface area, begin to gassify and get lost from the surface.

To see the effect of enhancement of surface area by some treatments, adsorption isotherms of phenol from the aqueous phase were determined on two activated carbons. These are plotted in Figure 1. It is seen that the amount of adsorption at each concentration rises appreciably with an increase in surface area, although not necessarily in the same proportion.

Transport Pores and Adsorption Pores. Adsorption in the case of activated carbons takes place mostly within the porous network of the particles. Distinction, however, has to be made between the larger transport pores (transitional pores) going through the whole particles and the finer pores (micropores) branching off from them and lying within the particles (4). The wider pores serve to transport adsorbate material from the external surface to the interior of the particles where the finer pores constituting the major internal surface adsorb them readily.

Table III. Effect of Treating Carbons with Ozonized Oxygen on Burn-offs and Surface Area

Carbon	Duration of Treatment (h)	Burn-off (%)	Surface Area (m^2/g)
C	0	Nil	832
	2	1.8	905
	4	3.3	948
	8	5.2	980
	12	7.1	874
	24	9.8	688
J	0	Nil	1011
	2	2.1	1082
	4	4.2	1154
	8	6.3	1108
	12	7.1	744
Sugar, charcoal,	0	Nil	550
outgassed at	6	1.2	780
1000° C	12	2.7	977
	24	3.9	1040

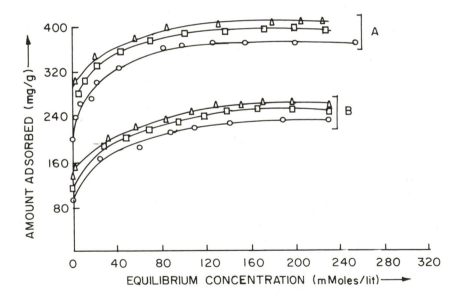

Figure 1. Adsorption isotherms of phenol on activated carbons A and B. Key: O, before any treatment; □, after deashing; △, after burn-off (10.4% in A and 9.6% in B).

In a good activated carbon, obviously, there should be, in general, some optimum relative proportion of the two types of pores so as to provide rapid transport of the adsorbate to the interior and also sufficiently large internal surface, for prompt and effective adsorption. This may, in fact, be another important parameter in characterizing carbons for their suitability in water treatment operations. It was thought of interest, therefore, to compare nitrogen surface areas of activated carbons with the values obtained from adsorption isotherms of some other gas or vapor with much larger molecular dimensions. Carbon tetrachloride with a molecular area of 36.4 Å^2 was selected. Adsorption of carbon tetrachloride vapor by activated carbons at ambient temperature is used in some quality control laboratories (5), although it is not known what it actually measures.

Adsorption isotherms (35° C) of carbon tetrachloride vapor on some activated carbons are given in Figure 2. Similar isotherms on a few typical carbon blacks are given in Figure 3 for comparison. Considering the isotherms on carbon blacks first, it is seen that sorption–desorption branches almost exactly superimpose showing that the adsorption is reversible. It is also seen that the adsorption, after a sharp initial rise, proceeds by exceedingly small amounts at successive increments in relative vapor pressure (r.v.p.) until a point of inflection is approached beyond which the isotherm starts rising rapidly, indicating completion of a monolayer and commencement of a subsequent layer.

The isotherms, evidently, are Type II of the well-known system of classification. The isotherms on activated carbons (Figure 2) also appear to be reversible, there being little or no hysteresis. This appears to be due to rather large molecular dimensions of carbon tetrachloride and hence its inaccessibility to finer micropores which are largely involved in the hysteresis phenomenon. All the isotherms, except the one on Carbon E, show a steep rise in the initial stage and a sharp break at such a low r.v.p. as 0.1. Beyond this there is a small, although a significant and continuous, pick up of the vapor, indicative of partial filling of transitional pores. The isotherms seem to lie between those of Types I and II. This finding is due to the presence of transitional porosity as a result of which the tendency for pore filling is greater than that for formation of a subsequent layer. The isotherm on the activated Carbon E, however, appears to be Type II and this indicates much less transitional porosity in this material.

Specific surface areas of the various carbons were calculated by applying the BET equation to these isotherms. The values are given in Table IV along with the corresponding nitrogen values. In the case of carbon blacks, the two sets of values are fairly close; in the case of activated carbons, an appreciable fraction of the nitrogen surface remains inaccessible to carbon tetrachloride vapor. Adsorption in such cases,

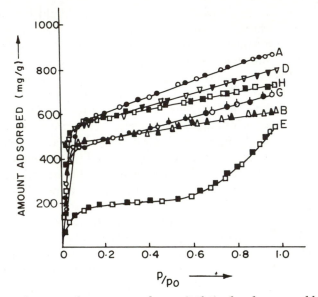

Figure 2. Adsorption-desorption isotherms (35°C) of carbon tetrachloride on activated carbons. Open points denote adsorption while solid points denote desorption values. (Reproduced, with permission, from Ref. 18. Copyright 1979, Indian Chemical Society.)

Table IV. Nitrogen and Carbon Tetrachloride Specific Surface Areas of Carbons

	Surface Area ($m^2 g$)		
Carbons	$N_2(78°K)$ (1)	$CCl_4(298°K)$ (2)	Ratio
Active Carbons			
A	1122	718	64
B	1046	616	59
D	922	701	76
E	549	252	46
G	912	565	62
H	901	677	75
Carbon Blacks			
ELF-O	171	165	96
Spheron-6	120	116	97
Mogul	301	258	86
Mogul-A	221	210	95
Graphon	86	85	99

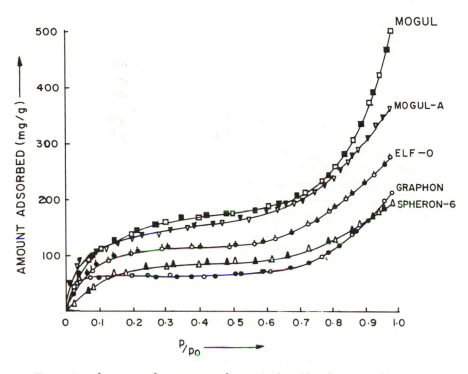

Figure 3. Adsorption–desorption isotherms (35°C) of carbon tetrachloride on a few carbon blacks. Open points denote adsorption while solid points denote desorption values. (Reproduced, with permission, from Ref. 18. Copyright 1979, Indian Chemical Society.)

therefore, appears to be subject to molecular sieve action. From the ratios of the two specific surface areas given in the last column of the table, it appears that while the frequency of micropores is only 24% in Carbon D, it is as high as 54% in Carbon E. Thus, Carbon D has much more transitional porosity than Carbon E. It is not possible at this stage to lay down any particular ratio of the two porosities as optimum for best performance of carbons in water treatment processes. For this purpose more work is needed particularly with carbons having nearly equal nitrogen areas but differing in carbon tetrachloride areas (compare Carbons D and G). In any case, adsorption of carbon tetrachloride vapor by activated carbons of known specific surface areas (nitrogen) does have some significance in characterizing carbons for their transitional porosity.

Modified Carbons

Acidic Sites for Adsorption of Ammonia and Other Alkaline Matter. Ammonia, generally found in some wastewaters, results from breakdown of urea, proteins, and other nitrogenous organic matter. Its presence is undesirable from several points of view. For example, chlorination of water in the presence of even a small amount of ammonia may result in the formation of chloramines which are toxic and pose a problem for elimination by granular activated-carbon treatment, as has been pointed out by Suidan et al. (6).

Activated Carbons A–H as such could not take up any noticeable amount of ammonia on shaking 1 g with 100 mL of a 0.075 M solution. The pH values of their aqueous suspensions were close to 8. But after giving them the various oxidation treatments, the pH values came down well within the acid range and with it the amount of ammonia neutralized rose considerably, as shown in Table V. This result is obviously due to development of the acidic carbon dioxide–surface complex. The efficiency of the various oxidation treatments to enhance neutralization of ammonia by a given carbon is seen to follow the order $HNO_3 > K_2S_2O_8 > H_2O_2 >$ moist air.

Neutralization of ammonia by HNO_3-treated carbons also was studied in flow experiments using flow rates of 80 and 150 mL/min. The results obtained by using Carbons A, B, and C, given in Table VI, show fairly high breakthrough times which increase with a decrease in flow rates, as expected. The beds could be regenerated to the extent of over 90% by treating them with a solution of 0.1 N HCl in a similar manner.

Table V. pH Values and Ammonia Neutralization Values of Oxidized Carbons

Oxidizing Agent	Carbon	pH	Ammonia Value (mEq/g)
HNO_3	A	4.15	2.2
	B	4.50	1.8
	D	4.58	1.8
	G	4.66	1.6
H_2O_2	A	5.05	0.9
	B	5.12	1.0
	D	4.95	0.9
	G	4.84	0.8
$K_2S_2O_8$	A	4.65	1.8
	B	4.74	1.5
	D	4.74	1.4
	G	4.85	1.4
Moist air	A	5.46	0.5
	B	5.38	0.6
	D	5.50	0.5
	G	5.48	0.4

**Table VI. Breakthrough Times for Flow of Aqueous Ammonia
Through HNO$_3$-Treated Carbon Columns**

Carbon	Flow Rate (mL/min)	Breakthrough Time (min)
A	80	14.1
	150	8.4
B	80	12.3
	150	7.5
C	80	12.8
	150	7.6

The oxidized carbons neutralized appreciably large amounts of aliphatic amines as well. These values are given in Table VII and are seen to be comparable with those for ammonia which are also reproduced in a separate column for easy reference. The amounts of secondary and tertiary amines neutralized, however, are relatively lower. It appears that the presence of more than one hydrocarbon chain introduces steric effects and, therefore, some acidic sites may not be available to the basic amino groups in the solution. The results, however, are clearly indicative of the effect of acidic surface oxides in neutralizing alkaline pollutants that might be present in wastewaters and waterways.

Carbonyl Quinonic Sites for Adsorption of Aromatic Species. Aniline and methylaniline are reported (7) to be among the organic compounds detected in wastewaters. These substances are weakly alkaline but their adsorption was found to be adversely affected by the

**Table VII. Neutralization Values of Oxidized Carbons for
Ammonia and Aliphatic Amines**

Oxidizing Agent	Carbon	Neutralization Values (mEq/g)				
		NH_3	$BuNH_2$	$(CH_3)_2NH$	$(C_2H_5)_2NH$	$(CH_3)_3N$
HNO$_3$	A	2.2	2.2	1.9	1.7	1.2
	B	1.8	1.7	1.5	1.4	1.1
	D	1.8	1.8	1.4	1.2	0.8
	G	1.6	1.5	1.2	1.0	0.6
H$_2$O$_2$	A	0.9	0.8	0.6	0.5	0.4
	B	1.0	1.0	0.7	0.6	0.4
	D	0.9	0.8	0.6	0.6	0.3
	G	0.8	0.8	0.5	0.5	0.2
K$_2$S$_2$O$_8$	A	1.8	1.8	1.4	1.2	0.8
	B	1.5	1.6	1.3	1.1	0.7
	D	1.4	1.3	1.0	0.9	0.6
	G	1.4	1.4	1.0	0.9	0.6

presence of acidic surface oxides. The results obtained by using HNO_3-treated Carbon A, before and after outgassing at 700° and 1000° C, for the adsorption of aniline and methylaniline are given in Figures 4 and 5, respectively. The results have been normalized with respect to surface area. The amount of adsorption of each substance is minimal in the case of the oxidized carbon but rises substantially as the acidic CO_2 complex is eliminated but the CO complex, arising from the presence of carbonyl and quinonic surface groups, is allowed to be retained on outgassing at 700° C (8). It is also significant that, when these groups are also eliminated and the carbon is rendered essentially oxygen free on outgassing at 1000° C, the amount of adsorption undergoes an appreciable fall at all relative concentrations. This effect is akin to that observed and reported earlier

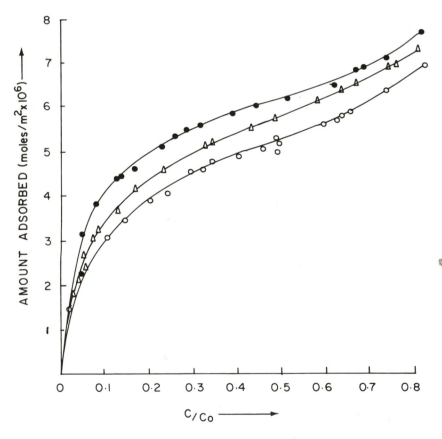

Figure 4. Adsorption isotherms of aniline from aqueous solution on oxidized carbon A. Key: O, before outgassing; ●, after outgassing at 700°C; and △ after outgassing at 1000°C.

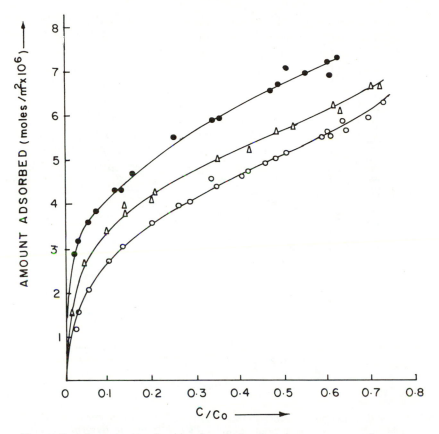

Figure 5. Adsorption isotherms of methylaniline from aqueous solution on oxidized carbon A. Key: ○, *before outgassing;* ●, *after outgassing at 700° C; and* △, *after outgassing at 1000°C.*

(9–11) in the case of adsorption of phenols and substituted phenols by carbons associated with carbonyl oxygen and may be attributed, as before, to the complexing of π-electrons of the benzene nucleus with a partial positive charge on the carbonyl groups.

This effect was checked more directly by studying adsorption isotherms of benzene (35° C) on activated carbons as had been reported earlier in the case of similar adsorption in carbon blacks (12). The results obtained in the case of one activated carbon are shown in Figure 6. As the acidic surface complex is eliminated at 700° C and the CO complex emerges as the only predominant surface complex, benzene sorption rises appreciably at all r.v.p.s. Furthermore, as this complex is eliminated and the carbon is rendered almost oxygen free on outgassing at 1000° C, the isotherm shifts again downward.

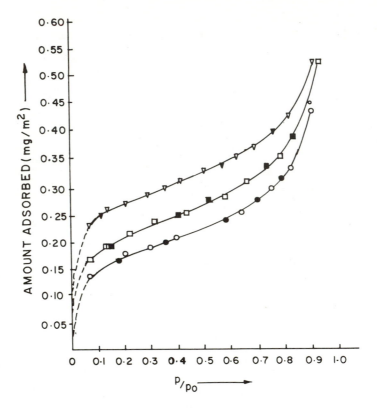

Figure 6. Adsorption–desorption isotherms (35° C) of benzene on carbon A;
open points denote adsorption while solid points denote desorption values.
Key: O-●, before outgassing; △-▲, after outgassing at 700° C; and □-■,
after outgassing at 1000° C.

Unsaturated Sites and Chemical Fixation of Chlorine and Other Species. Identification of certain highly reactive sites, referred to as unsaturated sites, generated on evacuating carbons around 700° C and estimated on reacting with aqueous bromine, has been reported in previous investigations (*13, 14*). It has also been shown (*15*) that concentrations of these sites can be enhanced, within limits, by surface oxidation followed by evacuation treatment. The concentration of such sites in our carbons, before and after this treatment, is recorded in Table VIII. There is, evidently, a considerable rise in the value after the treatment in each case. It was thought of interest to examine the use of these carbons in the adsorption of chlorine from the aqueous phase.

Chlorine is added to wastewaters as a disinfectant and sometimes also for removal of ammonia. This treatment, however, enhances the potential for the formation of trihalomethanes, chlorophenols, etc. There is need, evidently, for a fairly effective dechlorination treatment. The results obtained on interacting the various activated carbons with aqueous chlorine, given in Table IX, indicate chemical fixation of oxygen as well as chlorine. The values are seen to rise with an increase in surface unsaturation. The oxygen, in this case, originates, evidently, from hydrolysis of a part of the chlorine to hypochlorous acid reaction. More than 90% of the chlorine initially present in water could be eliminated in this manner by carbons associated with high concentrations of unsaturated sites. The total amount of oxygen and chlorine fixed by a carbon is in excess of the initial surface unsaturation. Besides this, a part of the unsaturation remains intact at the end of the process. This indicates that fixation of oxygen and chlorine takes place at some other active sites as well.

The carbons once used in the process could be regenerated to the extent of about 80% of the initial performance when treated in a current of hydrogen at 600° C.

The results of similar interactions with aqueous hydrogen sulfide are given in Table X. There is an appreciable fixation of sulfur in every case and the amount increases with a rise in surface unsaturation. Thus, Carbon A is seen to fix 4.74 milliequivalents (mEq) of hydrogen sulfide, corresponding to as much as 76 mg of sulfur per g carbon. It seems highly likely that activated carbons associated with high concentrations of unsaturated sites may be effective adsorbents for several other molecular species from the aqueous phase as well.

The possibility of formation of new compounds, not originally present in water, by some unspecified catalytic reactions taking place at the carbon surface was pointed out by Cookson (16). It may be mentioned that carbons associated with unsaturated sites, referred to above, have been found (17) to catalyze chlorination of benzene and toluene, both

Table VIII. Effect of Surface Oxidation Followed by Outgassing at 800° on Surface Area and Surface Unsaturation of Activated Carbons

	Surface Area (m²/g)			*Surface Unsaturation (mEq/g)*		
		After Treatment[a]			*After Treatment[a]*	
Carbon	*Before Treatment*	*A*	*B*	*Before Treatment*	*A*	*B*
A	1122	1160	1188	1.23	4.35	4.51
B	1046	1130	1174	1.55	3.65	3.44
C	832	888	923	1.39	4.01	3.20
D	922	959	1011	0.83	3.08	3.31
E	549	592	624	1.25	2.43	2.51

[a]Treatment A was with potassium persulfate followed by evacuation. Treatment B was with aqueous hydrogen peroxide followed by evacuation.

Table IX. Interaction of Activated Carbons with Aqueous Chlorine

Carbon[a]		Surface Unsaturation (mEq/g)	Oxygen Sorbal (mEq/g)	Chlorine Sorbal (mEq/g)	Surface Unsaturation After Treatment (mEq/g)
A	a	1.23	0.84	0.81	0.72
	b	4.51	2.05	2.40	1.52
B	a	1.55	1.45	1.25	0.36
	b	3.44	1.94	2.32	1.64
C	a	1.39	1.05	1.23	0.46
	b	3.20	1.66	2.45	0.72
D	a	0.83	0.70	0.64	0.42
	b	3.31	2.08	2.04	0.88

[a]a = Original, and b = oxidized and outgassed.

Table X. Fixation of Hydrogen Sulfide from Aqueous Phase by Carbons in Relation to Surface Unsaturation

Carbon	Surface Unsaturation (mEq/g)		H_2S Fixed (mEq/g)	
	Before[a]	After[b]	Before[a]	After[b]
A	1.23	4.35	1.47	4.74
B	1.55	3.65	1.72	3.42
C	1.39	4.01	1.32	4.18
D	0.83	3.08	1.05	3.38
E	1.25	2.43	1.62	2.84
Coconut charcoal outgassed at 750° C	3.50	—	3.37	—

[a]Carbons without any treatment.
[b]Carbons after surface oxidation following by outgassing.

reported to be present in wastewaters, yielding chlorinated derivatives which may be even more toxic. But it must be emphasized that such reactions are not at all likely to occur under the conditions in which carbons are used in water treatment processes.

Literature Cited

1. Laine, N. R.; Vastola, F. J.; Walker, Jr., P. L. *J. Phys. Chem.* **1963,** *67,* 3020–4.
2. Puri, B. R.; Singh, S.; Mahajan, O. P. *J. Indian Chem. Soc.* **1965,** *42,* 427–34.
3. Puri, B. R.; Sharma, G. K.; Sharma, S. K. *J. Indian Chem. Soc.* **1977,** *44,* 64–8.
4. Juntgen, H. *Carbon* **1977,** *15,* 273–83.
5. Smisek, M.; Cerny, S. "Active Carbon Manufacture, Properties and Applications"; Elsevier Publishing Co.: Amsterdam, 1970; p 321.
6. Suidan, M. T.; Kim, B. R.; Snoeyink, V. L. In "Activated Carbon Adsorption of Organics from the Aqueous Phase"; Suffet, I. H.; McGuire, M. J., Eds.; Ann Arbor Science: Ann Arbor, Mich. 1980; Volume 1, pp 397–421.
7. Cotruvo, J. A.; Chieh, W. In "Activated Carbon Adsorption of Organics from the Aqueous Phase"; Suffet, I. H.; McGuire, M. J., Eds.; Ann Arbor Science: Ann Arbor, Mich., 1980; Volume 1, pp 1–11.
8. Puri, B. R. In "Chemistry and Physics of Carbon"; Walker, P. L., Jr., Ed.; Marcel Dekker; New York, 1970; Volume 6, pp 191–282.
9. Mattson, J. S.; Mark, H. B., Jr.; Malbin, M. D.; Weber, W. J., Jr.; Crittender, J. C. *J. Colloid Interface Sci.* **1969,** *31* 116–24.
10. Puri, B. R.; Bhardwaj, S. S.; Mahajan, O. P. *J. Indian Chem. Soc.* **1975,** *52,* 26–9.
11. Puri, B. R. In "Activated Carbon Adsorption of Organics from the Aqueous Phase"; Suffet, I. H.; McGuire, M. J., Eds.; Ann Arbor Science: Ann Arbor, Mich., 1980; Volume 1, pp 353–78.
12. Puri, B. R.; Kaistha, B. C.; Mahajan, O. P. *Carbon* **1973,** *11,* 329–36.
13. Puri, B. R.; Sandle, N. K.; Mahajan, O. P. *J. Chem. Soc.* **1963,** 4880–4.
14. Puri, B. R.; Bansal, R. C. *Carbon* **1966,** *3,* 533–9.
15. Puri, B. R.; Mahajan, O. P. Gandhi, D. L. *J. Indian Chem. Soc.* **1972,** *10,* 848–9.
16. Cookson, J. T., Jr., In "Activated Carbon Adsorption of Organics from the Aqueous Phase"; Suffet, I. H.; McGuire, M. J., Eds.; Ann Arbor Science; Ann Arbor, Mich., 1980; Volume 1, 379–424.
17. Puri. B. R., Singh, D. D.; Kaura, N. C.; Verma, S. K. *Indian J. Chem.* **1980,** *19A,* 109–12.
18. Puri, B. R.; Arora, V. M.; Verma, S. K. *J. Indian Chem. Soc.* **1979,** *56,* 802–4.

RECEIVED for review August 3, 1981. ACCEPTED for publication September 16, 1982.

6

Effect of Surface Characteristics of Activated Carbon on the Adsorption of Chloroform from Aqueous Solution

CHANEL ISHIZAKI, IRIS MARTÍ, and MAGALY RUÍZ

Instituto Venezolano de Investigaciones Cientificas, Laboratorio de Ingenieria Ambiental, Apartado 1827, Caracas 1010-A, Venezuela

Chloroform adsorption on two activated carbons of different surface characteristics was investigated. The chemical and physical characteristics of the carbon surfaces were evaluated. The equilibrium and kinetic studies indicate different adsorptive behaviors for these systems. For the equilibrium concentration range of 10–200 µg/L, the data fit the Langmuir isotherm. The observed differences in adsorption isotherms cannot be explained by differences in pore size distributions, but can be explained by the chemical nature of the carbon. The carbon with the least amount of oxygen exhibits the highest affinity for chloroform. The relative capacities of the carbons to adsorb chloroform are a function of the solution equilibrium concentrations. For equilibrium concentrations below 270 µg/L, the carbon with the least amount of oxygen exhibits the highest capacity.

THE ABILITY OF ACTIVATED CARBONS to adsorb various organics from solution is recognized. However, the mechanisms by which these compounds adsorb are not clearly understood.

Adsorption is governed by the chemical nature of the aqueous and solid phases, and by the chemical nature of the adsorbing compounds. Any interpretation of the adsorptive behavior of activated carbons based only on surface area is incomplete. Carbons having equal weights and equal total surface areas, when prepared by different methods, exhibit different adsorptive characteristics. Some adsorptive properties can be explained by differences in relative pore size distributions, but a more important consideration is the difference in surface chemistry of the carbons. The adsorptive properties of carbons activated by oxidation depend primarily on the chemical nature and concentration of the oxidizing agent, the temperature of the reaction, the extent to which the activation is conducted, and the amount and kind of mineral ingredients in the char.

Activated carbons are structurally similar to turbostratic carbon, having microcrystallites only a few layers in thickness and less than 100 Å in width. The level of structural imperfections in activated-carbon microcrystallites is very high, which results in many possibilities for reactions of the edge carbons with their surroundings (1).

Oxygen combines with the carbon to form a physicochemical complex, C_xO_y, of variable composition. The oxygen functional groups suggested as being present on the surface of activated carbon include carboxyl groups, phenolic hydroxyl groups, quinone-type carbonyl groups, lactones, carboxylic acid anhydrides, and cyclic peroxides. Carboxylic, lactone, and phenolic groups are the acidic surface oxides (2). The nature of the basic surface oxides is less understood, and no fully convincing formulation has been proposed so far.

Besides oxygen, hydrogen is present in most carbon surfaces. The hydrogen not only forms part of the oxygen functional groups, but it is also directly combined with carbon atoms. Hydrogen is more strongly chemisorbed than oxygen and is difficult to remove from the surface (3).

It is evident that the physical and chemical properties of the carbons depend on the activation procedure employed in the manufacturing or regeneration process. Activation procedures should be controlled to optimize the adsorption of specific compounds from water and maintain maximum capacity during regeneration. Implementation of proper activation and regeneration procedures requires basic knowledge of the mechanisms by which particular adsorbates or groups of adsorbates of the same nature adsorb.

The adsorption of many organics, especially phenol and its derivatives, on different carbons has been studied extensively, but the biggest difficulty in drawing conclusions about the adsorption mechanisms arises from lack of information on the surface characteristics of the carbons used in the different studies.

Mullins et al. (4) clearly showed that the variation in adsorptive characteristics of different carbons for the removal of trihalomethanes is typically in the 200–300% range, indicating a strong influence of the carbon's characteristics on adsorptive capacities.

The present study tries to correlate the effectiveness of different carbons for the removal of a specific adsorbate of major concern, chloroform (5), with the physical and chemical properties of the carbon surface.

Experimental

Carbons. The activated carbon used was Filtrasorb 200 (F200), granular carbon produced from coal by high temperature steam activation. The mean particle diameter of the carbon used was 606 μm. The amount of oxygen surface groups on the original carbon was reduced by heat treating the carbon at 1000°C for 17 h in an inert atmosphere. The carbon was then allowed to cool to room temperature under inert atmosphere and kept under these conditions until used for its characterization and adsorption studies. This carbon will be identified as OG.

The F200 carbon was washed several times with distilled water and oven-dried in thin layers at 105°C for 24 h. The OG carbon was used without further treatment.

The physical and chemical characteristics of the surface of the two carbons were evaluated as follows. Specific surface area measurements were made by the BET technique, and pore size distributions were made using a Monosorb surface area analyzer model MS-4 and Quantasorb adsorption system model QS-7, respectively. Direct transmission infrared spectroscopic studies of 0.25% in weight KBr tablets coupled with standard neutralization techniques were used to investigate the nature and distribution of the surface functional groups on the different carbons. The technique used for the preparation of the KBr pellets has been described previously (6).

Table I summarizes the specific area and average pore size of the different carbons. Figure 1 shows the pore size distribution for these carbons in the 10–300 Å pore radius range.

The neutralization capacities of these carbons for different neutralizing solutions are given in Table II, and the acidic surface group distribution according to Boehm (2) are presented in Table III. Figure 2 shows the infrared direct transmission spectra of the carbons. A thorough discussion of the structure of the surface oxides present on the different carbons has been presented elsewhere (6,7). The spectra as well as the neutralization capacities indicate that there has been a reduction in surface oxides for the OG carbon, as expected.

Table I. Specific Surface Area and Average Pore Size

Carbon	Specific Surface Area (m^2/g)	Average Pore Size in the 10–300 Å Range
F200	655	23
OG	647	20

Table II. Neutralization Capacities for Different Neutralizing Solutions (in mEq/g)

Neutralizing Solutions (~0.25 N)

Carbon	$NaHCO_3$	Na_2CO_3	$NaOH$	HCl
F200	0.07	0.26	0.74	0.50
OG	0	0.05	0.42	0.52

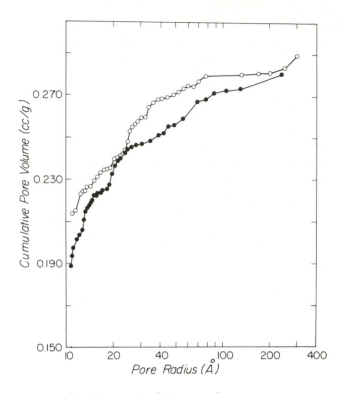

Figure 1. Pore size distributions for different carbons. Key ○, F200; and ●, OG.

Adsorption Studies. All adsorption studies were conducted in batch systems consisting of 125-mL glass vials with "Mininert" Teflon sealing valve caps. The initial chloroform concentration was 300 μg/L for the equilibrium and kinetic studies. The carbon dosages varied between 0.2 and 1.0 g/L. The stock chloroform solution was prepared with distilled water freed from volatile organics and gases by nitrogen gas striping. The pH of the solutions was not buffered to avoid the introduction of additional species into the water–chloroform–activated-carbon systems.

A 48-h contact time was selected for the equilibrium studies based on the adsorption kinetics. A Burrell wrist-action shaker was used for agitating the samples. Controls without carbon were run in all cases.

Analytical Procedure. A model 3700 Varian gas chromatograph with a CDS 111 Varian data system was used for chloroform analysis. This chromatograph was equipped with a flame-ionization detector and temperature programming. A 1.8-m (6-ft) long 0.3-cm (1/8-in.) diameter stainless steel column filled with 100–120-mesh Chromosorb 101 and 5% Carbowax was used. The samples were concentrated using a Teckmar model LS C-2 liquid concentrator.

Table III. Distribution of Acidic Functional Groups According to Boehm's Selective Neutralization Capacities

Carbon	Phenolic Groups (%)	f-Lactone Groups (%)	Carboxylic Groups (%)
F200	65	26	9
OG	88	12	0

Results and Discussion

Equilibrium Studies. The equilibrium isotherms for the adsorption of chloroform on the two carbons investigated are presented in Figure 3 (on a weight basis) and in Figure 4 (normalized for the corresponding surface areas). It is observed in these figures that, for the equilibrium concentration range under consideration (10–200 μg/L), the OG carbon exhibits a stronger affinity and larger capacity than the F200 original carbon. The equilibrium data are plotted according to the Langmuir equation in Figure 5. The equilibrium constants (K) and monolayer coverages (Z) obtained from these plots are shown in Table IV.

Considering the relationship between the equilibrium constant (K) and the adsorption energy given by the equation:

$$K \propto e^{-\Delta H°/RT} \tag{1}$$

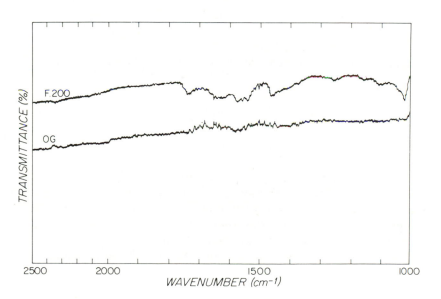

Figure 2. IR direct transmission spectra of different carbons, top trace (F200) and bottom trace (OG).

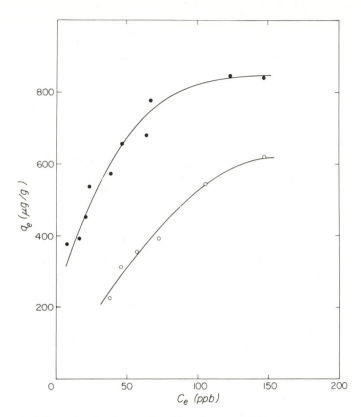

Figure 3. Adsorption isotherms for chloroform on different carbons on mass basis. Key: ○, *F200; and* ●, *OG.*

and using the experimentally obtained values of K (Table IV), it follows that:

$$\Delta H^{\circ}_{OG} = 2.6 \; \Delta H^{\circ}_{F200} \qquad (2)$$

According to the Polanyi approach to model adsorption, the distribution of adsorption energies follows the pore size distribution, and it is limited to London force adsorption. Rozwadowski et al. (8) in studying the adsorption of methanol on different active carbons reported that the experimental curves differed markedly from the theoretical ones calculated from benzene isotherms assuming that the theory of volume filling is fully satisfied.

As has been recognized by Manes (9), even though the theory of volume filling has been used successfully, for a variety of compounds it does not always work, especially because the model does not take into consideration the nature of the adsorbate (e.g., polarizability) or the chemical characteristics of the adsorbent surface. The present study is

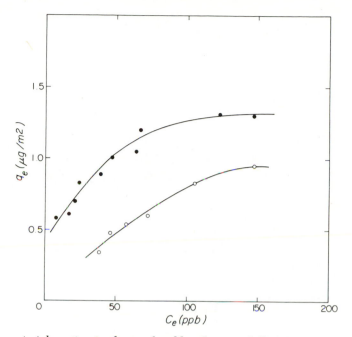

Figure 4. Adsorption isotherms for chloroform on different carbons on area basis. Key: ○, F200; and ●, OG.

another example where the Polanyi model would fail to predict the actual behavior of the systems.

As seen in Figure 1, the pore size distributions for the OG and F200 carbons are similar, but the adsorptive behavior of these two carbons for chloroform (Figures 3–5) are different, especially with respect to the adsorption energies. Therefore, it is evident from these results that the explanation must be other than merely differences in pore size distribution for the adsorbate under consideration.

Consider now the net adsorption energy concept presented by McGuire and Suffet (*10*) and the chemical characteristics of the different carbons to explain the observed results. According to these authors, the net adsorption energy is given by:

$$E_T^A = E_{js}^A - E_{is}^A - E_{ji} \tag{3}$$

Table IV. Langmuir Parameters for the Adsorption of Chloroform on Different Carbons

Carbon	Equilibrium Constant (L/mg)	Monolayer Coverage (mg/g)	(mg/100 m²)
F200	4.1	1.75	0.27
OG	39.6	1.00	0.15

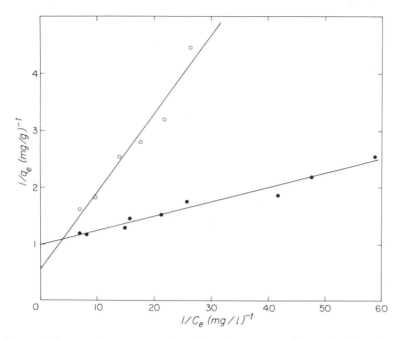

Figure 5. Langmuir representation of the adsorption isotherms for chloroform on different carbons. Key: ○, *F200; and* ●, *OG.*

where E_T^A is the net adsorption energy, $E_{j\,s}^A$ is the energy of affinity of the solute for the surface phase, $E_{i\,s}^A$ is the energy of affinity of the solvent for the surface phase, and E_{ji} is the energy associated with the affinity of the solute with the solvent phase.

For the systems under consideration in the present study, the term E_{ji} is the same in both cases, but the terms $E_{j\,s}^A$ and $E_{i\,s}^A$ are different due to the chemical nature of the carbons used. The differences in the $E_{i\,s}^A$ components for the water carbons will be primarily due to the difference in total amount of oxygen present on the carbon surface, and therefore E_{is}^A for the F200 carbon should be larger than the E_{is}^A for the OG carbon. Since this energy term should be subtracted from the energy of affinity of the solute for the surface phase, as a result, even if the two $E_{j\,s}^A$ terms would be the same, the E_T^A value for the OG carbon would be larger than for the F200 carbon, as it is actually observed experimentaly $[E_{T(OG)}^A = 2.6\ E_{T(F200)}^A]$.

Consider now the $E_{j\,s}^A$ values themselves. According to Rozwadowski et al. (8) when considering the adsorption of polar substances, the participation of at least the following types of interactions should be considered: disperse interactions of adsorbate–adsorbent, specific interactions of adsorbate–adsorbent, and associations of adsorbate–adsorbate. The relative share of these interactions in every case leads to very different E_{js}^A values, the greater the share of specific interactions, the greater the expected adsorption energy.

With the information available, it is not possible to draw conclusions about the relative values of the E_{js}^A terms for the two carbons. It appears from the isotherms that the adsorptive sites on the OG carbon are more limited, as expected, but more specific for the adsorption of this particular adsorbate relative to the solvent. For low adsorbate concentrations, the contribution of the energy of affinity of the solvent for the surface phase plays a more important role in defining the differences in net adsorption energies than the energies of affinity of the solute for the surface phase.

The following general comments should be made on adsorption capacities: From Figures 3 and 4, within the range of measured equilibrium concentrations, the OG carbon exhibits the highest capacity both on a mass and area basis. When the Langmuir equation is used and extrapolation is made to obtain the intercept from which the monolayer coverage is obtained, the corresponding equations for the different carbons cross each other indicating that the F200 carbon has a greater capacity than the OG carbon at equilibrium concentrations greater than 270 μg/L. The isotherms indicate a greater dependency of availability of surface sites on solute equilibrium concentration for the F200 carbon than for the OG carbon; therefore, for high solute concentrations, other adsorptive sites, present on the surface of the F200 carbon and not on the OG carbon, would be filled.

The results presented by Chudyk et al. (*11*) on the adsorption of chloroform by different carbons and a carbonaceous resin support the findings of this study in the sense that the pore size distribution of the adsorbing material is not relevant for the adsorption of chloroform and that the carbonaceous resin can perform as well as the activated carbon. The decrease in capacity observed by these authors after baking the carbons at 450°C cannot be explained with the information provided.

The results presented by Puri (*12*) for the adsorption of phenol on different carbons with different chemical surface characteristics also indicate that the behavior of the different carbons, so far as capacity is concerned, changes with the equilibrium concentration.

Further investigations are being conducted to gain better insight on the effect of specific sites for the adsorption of chloroform and the dependency on equilibrium concentration.

Kinetic Studies. The adsorption of chloroform as a function of time for the different carbons is shown in Figure 6. The carbon dosage used in these experiments was 1.0 g/L, and the initial chloroform concentration was 300 μg/L.

A trend similar to the equilibrium results is observed for the rates of adsorption. The OG carbon exhibits a stronger affinity for chloroform and a faster adsorption rate.

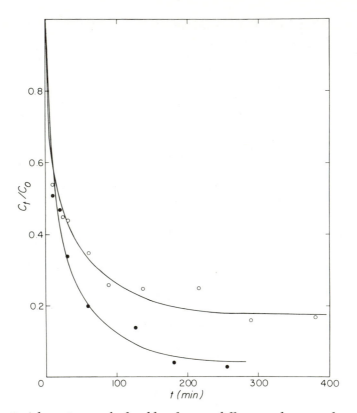

Figure 6. Adsorption uptake for chloroform on different carbons as a function of time. Key: O, F200; and ●, OG.

Since the solutions were not buffered (as indicated in the experimental procedures) to avoid competition of the buffer species for sites on the carbon surface, the pH values of 1.0 g/L carbon solutions after 48 h of contact time are given in Table V. As indicated, the pH for the OG and F200 carbon solutions are similar, and the observed different adsorption behaviors are not due to differences in pH.

The main purpose of the kinetic studies was to establish correlations between the adsorption rates and the chemical and physical characteristics of the carbon surfaces. Again, the pore size distribution did not play an important role in controlling chloroform adsorption, and the main explanation for the observed increased rate shown by the OG carbon should be regarded as related to the chemical characteristics.

Contrary to what was observed previously for the adsorption of butyl disulfide and decane (*13*) for which the adsorption rates were drastically influenced by the presence of oxygen on the surface, for chloroform (this study) and *p*-hydroxybenzaldehyde (*13*), the influence is less pronounced.

Table V. pH of 1.0-g/L Carbon Solutions after 48-h Contact Time

Carbon	pH
F200	6.1
OG	6.8

Summary

Chloroform adsorption was studied using two different carbons, for which the chemical and physical characteristics of its surfaces were evaluated (Figures 1 and 2 and Tables I–III).

The equilibrium studies indicate different adsorptive behaviors for these systems. For equilibrium concentrations in the 10–200 μg/L range, the observed differences in adsorption isotherms (Figures 3–5) cannot be explained by differences in pore size distributions, but they can be explained by the chemical nature of the carbon surface. In this regard, the net adsorption energy concept presents a more realistic approach for the modeling of adsorption systems than the theory of volume filling.

The carbon with the least amount of oxygen (OG) exhibits the highest affinity for chloroform ($K = 39.6$ L/mg).

The relative capacities of the carbons to adsorb chloroform are a function of the solution equilibrium concentrations (Figure 5), showing that, for solution equilibrium concentrations below 270 μg/L, the OG carbon is capable of adsorbing more chloroform than the F200 carbon. These findings question the validity of the extrapolation made using the Langmuir equation to estimate the monolayer coverage and the comparison made using these values for different carbons when trying to adsorb organics present in water at very low concentrations.

As a general conclusion, the adsorption of organics from the aqueous phase onto activated carbon, even when dealing with simple systems as the ones presented here, is a very complex and dynamic process, and even though at present we do not have enough knowledge to be able to model or predict to a great extent the behavior of real systems, basic research should be encouraged to fill in the gaps.

Literature Cited

1. Mattson, J. S.; Mark, H. B., Jr. "Activated Carbon Surface Chemistry and Adsorption from Solution"; Marcel Dekker: New York, 1971.
2. Boehm, H. P. In "Advances in Catalysis", Eley, D. D.; Pines, H.; Weisz, P. B., Eds.; Academic: New York, 1966; Vol. 16, p. 179.
3. Puri, B. R. In "Chemistry and Physics of Carbon", Walker, P. L., Jr., Ed.: Marcel Dekker: New York, 1970; Vol. 6, p. 191.
4. Mullins, R. L., Jr.; Zogorski, J. S.; Hubbs, S. A.; Allgeier, G. D. In "Activated Carbon Adsorption of Organics from the Aqueous Phase"; Suffet, I. H.; McGuire, M. J., Eds.; Ann Arbor Science: Ann Arbor, Mich. 1980; Vol. 1, p. 273.

5. Ishizaki, C.; Van der Biest, L. *Interciencia*, **1979**, *4*, 333.
6. Ishizaki, C.; Martí, I. *Carbon*, **1981**, *19*, 409.
7. Marti, I., "Estudio de Modificaciones Químicas y Físicas en la Superficie de un Carbón Activado", Trabajo de Grado, Universidad Somón Bolír, Venezuela, 1979.
8. Rozwadowski, M.; Siedlewski, J.; Wojsz, R., *Carbon*, **1979**, *17*, 411.
9. Manes, M. In "Activated Carbon Adsorption of Organics from the Aqueous Phase"; Suffet, I. H.; McGuire, M. J., Eds.; Ann Arbor Science: Ann Arbor, Mich., 1980; Vol. 1, p. 43.
10. McGuire, M. J.; Suffet, I. H. In "Activated Carbon Adsorption of Organics from the Aqueous Phase"; Suffet, I. H.; McGuire, M. J. Eds.; Ann Arbor Science: Ann Arbor, Mich. 1980; Vol. 1, p. 91.
11. Chudyk, W. A.; Snoeyink, V. L.; Beckmann, D.; Temperly, T. *J. Am. Water Works Assoc.*, **1979**, *71*, 529.
12. Puri, B. R. In "Activated Carbon Adsorption of Organics from the Aqueous Phase"; Suffet, I. H.; McGuire, M. J., Eds.; Ann Arbor Science: Ann Arbor, Mich. 1980; Vol. 1, p. 353.
13. Ishizaki, C.; Cookson, J. T., Jr. In "Chemistry of Water Supply, Treatment, and Distribution"; Rubin, A. J.; Ed.; Ann Arbor Science: Ann Arbor, Mich. 1974; p. 201.

RECEIVED for review August 3, 1981. ACCEPTED for publication May 4, 1982.

Discussion I
Theoretical Approaches

Participants

Katherine Alben, New York State Department of Health
W. Brian Arbuckle, University of Florida
Georges Belfort, Rensselaer Polytechnic Institute
Andrew Benedek, McMaster University
Francis A. DiGiano, University of North Carolina
Milton Manes, Kent State University
Michael J. McGuire, The Metropolitan Water District of Southern California
Alan L. Myers, University of Pennsylvania
Paul V. Roberts, Stanford University
Michael R. Rosene, Calgon Corporation
Irwin H. Suffet, Drexel University
Makram T. Suidan, University of Illinois
Walter J. Weber, Jr., University of Michigan

SUFFET: What are the similarities and what are the differences that we see developing from the different theoretical approaches that people are taking? I asked this two years ago, and two years ago we didn't have many good answers. Now I think we're getting there.

MANES: Since the ideal adsorbed solution theory—at least for liquids—and the Polanyi theory give pretty much the same results, one would expect them to be related.

As a matter of fact, in the Grant–Manes papers of 1962 and 1964[1], where we did multicomponent adsorption of gases, we pointed out the relationship between the Polanyi model and the Myers–Prausnitz model. What we proceeded to do was to evaluate adsorbates at equal total volume. The Myers–Prausnitz model would evaluate them at equal spreading pressure.

[1]Grant, R. J., M. Manes, and S. B. Smith, "Adsorption of Normal Paraffins and Sulfur Compounds on Activated Carbon," *Am. Inst. Chem. Eng. J.* 8:403 (1962); and Grant, R. J., and M. Manes, "Correlation of Some Gas Adsorption Data Extending to Low Pressures and Supercritical Temperatures," *Ind. Eng. Chem. Fundam.* 3:221 (1964).

0065-2393/83/0202-0107$06.00/0
© 1983 American Chemical Society

This spreading pressure turns out to be equal to that area I showed you under the correlation curve, and if you have liquids of what I would call approximate scale—the same scale factor—you get exactly the same results. So the two do come together.

Now, as I said, there are places you can find systems where the IAS theory breaks down, and one would say in fairness that the IAS theory was claimed to be accurate only in dilute solutions. For some liquids, actually the IAS theory is probably better than the authors think because it can go up to high concentrations, and it doesn't break down.

So, again, it depends on what you're interested in. From our physical chemists' point of view, we like the Polanyi model better because we can think in it and because it yields many qualitative observations that are worthwhile, some of which I will get a chance to list for you. But there is a strong relation in the mathematics. If we now consider the net adsorption energy concept, of course, the Polanyi model is deeply involved with adsorption energy.

The solubility parameters that McGuire and Suffet use are deeply related, certainly, for nonpolar materials. The solubility parameters are again deeply related to the refractive index. So there, too—at least for the multicomponent liquids—we're really saying pretty much the same thing, and the question is which is better.

Now, let's consider a correlation between these theories and the octanol–water partition coefficient. Dr. Chiou, Oregon State, is interested in octanol-water partition coefficients, and the environmental people are interested in these too.

It turns out that the octanol–water partition coefficient correlates very, very well with solubility, and the reason is this: Most solutes dissolve almost ideally in octanol, and therefore you may have an activity coefficient of two or three at the very most, which represents the incompatibility.

Then the rest of the partition coefficient is nothing but water solubility. Now, I think here we have to be kind of careful in what the statisticians call correlated variables. There are certainly dozens of variables you can think of, and what we are trying to do is not to find more variables, but to see whether we can reduce them to what we consider to be essential. In many cases, for example, a surface area correlates with molecular volume, which is going to correlate with molar polarizability.

The question is which one you like. Any one of them is going to correlate with strength of adsorption, and our job is to see if we can find two or three variables that leave us nothing else to do.

Branching does not fit comfortably into the Polanyi model, although there are ways one can make it do so. In the Hofer–Manes paper of 1969[2],

[2]Hofer, L. J. E., and M. Manes, "Adsorption on Activated Carbon from Solvents of Different Refractive Index," *Chem. Eng. Prog.* 65:84–8 (1969).

we had recognized that some substances adsorb anomalously weakly and could not be made to fit on a plane surface (carbon tetrachloride, for example, was one of them); and at that time we had guessed that this was because they were far away from a plane surface.

Then in the Chiou paper of 1972[3], we tried to see just how bad we could make branching. Are you familiar with this paper, where we adsorbed some square planar and some octahedral complexes of copper, and we figured that an octahedral complex was about as miserable a steric fit as one could think of to a plane surface? There we picked up some major effects.

Now, I think one good way if one is going to investigate the effects of branching (we haven't, just because of the pressure of other things)—if, for example, one were to experimentally find correlations between branching and what I call the scale factor—one could very well find group contributions which would be relatively constant, and so I think it should become possible to fit branching comfortably into the Polanyi model.

BELFORT: I think that there should always be a diversion of attempts and direction, and that maybe now and then we need to be reminded to get back to simplistic models and simplify everything. But I don't think there's anything wrong with attempting to look in detail at all the interactions and trying to incorporate an overall general model.

After all, what are we doing at universities if we're not looking for the truth? On the other hand, we have the pressure from reality, and that is to get funding, and funding often requires us to become less complicated and move in the direction of simplified models. So I don't think there's anything wrong with diverting in the two directions.

I would like to make a point, and that is that what we've discussed up till now are equilibrium models, and often the limitations of adsorption of species are determined kinetically or dynamically by mass-transfer co-efficients across mass-transfer films and by internal diffusion with a pore or surface diffusion. So we shouldn't forget that maybe we can correlate something in some idealized laboratory and get great fits, or let's say we can fit adsorption isotherms by using various indices to collapse data.

But in the end it may also be kinetically determined, and I think there are some papers in this conference that will discuss the kinetics of slow and fast adsorption, the MADAM model from Professor Weber, et cetera.

WEBER: I would ask this question to the panel: Doesn't the relative utility of any one of the different techniques that you discussed in this session depend upon one's objectives in examining adsorption data?

If you're simply interested in roughly screening the relative ad-sorbability of a long list of compounds, and you don't want to do a host of

[3]Chiou, C. T., and M. Manes, "Application of the Polanyi Adsorption Potential Theory to Adsorption from Solution on Activated Carbon IV. Steric Factors, as Illustrated by the Adsorption of Planar and Octahedral Metal Acetylacetonates," *J. Phys. Chem.* 77:809 (1972).

measurements, wouldn't reference to a lumped-parameter physical/chemical property like the water–octanol partition coefficient suffice? Between that and the ultimate application of such information—namely, as input to the design of an adsorption system—there's a range of different levels of sophistication. Clearly the octanol–water partition coefficient does not say anything about the adsorbent, and so the next step in the process would be to go to Professor Myers' type of analysis of relative energies of immersion.

But in the design of an adsorption system where one applies a model like MADAM, it is necessary to have a quantitative description of the equilibrium relationship between adsorption capacity and concentration remaining in solution for the particular system you are designing an adsorber for. My experience is that none of these predictive equilibrium models are really sufficient, because a given solute in two different waters can behave remarkably differently because of the effects of background constituents which are not identified.

There may be humic acids and tannins and lignins. We know nothing about them. We know nothing about their physical/chemical properties. How do we characterize the adsorption relationship other than by doing an adsorption isotherm?

And if we have to do that, then, is not something like the Freundlich equation as good as anything? We need an equation we can use in a mathematical model to do the designs, to do the predictions, and we don't have a sufficient level of information on the physical/chemical parameters of the substances in the solution to induce that type of equation. We have got to deduce rather than induce, I'm afraid.

ALBEN: Professor Manes, your conceptual model of a centrifuge tube seems to me to inspire a concept of liquid films in the micropores. I'm wondering, can you extend your theory to deal with an adsorption capacity of a liquid phase which develops in time as opposed to the adsorptive capacity of the surface? Or do you just ignore it and shut down your carbon filters?

MANES: The Polanyi model is really a capillary condensation model. The capillary condensation model was discredited for a while because it ran into some problems. Now it is well adapted to activated carbon, where practically all of the surface area is in pores, so that conceptually we think of the adsorption being in pores rather than on a surface.

That is one of the reasons why the model says nothing about surface area. A geographical analogy would be something like this: You might be interested in how many small harbors and how many big harbors there are on the Atlantic coast and what their boat-carrying capacity might be, in which case if you knew the length of the Atlantic coastline, it might not help you too much.

So essentially what we're now saying—and there are some contradictions in the model which I could take up with you later—what we're really saying is that the pore is so small that the energies of adsorption may get to be twice the heat of vaporization and yet sufficiently large that the interactions between the molecules in that pore are not too far removed from what they are in bulk.

Now, so far that approximation has served us in good stead, but we run into trouble with it when we start to adsorb solids. Then you find that an adsorbed solid is going to have its own crystal structure, and if that crystal structure is not congenial with the pore structure then you're going to get some voids.

So there is one limitation where we get into trouble. But essentially on activated carbon we think of an adsorption space rather than a surface, and in terms of pore volumes rather than surfaces, and we find that helps a lot.

Again, you see, you have to shuck off a number of things you've been brought up to believe.

ALBEN: One other thing which a couple of speakers touched on was heterogeneity of the carbon surface, and I think that's very interesting to consider. Professor Myers commented that you could treat them independently, and I think experimentally that's interesting too, because if you look at compounds that are in very different chemical classes rather than essentially the same chemical class, you will see their adsorption, but it's almost independent, and they are not really competing. But it is a heteogeneous adsorbent.

MYERS: The nice thing about the IAS approach is that it leads to a distribution which you can use to do other things with. One of the things we're trying to do is to find this distribution by another method so that instead of just being a parameter of the model, it's something we can verify.

ALBEN: Why do you want to have a model that has heterogeneity in it?

MYERS: Because I can use it to make predictions of things that right now we can't predict.

ALBEN: For what compounds?

MYERS: Well, for example, for adsorption of liquid mixtures. Right now you can't use the theory of ideal· adsorbed solution to predict adsorption of liquid mixtures, and I believe that the reason for this is that we're not taking into account the heterogeneity of the surface. I think that's by far more important than any nonidealities due to the interactions of the liquids themselves.

ALBEN: Okay, but would you say then that the surface has extra capacity because it's heterogeneous, and that the molecules are not actually competing because they can find different sites?

MYERS: I'm saying that no matter how much the molecules compete,

it's unimportant relative to the heterogeneity of the surface. The heterogeneity of the surface completely dominates the picture, so that any
molecular interactions, between either like or unlike species, are absolutely unimportant, at least on carbon.

MANES: What we're saying is, you know what the energy of adsorption is if you put a molecule on a block of carbon. If you put it in a
sandwich it's going to have twice as much energy. Essentially, about the
highest energies of adsorption that you get on carbon will be about twice
the energy that you calculate for a model-maker.

So, therefore, if one says that in the fine pores—and this, of course, is
part of the Polanyi model—that in the fine pores your energy of adsorption is bigger simply because you are closer to more carbon, this, in
fact, is what we mean by the heterogeneity from our point of view.

The heterogeneity of the carbon surface means you have fine pores
and you have coarse pores.

ALBEN: I'm curious. What do you people mean by heterogeneity?

MYERS: I don't have a molecular model. I have an energy distribution which we can fit by a function. This is a forthcoming publication
which will give a numerical adsorption isotherm rather than a Polanyi
potential.

I have an analytical distribution function. I don't yet have a molecular
model for calculating that distribution. That's what we're working on.
That's more complicated.

MANES: I think this is an interesting point.

SUIDAN: The energy of distribution has to be a function of the
extent of surface coverage. When you look at different molecules that are
being adsorbed in single-solute systems, and you come back to reduce
them all to one expression, how is the effect of that molecule—which has
been adsorbed and is reflecting the energy of distribution of the remaining sites—taken into account?

MYERS: Take a pair of sites. Different solutes will have different
energies on different pairs of sites, but the ratio will be the same.
Therefore, in terms of theta the distribution will be the same for both
solutes.

SUIDAN: Wouldn't it be affected differently by different extents of
coverage by different molecules?

MYERS: Different sites will have different ratios. Pick any pair of
sites, pick any pair of solutes, and the ratio of energies will be the same.

SUIDAN: I would have thought that a covered site is more influenced by the new molecule than it is by the original carbon surface, and
I was wondering whether the type of molecule that is adsorbed will have
more of an effect.

MYERS: I'm just saying that the distribution function will look the
same if you plot it versus theta.

BENEDEK: First of all, I agree with Walt Weber; I'm not about to give up on measuring an isotherm. When it comes to doing a modeling or engineering type of work on wastewater, or on drinking water, I would like to have an isotherm.

However, I think that there's a wonderful application for all this theoretical work that we heard today—and I'd like to have the panel comment on it—and that is, once you have a theory you have a tool which you can turn around, and once you understand how adsorption occurs you can try to use the tool that you now have and figure out how to make better carbons or adsorbents.

Can you comment on that?

BELFORT: It's still too early for us to comment relative to the use of a solvophobic approach. However, incorporated in the approach—and this is included in our paper—is that one can without any trouble look at different adsorbents as a function of a particular chosen model group of components—say, some homologous series.

One could then look at and determine why and wherefore different solutes adsorb differently by different adsorbents. So theoretically we have a way of groping at that problem. We have not approached it yet, and we hope to as soon as some of the manufacturers approach me and tell me they will give me some samples. We will be very happy to try to look.

ROSENE: That was a very interesting point Andy brought out. In point of fact, we do use Polanyi theory as a tool in our research and development to characterize adsorbents and to assist us in developing new adsorbents for different applications. And the pore-size distribution or the characteristic curve—or however you want to describe it—which de-scribes the heterogeneous nature of the pore distribution in the material is indeed very important in how a particular material will work in any given application. We certainly do use that as a tool, without doubt.

ARBUCKLE: I have a question for you then, Mike. If you char-acterize adsorbents that way, do you characterize them as they go into regeneration? Because whatever we do with virgin carbon is far different than what we're going to get with regenerated carbon.

ROSENE: Absolutely. We look at the material on a cyclic basis, obviously, depending on how well or how poorly you regenerate; you change the pore-size distribution, and you can go a long way towards explaining why some reactivated materials don't work very well by looking at what you've done to the pore structure after you've regenerated it. You can take a perfectly good virgin carbon and make it into something that's not very good by improperly regenerating it.

ARBUCKLE: Is this all proprietary, or do you plan on publishing that?

ROSENE: I think it's pretty well known that if you don't regenerate the adsorbent very well you change the pore structure.

ARBUCKLE: You mentioned factors which affect the regeneration.

ROSENE: It depends on the application, obviously, which conditions you're going to use to regenerate. Industrial waste applications are a much more difficult problem than, say, drinking water applications because of the type and the amount of adsorbed materials that you put on the carbon. The pore-structure changes that occur can materially affect the performance of the material.

So it's important to look at the material on a cyclical basis. But we can do that using a pore-size characterization, essentially, to look at the change in the adsorbent, and the Polanyi theory obviously is a very integral part of looking at it.

The Polanyi model is obviously an equilibrium model, and if you're looking at adsorption kinetics, changes in the pore structure there will affect the intraparticle effective diffusivity, which can materially alter the performance, again, in an actual column system.

ROBERTS: I have a two-part question about pore-size measurement and the effect of pore-size distribution. I'm surprised that there's such optimism about being able to measure pore sizes in a meaningful way in activated carbon.

MANES: I'm not optimistic at all. I was going to say something about that, but thanks for saying it first.

ROBERTS: But if we could measure it, and if we found that a substantial portion of the adsorption capacity is present in pores that are of the same order of magnitude and size as the adsorbing molecules, what would that say to us about the applicability of the alternative models, about which one would be better suited for that situation?

MANES: First of all, I agree with you about pore-size distribution. Pore-size distributions are based on the use of the Kelvin equation, and the Kelvin equation works fine. But when you get down to the size pores that you're interested in, the Kelvin equation has something called the surface tension, and who the hell is to know what a surface tension is when you are three molecules or four molecules thick? So the Kelvin equation really falls apart.

Secondly, you have to make an assumption that the pores are going to be slits or circles, and you really have no way of knowing what shape they are. Actually, one of the ways the Polanyi model gets around this is to characterize the pores by the energy of adsorption, which is what you're interested in.

Now, to a certain extent the specifications in the carbon business— maybe in some others—have been dominated by some former great workers in the field, and because of this we keep thinking in terms of specifying pore-size distributions in surface areas. But you give a user a pore-size distribution and a surface area, and he does not know how to get an isotherm for something he's interested in. If you now specify in terms of a Polanyi correlation curve or the adsorption isotherm of some substance that you can measure over a wide range of capacities, then the customer can pick this up and estimate an adsorption isotherm for many things.

Now, to get back to the size. We have—and I'm sure other people have—observed molecular sieving effects, so if, for example, you get to something like octane or decane, you may find that the calculated energy of adsorption is not as big as you might think it is or the adsorption isotherm drops from where it ought to be.

For example, this could be in terms of pore-size as in molecular sieves, but I rather suspect it's like pore depth. In other words, if you're a snake and you go into a shallow hole, your head and your midsection may be in there, but your tail keeps sticking out, and that doesn't help any as far as your energy of adsorption is concerned.

So one would imagine that on carbons when one does get molecular sieving effects, usually what happens is, the kinds of big molecules that give you these big effects are so insoluble in water anyway that you get very powerful adsorption.

But I have no doubt that when you get to big enough—when you get to long enough—molecules, you do run into molecular sieving effects.

DiGIANO: Dubinin's work covers both micropore and macropore models, and the adsorption energy epsilon or whatever we're calling it can be used as either epsilon or epsilon squared.

I wondered why in your work you continue to use Dubinin's macropore model, which gives a nonlinear characteristic curve, if you will. And I guess the question I have specifically is, have you tried to interpret your work by a micropore model, and would you get linear characteristic curves?

MANES: To put it bluntly, we are not really interested in linear characteristic curves. For one thing, if you did have a linear characteristic curve and you reactivated the carbon, you'd have a different curve. So really we have figured that trying to linearize a curve which goes very nicely with three parameters in a polynomial into a computer is really not where we want to spend our limited time, essentially.

The other point is the distinction between macropores and micropores, which Dubinin makes a big deal about because he's still trying to straighten out the Dubinin–Radiskovitch equation after 20 or 30 years. We really have not been interested in it, and I don't think it's a good way to go.

MYERS: The Dubinin–Radiskovitch equation, as many people here know, blows up in the region of low coverage. However, that's a very important region because that's precisely where you get the Henry's constant. It seems to me if you're going to follow up on the kind of thing that Professor Weber was discussing, getting good molecular information, the first thing you want is the Henry's constant.

So what we've been trying to do is to correct the Dubinin–Radiskovitch equation so that it does give a Henry's constant. I think this would be a valuable contribution. Also, it would be a useful way of correlating data.

BELFORT: I have a question to three people on the table here and the panel. It's a dual question because it concerns two theories that were described today.

One of them—if I could direct this to Professor Manes—is in his chapter[4] in the book that was published of the Miami meeting, where Professor Manes mentions very clearly that one of the parameters that we really don't know much about, or how to estimate, is the density. My question is, do we know more after these few years on how to estimate the density of the adsorbed film? Because evidently one of the discreditations of the capillary condensation model in the 1920's—in fact, Polanyi himself discredited the model—was that we can't estimate across that film or that adsorbed layer; we have difficulty estimating the three-dimensional properties of that adsorbed layer because we have enormous potential differences or pressures, if you will.

So that's the first question: Can we estimate that a priori?

The next question I have is about the adsorption energy theory. I notice that in the table that was given—which was taken out of McGuire's thesis[5]—is an arbitrary choice of allocating deltas for hydrogen bonding. If you look at the table that Dr. Arbuckle used, in fact, it goes 0.51, 0.65 for the different complexes.

I ask the question, do we now know how to choose that parameter without an arbitrary choice, or do we need to use that parameter also to fit the model?

SUFFET: I'll answer the second one. The answer is no, and the reason that it was picked was that there was no other way to go to make that estimate at that time.

Now, we'd like to see—and I think everybody would agree—improvements in all of these theories, and I think the advantage of developing theories like this is to get the fundamentals down so that we can go into the field with an understanding.

We don't know how to measure and see things on a molecular level on a carbon surface. We haven't done that well. It is an open research area. We haven't heard anything about it today; nobody seems to be doing it. We're developing surface techniques to look at molecular surfaces, and

[4]Manes, M., "The Polanyi Adsorption Potential Theory and Its Applications to Adsorption from Water Solution onto Activated Carbon," in *Activated Carbon Adsorption of Organics from the Aqueous Phase*, Vol. 1 (I. H. Suffet and M. J. McGuire, eds.), Ann Arbor Mich.: Ann Arbor Science Publishers, Inc., 1980.

[5]McGuire, M. J., "The Optimization of Water Treatment Unit Processes for the Removal of Trace Organic Compounds with an Emphasis on the Adsorption Mechanism," Ph.D. Thesis, Drexel University (1977).

this is the hope, that the next time a carbon symposium is held we see something on this.

And I think that that's where the problem is here—that we're either looking at the molecular level which we're trying to estimate because we don't have a good feeling of what's happening there or we're looking at complex situations in real cases where we have hundreds of compounds competing at one time and we don't have an estimate, and we're using theories to develop an understanding of real cases. And that's what theories are for.

McGUIRE: The table that Georges refers to is the empirical part of the whole process, and that was done because, again, as Brian Arbuckle has stated, I was an engineer sitting at my desk trying to make some decent adsorption predictions.

Unfortunately, when you look at the solubility parameter things fall apart in polar systems, particularly in water, and you have difficulty in predicting the acid and base components of those reactions. Therefore, if you're going to handle that part of the interaction at all you have to find something empirical that will work, and that's what happened.

It was an empirical decision, and it is arbitrary; that's true. But the point is that for what we were looking at it worked.

MANES: I'll take the first point—that for most liquids this is not a problem. It's a problem for either big liquids or solids, so for many liquids if you ignore it you're in pretty good shape.

But the other part is this: Experimentally we ran adsorption isotherms of a number of solids, and we did find that the limiting volume that we got was less than the limiting volume for gases—in some cases half as much; in some cases one-tenth as much. In the cases where it was sensational, like one-tenth as much for some amino acids, we thought that the solid was coming down as a hydrate, and when we switched to methanol as a solvent we got the different results which were in that direction.

No, we do not yet know how to predict what I call the adsorbate density of solids, but, yes, we have a lot of independent evidence to show that this is not simply a parameter that one fits to make one's equations come out right, but that this is a real phenomenon. And there are a number of qualitative observations, some of which will come out in a second paper, that I will tell you which fit it.

Qualitative observation: If you happen to have a strongly adsorbed solid and you now wish to adsorb on it a weakly adsorbed liquid, the liquid will adsorb and appear to go into the supposed interstices of the solid. On the other hand, if we have a weakly adsorbed solid and a strongly adsorbed liquid, then we would expect that the solid would be completely excluded. In other words, if you happen to have an iron sponge floating on mercury, you would expect it to be completely excluded. So these turn out to be real effects.

Another case: These were not handled in an arbitrary fashion just to make equations work out. In Rosene's dissertation[6], where he did work on competitive adsorption of solids, this adsorbate density, which was not thrown in as an arbitrary parameter but which was in fact determined experimentally, did fit the adsorption data better than other numbers did.

So there appears to be this complexity that we have when we're dealing with solids—and particularly when we're dealing with solids and liquids—that seems to be a real complexity that we will come to grips with, and I don't think we're going to get at it by arbitrarily fooling around with a number of variables that are very far removed from what's going on.

Therefore, we're going to continue along this line. If somebody has a good idea for calculating what we call this degree of incompatibility, fine and dandy. We haven't yet. We *have* checked it to make sure or to convince ourselves that we are dealing with a real phenomenon.

[6]Rosene, M. R., "Application of a Polanyi-Based Adsorption Model to the Competitive Adsorption of Solids from Water Solution onto Activated Carbon," Ph.D. Thesis, Kent State University, Oh. (1977).

MODELING AND COMPETITIVE
ADSORPTION ASPECTS

7

Removal of Carbon Tetrachloride from Water by Activated Carbon

WALTER J. WEBER, JR. and MASSOUD PIRBAZARI[1]

University of Michigan, Department of Civil Engineering, Ann Arbor, MI 48109

Significant parameters associated with carbon tetrachloride adsorption by activated carbon were investigated to provide a basis for a more quantitative description of the process. Adsorption–desorption equilibrium studies suggested that the uptake of carbon tetrachloride by activated carbon is readily reversible. The Michigan Adsorption Design and Applications model (MADAM) was generally able to simulate and predict the performance of fixed-bed adsorbers for carbon tetrachloride removal from water. The effectiveness of activated carbon for carbon tetrachloride removal is likely to be adversely affected by the competitive adsorption and displacement effects of background organic substances. The configuration of the adsorber system is another factor that should be taken into consideration.

CARBON TETRACHLORIDE (CCl_4) is a contaminant of many raw water supplies and finished drinking waters (*1–5*). Because it has induced cancer in laboratory animals and is strongly suspected as a human carcinogen (*6*), its presence in potable waters is of concern (*5*).

Activated carbon is capable of adsorbing CCl_4, but there is little known about the quantitative aspects of this adsorption and of factors that govern its effectiveness in water supply applications. The work reported here was designed to identify significant parameters associated with CCl_4 adsorption by activated carbon and to provide a basis for more quantitative description of the process.

[1]Current address: University of Southern California, Los Angeles, CA 90007.

0065-2393/83/0202-0121$12.50/0

Experimental

Solute. Baker Instra-Analyzed CCl₄ (Baker Chemical Co.; Phillipsburg, N.J.) was used. Experimental solutions were prepared in organic-free water (OFW) [deionized distilled water (DDW) passed through a Milli-Q water purifier (Millipore Corp., Bedford, MA)] or a solution of humic acids (Technical Grade, Aldrich Chemical, Milwaukee, WI) prepared in OFW.

Adsorbent. Filtrasorb 400 granular activated carbon (Calgon Corp. Pittsburgh, PA) was sieved using a set of U.S. standard sieves, and selected size fractions were washed with DDW to remove leachable materials and carbon fines. The carbon was then dried to constant weight at 105°C for 12 h and immediately desiccated. After reaching room temperature, the carbon was stored in air-tight glass bottles.

Analysis. Several important points were considered in the selection of the most suitable sampling, concentration, and analytical methods including: specificity, sensitivity, sample size, reproducibility, a detection limit requirement of < 1 μg/L, and quickness and ease (7).

A Varian 2700 equipped with a scandium tritide electron-capture detector was employed for gas chromatographic (GC) analysis of samples in which CCl₄ was present in very low concentration. The chromatograph was equipped with a 1.8-m × 0.3-cm (6-ft. × ⅛-in.) stainless steel column packed with 20% SP 2100/ 0.1% Carbowax 1500 on 100–120 Supelcoport (Supelco, Inc.). To eliminate the problems associated with the characteristically small linear range of response of the scandium tritide electron-capture detector, more concentrated samples were analyzed using a Varian 1700 equipped with a similar column and a Tracor Instruments Model 700 Hall electrolytic conductivity detector, in the chlorine detection mode.

Sample Collection and Concentration. DYNAMIC HEADSPACE. The stripping and preadsorption technique developed by Bellar and Lichtenberg (8) for volatile halogenated compounds was investigated as a potential technique for the analysis of CCl₄ at ≤ 1 μg/L. This technique, although highly efficient and reproducible, required special equipment for the stripping, adsorption, and desorption procedures and demanded considerable time for the large number of samples to be analyzed.

STATIC HEADSPACE. The concentration of a gas dissolved in a liquid is, pursuant to Henry's law for dilute solutions, proportional at equilibrium to the partial pressure of that gas in the vapor phase, or headspace gas. Static headspace analysis was reported to be a sensitive technique for halomethane detection (9). As discussed by Weber et al. (10), the procedure is efficient and reproducible but highly temperature sensitive. Furthermore, the fact that solid, liquid, and vapor phases were all present in the adsorption experiments described here would have required a total mass balance estimation necessitating preparation of two sets of calibration curves.

AQUEOUS INJECTION. Nicholson and Meresz (11) developed a technique whereby aqueous samples are directly injected into the chromatograph using a Sc³H electron-capture detector. In the present work, however, aqueous injection was detrimental to the detector, substantially shortening its useful life.

LIQUID–LIQUID EXTRACTION. Liquid–liquid extraction techniques have attracted interest since the development of specific modifications for the analysis of haloforms (e.g., 12, 13). In general, these techniques involve extraction into an organic solvent followed by GC measurement using an electron-capture detector; a detection limit of less than 1 μg/L can be obtained.

The liquid–liquid extraction technique was the most attractive alternative for the present work, particularly with respect to compatibility with the adsorption experiments. In this procedure, 5–10-mL samples were mixed with 1–5 mL of Phillips 66 pure grade *n*-hexane and 0.5–1 g of Baker Analyzed sodium chloride in 10–20 mL vials fitted with Teflon-coated screw-on caps. The samples were shaken vigorously for 1 min, the hexane was allowed to separate, and solute analysis was performed by injecting a few microliters of the extract into a gas chromatograph.

Adsorption Methodologies. EQUILIBRIUM STUDIES. Adsorption equilibrium experiments for CCl_4 were using both bottle-point techniques and a mini-adsorber-column procedure.

Two distinct methods of preparing and analyzing the isotherms were evaluated in the bottle-point experiments; these were the static-headspace and liquid–liquid extraction techniques already described. Each method yielded results that were reproduced easily by the other method.

The static-headspace method consisted of placing different amounts of granular activated carbon in each of a series of 150-mL air-tight Hypo-Vials which had previously been cleaned carefully and heated in a muffle furnace at 400°C for 2 h. The vials were then filled with 100 mL of background solution spiked with CCl_4, quickly sealed using Teflon-coated septa and crimped-on aluminum caps, and agitated at room temperature for 4 days to achieve equilibrium.

Equilibrium vapor phase concentrations were determined by analysis of 0.1–0.5 mL of the overhead gas from each vial and by comparison with a calibration curve prepared using known weights of compound dissolved in hexane. The vapor phase concentration of each system containing carbon was then compared with the vapor phase concentrations of appropriate carbon-free standards to yield, after correction for vapor solution equilibria, the corresponding residual aqueous phase concentration.

The liquid–liquid extraction procedure was employed in adsorption experiments conducted in a manner similar to that already discussed, except that there were no vapor phase headspaces in the Hypo-Vials used to bring the activated carbon and CCl_4 solution to equilibrium. After equilibrium had been achieved, a 10-mL aliquot of each sample was subjected to a liquid–liquid extraction with the solvent system described earlier.

Four sets of minicolumn studies were conducted to evaluate dynamic adsorption capacities of activated carbon with respect to four different influent concentrations of CCl_4. These concentrations were chosen to cover a broad range, i.e., 50, 200, 600, and 2,000 μg/L. Feed solutions of CCl_4 were prepared in 12-gallon (45-L) glass carboys using OFW. A second carboy of 5-gallon (19-L) capacity was used in each experiment to attenuate variations in influent concentration. A 0.5-mm (ID) capillary column was pierced through the Teflon-coated stopper of the larger carboy to prevent formation of a vacuum as solution was displaced. A precision volumetric pump (Flow Metering, Inc., Model #RP-SY) was employed to transfer the solution through Teflon tubing to a 0.3-m (1-ft) long glass adsorber having an internal diameter of 1 cm and containing 1 g of 50/60 U.S. standard sieve size activated carbon. A flow rate of 1.6 gpm/ft² (65 L/m²/min) was maintained in all column runs. The experimental apparatus and arrangement is illustrated in Figure 1. Effluent and influent samples were collected in glass vials and subjected to liquid–liquid extraction and GC analysis.

RATE STUDIES. Completely mixed batch reactor (CMBR) adsorption rate experiments were conducted for CCl_4—individually and in the presence of humic acids—in carefully sealed 2.6-L glass reactors of the type illustrated in Figure 2. Weighed and prewetted quantities of granular activated carbon were added to

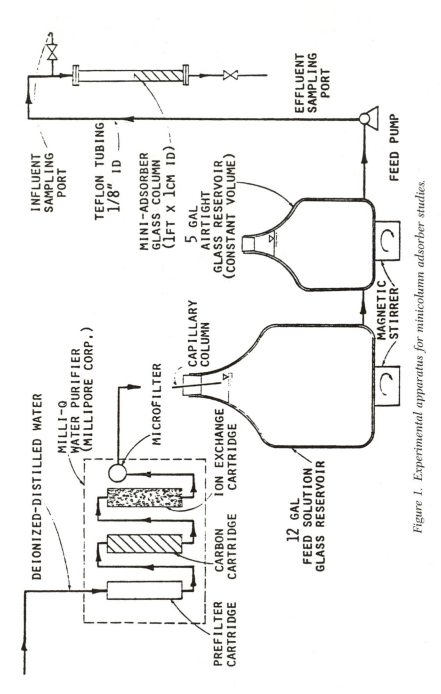

Figure 1. Experimental apparatus for minicolumn adsorber studies.

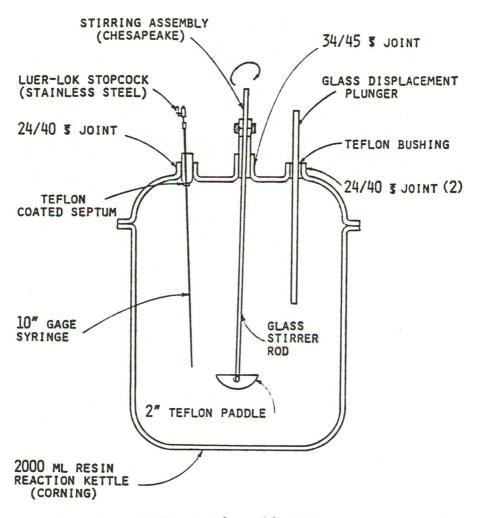

STIRRING ASSEMBLY
(CHESAPEAKE)

34/45 ꙅ JOINT

LUER-LOK STOPCOCK
(STAINLESS STEEL)

GLASS DISPLACEMENT
PLUNGER

24/40 ꙅ JOINT

TEFLON BUSHING

24/40 ꙅ JOINT (2)

TEFLON
COATED SEPTUM

10" GAGE
SYRINGE

GLASS
STIRRER
ROD

2" TEFLON PADDLE

2000 ML RESIN
REACTION KETTLE
(CORNING)

Figure 2. Scheme of the CMBR.

appropriate experimental solutions (pH 7; OFW or 5 mg/L humic acid prepared in OFW background solution) in the vapor phase free reactors. The carbon was dispersed by a motor-driven glass stirrer and 5-mL samples withdrawn at fixed time intervals using a 10-mL hypodermic syringe. A displacement plunger built into the reactor eliminated introduction of headspace by sample volume displacement.

Both dynamic headspace and liquid–liquid extraction techniques were compatible with the methods employed in the CMBR adsorption rate studies with CCl₄. The liquid–liquid extraction procedure was used for the majority of the studies because it involved less time.

The reactor systems and experimental procedures devised for these rate studies were particularly successful. The reactor design eliminated potential solute loss by volatilization and evaporation. Furthermore, because the solution in the CMBR was in contact with only with glass and Teflon, no measurable sorbate loss to containing surfaces was encountered. Elimination of these losses is important when rate studies are performed with very dilute solutions of volatile compounds.

COLUMN STUDIES. Column studies were conducted for evaluation of break-through profiles for CCl_4. Saturated solutions of CCl_4 were prepared in 200-L airtight stainless steel feed tanks. After reaching saturation, the experimental aqueous feed solutions were kept in constant contact with CCl_4 vapor.

Saturated solution was withdrawn by high precision volumetric pumps, diluted to appropriate concentration with high purity water, and then passed through glass mixing chambers to the adsorbers. All transfer tubing was stainless steel and/or Teflon. Each adsorber was comprised of a 1.8-m (6-ft) long glass column with an internal diameter of 3 cm containing 250 g of 16/20 U.S. standard sieve size activated carbon. Influent and effluent samples were collected in 20-mL vials. Care was taken during sample transfer to minimize CCl_4 loss by volatilization. The samples were then concentrated using the liquid–liquid extraction technique and analyzed by GC. Figure 3 illustrates the experimental apparatus and arrangement.

Two major problems were encountered during the experiments. First, the influent concentration fluctuated considerably, perhaps due to blockage of the carbon tetrachloride solution transfer tubing. This problem necessitated frequent maintenance and readjustment. Second, biological growth on carbon particles in the adsorber generated considerable headloss, requiring periodic air-scouring and backwashing. Various inhibitors for preventing biological growth were evaluated to permit singular determination of appropriate adsorption parameters, with only marginal success.

Results and Discussion

Adsorption Equilibria. Several different isotherm models were tested, and the Freundlich equation was found to provide the most straightforward and generally satisfactory method for fitting and describing the experimental data.

Equilibrium data derived from both bottle-point and minicolumn experiments are presented in Figure 4; some small differences in adsorption capacities obtained by the two methods may be observed. Figure 5 illustrates that pH has no significant effect on the equilibrium adsorption characteristics of CCl_4. No dependency of equilibrium capacity on carbon particle size was experienced (Figure 6). Figure 7 demonstrates that initial concentration also has no significant effect on the position of the isotherm.

Figure 8 compares isotherm data for CCl_4 in OFW background solution and in the presence of 5 mg/L humic acid in OFW. Comparative isotherm data for OFW background solutions and Ann Arbor tap water are presented in Figure 9. While the equilibrium capacities for adsorption of CCl_4 appear from the data presented in these figures to be little affected by background constituents characteristic of raw water sources and finished drinking water supplies, these data pertain only to equilibrium conditions and do not address potential competitive effects relating to adsorption rates. Moreover, they pertain only to systems in which initial

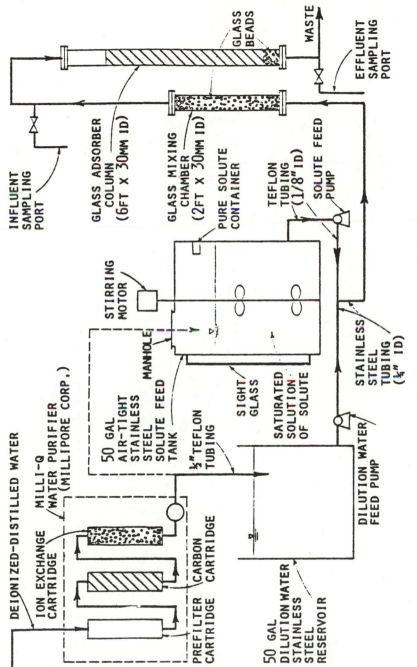

Figure 3. Experimental apparatus for fixed-bed adsorber studies.

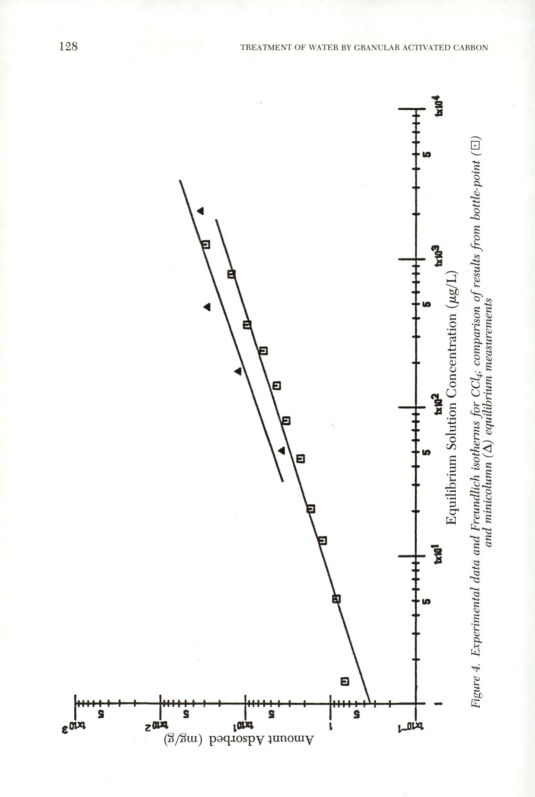

Figure 4. Experimental data and Freundlich isotherms for CCl₄; comparison of results from bottle-point (□) and minicolumn (△) equilibrium measurements

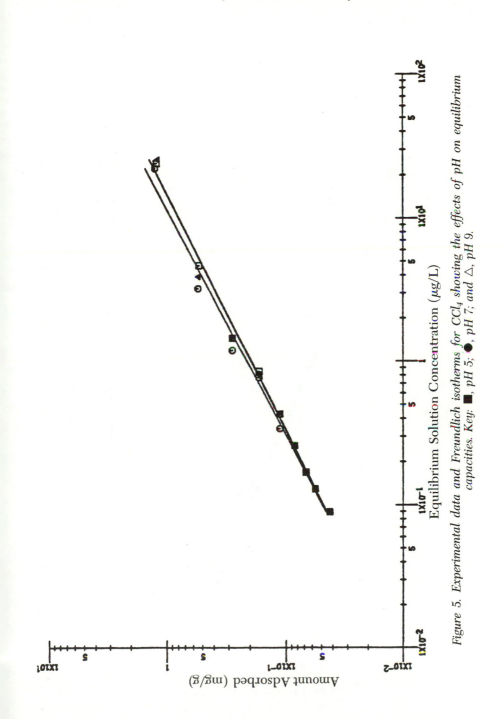

Figure 5. Experimental data and Freundlich isotherms for CCl_4 showing the effects of pH on equilibrium capacities. Key: ■, pH 5; ●, pH 7; and △, pH 9.

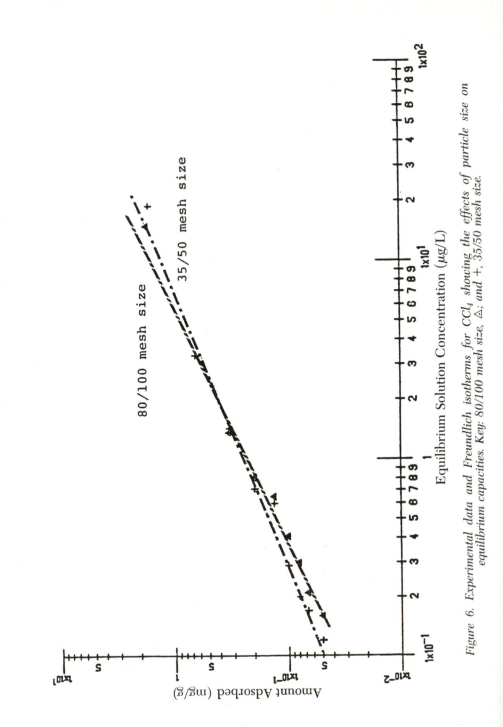

Figure 6. Experimental data and Freundlich isotherms for CCl₄ showing the effects of particle size on equilibrium capacities. Key: 80/100 mesh size, △; and +, 35/50 mesh size.

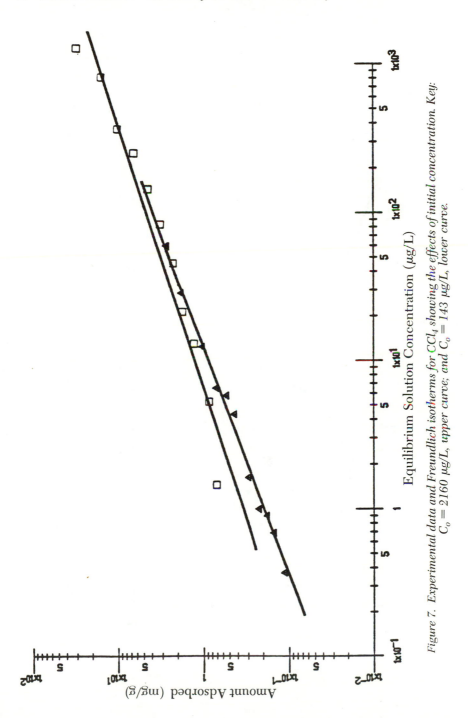

Figure 7. Experimental data and Freundlich isotherms for CCl$_4$ showing the effects of initial concentration. Key: C$_o$ = 2160 µg/L, upper curve; and C$_o$ = 143 µg/L, lower curve.

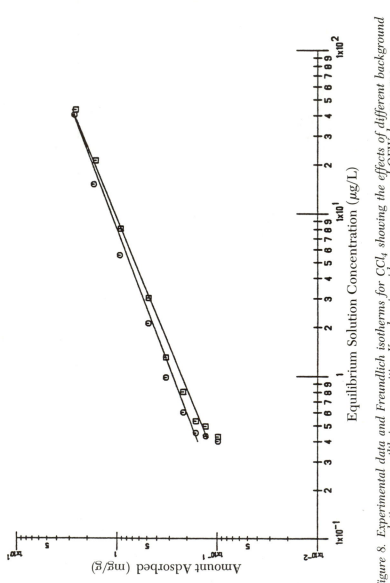
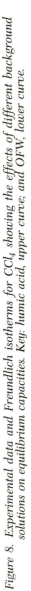

Figure 8. Experimental data and Freundlich isotherms for CCl_4 showing the effects of different background solutions on equilibrium capacities. Key: humic acid, upper curve; and OFW, lower curve.

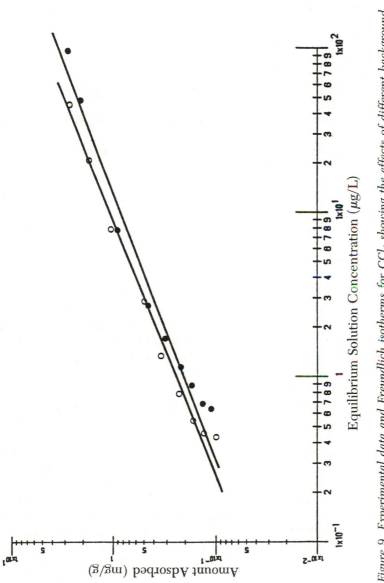

Figure 9. Experimental data and Freundlich isotherms for CCl₄ showing the effects of different background solutions on equilibrium capacities. Key: OFW, upper curve; and tap water, lower curve.

adsorption occurs simultaneously for both CCl_4 and potential competitive substances.

To test the possibility that the absence of competitive effects on equilibrium capacities may be due to more rapid equilibrium of the carbon by the CCl_4 relative to potentially competitive substances, additional equilibrium studies were conducted with carbon that had been previously equilibrated with humic acids. Figure 10 presents the experimental data and the Freundlich isotherms for this system and for the control. It may be observed that the presaturated carbon demonstrated a significantly lower adsorption capacity for CCl_4.

Isotherm parameters for all equilibrium studies are summarized in Table I.

Adsorption Rates. Data from CMBR experiments on the effects of carbon particle size on adsorption rates of CCl_4 are presented in Figure 11. Two mass transport steps—external and intraparticle diffusion—are usually involved in the uptake of organic compounds from aqueous solution by activated carbon, either one of which may be rate limiting (14). It has been observed for intraparticle-controlled systems that incompletely exhausted carbon can achieve some degree of enhanced rate of uptake of organic matter upon standing because of continued diffusion of the organic matter into the capillaries of the carbon during this period, thus freeing adsorption volume for greater rate of uptake upon remixing of the carbon with the solution. The results of similar experiments with CCl_4 (Figure 12) indicate no uptake during a 16-h quiescent period during which mixing of the reactor was suspended and no significant change in the rate of uptake after resumption of stirring. This finding suggests film or external diffusion control, rather than intraparticle control, for CCl_4 adsorption in the CMBR.

A simulation–modeling analysis was performed for adsorption of CCl_4 from OFW to evaluate appropriate rate coefficients, k_f and D_s, for the two-resistance mass transport model, MADAM, used for quantification of the adsorption data and adsorber performance (15–18). A preliminary value for k_f was estimated using a method similar to that discussed by Pirbazari (19) and Pirbazari and Weber (20). The estimated coefficients, $k_f = 4.1 \times 10^{-3}$ cm/s and $D_s = 3.0 \times 10^{-10}$ cm^2/s, were used to predict the experimental profile represented by the solid line in Figure 13. Figures 14 and 15 demonstrate the results of sensitivity analyses conducted for k_f and D_s, respectively. It may be observed from Figure 14 that variations of $\pm 25\%$ in the film transfer coefficient, k_f, result in substantially different predicted profiles, while, as noted in Figure 15, the system is relatively insensitive to changes in D_s. This finding accords with the observation that external mass transfer most likely controlled rates of adsorption of CCl_4 by activated carbon in the experimental CMBR systems.

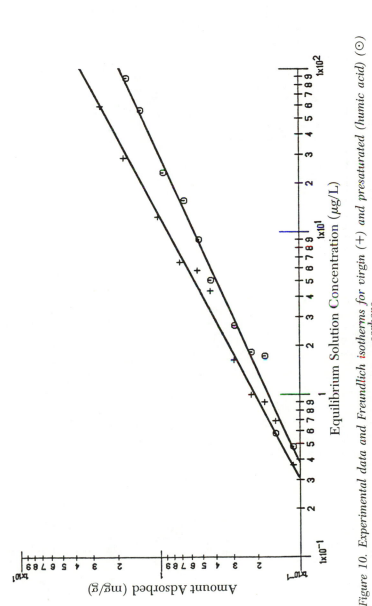

Figure 10. Experimental data and Freundlich isotherms for virgin (+) and presaturated (humic acid) (⊙) carbons.

Table I. Summary of Adsorption Equilibrium Studies for Carbon Tetrachloride

Figure	Background Solution	Carbon Size, U.S. Standard Mesh Size	Initial Concentration (μg/L)	Buffer KH_2PO_4	Freundlich Isotherm Constants[a]			Remarks
					K_f	$1/n$	r[b]	
4 & 7	OFW[c]	30/40	2,160	—	0.344	0.551	0.975	minicolumn
4	OFW	50/60	—	—	0.497	0.585	0.942	
5	OFW (pH 5)	200/325	100	$1 \times 10^{-3}M$	0.209	0.603	0.998	pH effect
5	OFW (pH 7)	200/325	100	$1 \times 10^{-3}M$	0.224	0.630	0.993	
5	OFW (pH 9)	200/325	100	$1 \times 10^{-3}M$	0.208	0.601	0.995	
6	OFW	35/50	100	$1 \times 10^{-3}M$	0.243	0.818	0.954	particle size effect
6	OFW	80/100	100	$1 \times 10^{-3}M$	0.234	0.700	0.995	
8	OFW	16/20	125	—	0.209	0.678	0.965	background solution effect
8	OFW + 5 ppm humic acid	16/20	125	—	0.262	0.674	0.987	
9	Tap water	30/40	178	—	0.195	0.594	0.970	
9	OFW	30/40	125	—	0.215	0.610	0.985	
10	OFW	30/40	142	—	0.169	0.517	0.992	presaturated carbon
10	OFW	30/40	143	—	0.202	0.633	0.996	virgin carbon

Note: All studies were conducted at pH 7 unless otherwise noted.
[a] Based on C_e and q_e expressed in μg and mg/g, respectively.
[b] r = Correlation coefficient for fit of the isotherm to the experimental data.
[c] Organic-free water.

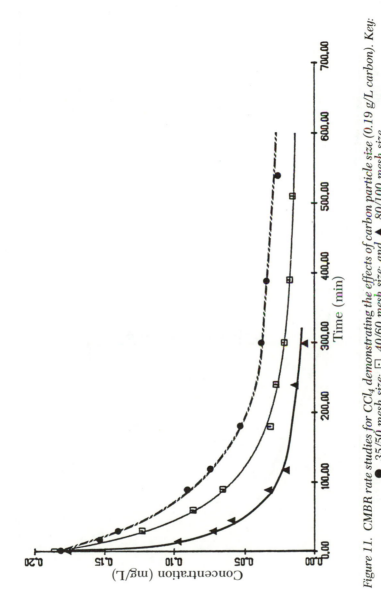

Figure 11. CMBR rate studies for CCl_4 demonstrating the effects of carbon particle size (0.19 g/L carbon). Key: ●, 35/50 mesh size; ⊡, 40/60 mesh size; and ▲, 80/100 mesh size.

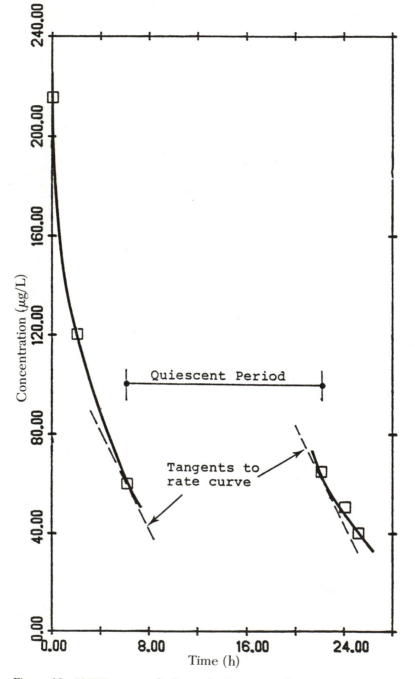

Figure 12. CMBR rate study for CCl$_4$ illustrating the effect of interrupted stirring (0.19 g/L of 16/20 U.S. sieve size carbon).

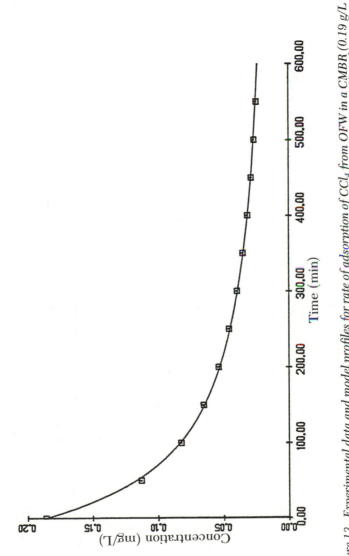

Figure 13. *Experimental data and model profiles for rate of adsorption of CCl_4 from OFW in a CMBR (0.19 g/L of 16/20 U.S. sieve size carbon). Conditions $k_f = 4.1 \times 10^{-3}$ cm/s, and $D_s = 3.0 \times 10^{-10}$ cm^2/s.*

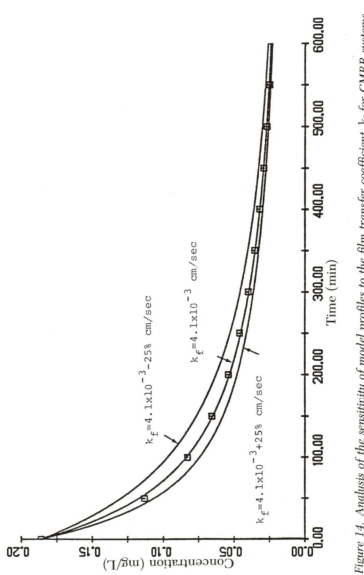

Figure 14. Analysis of the sensitivity of model profiles to the film transfer coefficient, k_f for CMBR systems.

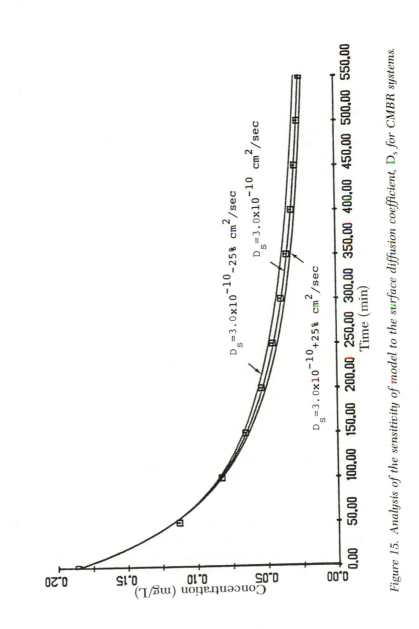

Figure 15. Analysis of the sensitivity of model to the surface diffusion coefficient, D_s for CMBR systems.

A similar simulation procedure was employed for analysis of data from CMBR adsorption rate studies with CCl_4 in the presence of 5 mg/L humic acid. Figure 16 presents the experimental data and the predicted profile (solid line) based on the estimated parameter values of $k_f = 3.8 \times 10^{-3}$ cm/s and $D_s = 1.9 \times 10^{-10}$ cm^2/s. The intraparticle diffusion coefficient for the system containing humic acids was lower than that for the organic-free system, suggesting that the humic acid molecules retard the diffusion of CCl_4 within the carbon pores. Experimental conditions and associated mass transfer coefficients for the systems depicted in Figures 13 and 16 are summarized in Table II.

Adsorption Columns. Experimental fixed-bed column studies for CCl_4 were designed and executed in two phases. Phase I studies consisted of minicolumn tests to obtain preliminary data for predictive model calibration. Phase II studies comprised breakthrough runs with large fixed-bed adsorbers to develop data for model verification. Examples of column data and best fit profiles are given in Figures 17 and 18 for the Phase I and Phase II studies, respectively.

It was assumed that properly designed minicolumn studies would yield necessary information for predictive model calibration. Experiments with the larger columns, which were devised to simulate more realistic adsorber systems and more closely approximate removal efficiencies, encountered operational problems stemming largely from biological growth on the carbon.

Table III summarizes the physical and chemical parameters associated with selected adsorber studies, including values for the surface diffusion coefficients, D_s, and the film transfer coefficient, k_f. Surface diffusion coefficients were estimated from the CMBR rate tests and film transfer coefficients by a correlation technique (21).

MODEL VERIFICATION. Experimental data and predictive breakthrough profiles for typical fixed-bed adsorber runs are presented in Figures 19–21. Good agreement between the experimental data and the predicted breakthrough profiles for Runs I and II is observed (Figures 19 and 20). The experimental data and predicted breakthrough profile associated with adsorber Run III do not agree as well (Figure 21). Two reasons can be advanced to explain this. First, it was necessary to air-scour and backwash the column twice during operation due to air entrapment within the carbon bed and/or excessive headloss caused by biological growth. Air-scouring and backwashing may have disturbed the original bed stratification and/or caused some elution or desorption. Second, the influent concentration fluctuated drastically during the run, imposing marked re-equilibration phenomena in the adsorber bed.

MODEL SENSITIVITY. Model sensitivity analyses with respect to k_f and D_s were conducted for the minicolumn fixed-bed Run I. Figures 22 and 23 demonstrate deviations in breakthrough profiles for variations of $k_f \pm 25\%$

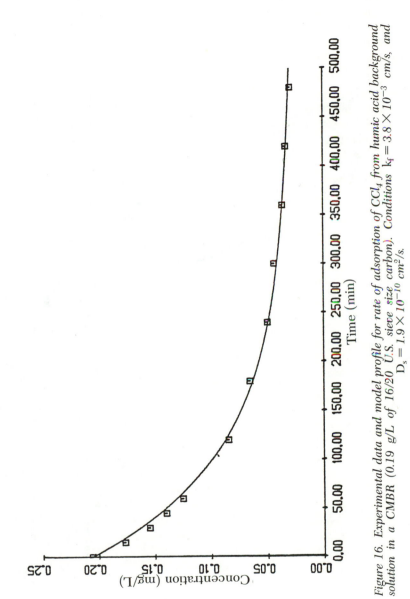

Figure 16. Experimental data and model profile for rate of adsorption of CCl$_4$ from humic acid background solution in a CMBR (0.19 g/L of 16/20 U.S. sieve size carbon). Conditions $k_f = 3.8 \times 10^{-3}$ cm/s, and $D_s = 1.9 \times 10^{-10}$ cm^2/s.

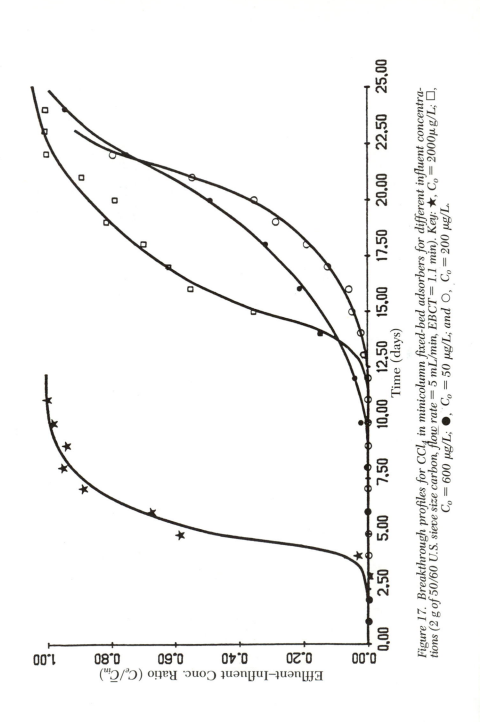

Figure 17. Breakthrough profiles for CCl$_4$ in minicolumn fixed-bed adsorbers for different influent concentrations (2 g of 50/60 U.S. sieve size carbon, flow rate = 5 mL/min, EBCT = 1.1 min). Key: ★, C$_o$ = 2000μg/L; □, C$_o$ = 600 μg/L; ●, C$_o$ = 50 μg/L; and ○, C$_o$ = 200 μg/L.

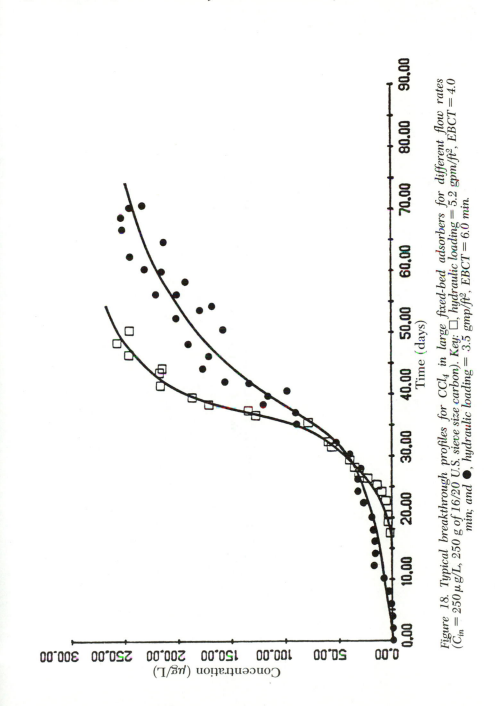

Figure 18. Typical breakthrough profiles for CCl_4 in large fixed-bed adsorbers for different flow rates ($C_{in} = 250 \, \mu g/L$, 250 g of 16/20 U.S. sieve size carbon). Key: □, hydraulic loading = 5.2 gpm/ft², EBCT = 4.0 min; and ●, hydraulic loading = 3.5 gmp/ft², EBCT = 6.0 min.

Table II. Estimated Adsorption Rate Mass Transfer Coefficients for CCl_4 in CMBR's

Figure	Background Solution	C_0 ($\mu g/L$)	Carbon Dosage (g/L)	K_f cm/s ($\times 10^3$)	D_s, cm^2/s ($\times 10^{10}$)
13	OFW	186	0.19	4.1	3.0
16	OFW + 5 mg/L humic acids	200	0.19	3.8	1.9

Table III. Parameters Used for Fixed-Bed Adsorber Predictive Breakthrough Profiles

Adsorber Runs	$\bar{C}_{in}$ ($\mu g/L$)	d (cm)	L (cm)	γ	v (cm/s)	k_f (cm/s)	D_s (cm^2/s)
I (Figure 19)	177	0.03	7.0	0.46	0.23	5.5×10^{-3}	3.0×10^{-10}
II (Figure 20)	570	0.1	84.5	0.37	0.63	3.0×10^{-3}	3.0×10^{-10}
III (Figure 21)	267	0.1	84.0	0.37	0.94	3.4×10^{-3}	3.0×10^{-10}

Note: $\bar{C}_{in}$ = average influent concentration; d = carbon particle diameter; L = bed length; γ = bed porosity, dimensionless; v = interstitial fluid velocity; k_f = film transfer coefficient and D_s = surface diffusion coefficient.

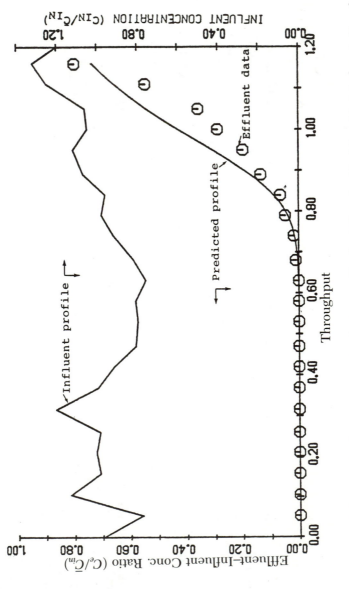

Figure 19. Influent and effluent concentration–history profiles for CCl$_4$ in fixed-bed Run I (minicolumn), (2 g of 50/60 U.S. sieve size carbon, C$_{in}$ = 177 μg/L, EBCT = 1.1 min).

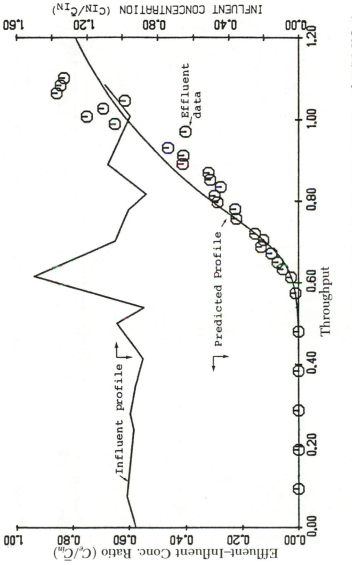

Figure 20. Influent and effluent concentration profiles for CCl$_4$ in fixed-bed Run II (250 g of 16/20 U.S. sieve size carbon, C$_{in}$ 570 µg/L, EBCT = 6.0 min).

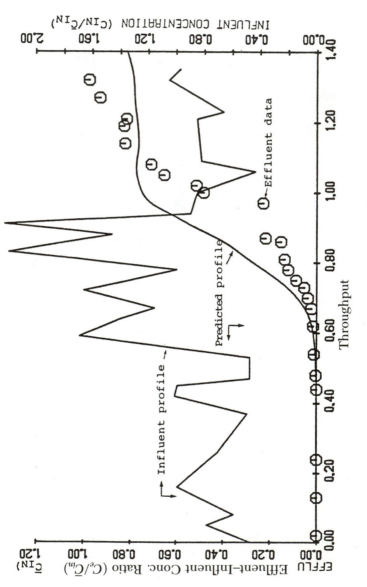

Figure 21. Influent and effluent concentration history profiles for CCl₄ in fixed-bed Run III (250 g of 16/20 U.S. sieve size carbon, $C_{in} = 267$ μg/L, EBCT = 4.0 min).

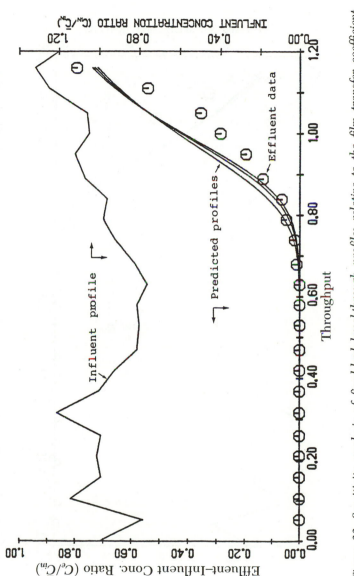

Figure 22. Sensitivity analysis of fixed-bed breakthrough profiles relative to the film transfer coefficient, $k_f \pm 25\%$, for CCl_4 (Run I).

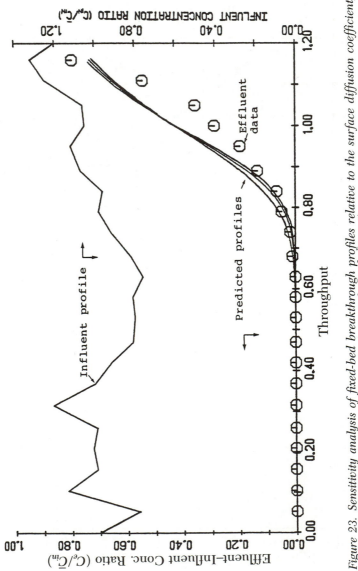

Figure 23. Sensitivity analysis of fixed-bed breakthrough profiles relative to the surface diffusion coefficient, $D_s \pm 25\%$, for CCl_4 (Run 1).

and $D_s \pm 25\%$, respectively. The predictive model displays similar sensitivity to k_f and D_s, suggesting that both film diffusion and intraparticle diffusion present important mass transport rate limits for CCl_4 in this particular minicolumn fixed-bed adsorber system.

Figures 24–27 present similar sensitivity studies for the larger column fixed-bed Runs II and III. It is evident that film diffusion appears to be the principal control in adsorber Run II while surface diffusion appears to dominate mass transport control in adsorber Run III.

For the large fixed-bed adsorbers (Runs II and III), increasing the interstitial fluid velocity from 0.63 cm/s to 0.94 cm/s appears to result in a change in the rate-limiting mass transport mechanism from film diffusion to intraparticle diffusion. It was demonstrated that in the minicolumn study (Run I), for which both lower interstitial velocity (0.23 cm/s) and smaller carbon particle size were used, both film diffusion and intraparticle diffusion were important. The controlling mass transport mechanism for CCl_4 thus appears to be particularly sensitive to the physical characteristics of the system (e.g., flow velocity and particle size). Further study is required to refine this conclusion.

Desorption and Chromatographic Effects. Equilibrium adsorption–desorption experiments were performed using the described bottle-point technique. When equilibrium was achieved, the supernatant aqueous solution was discarded, and the Hypo-Vials containing the carbon were refilled with OFW, sealed, and agitated until equilibrium was reestablished. Samples were withdrawn from each vial, concentrated by liquid–liquid extraction and analyzed by electron-capture GC. The experimental data for the adsorption–desorption studies (Figure 28) indicate no significant equilibrium hysteresis.

Desorption rates also were examined in the CMBR adsorption rate studies described. After equilibrium had been attained, the supernatant solution was discarded, and the reactor was filled with OFW. The rate of release of CCl_4 from the carbon was then measured, using the same procedure employed for evaluating the uptake rate. The approach to equilibrium in the desorption experiments was faster than in the adsorption tests. Data for a typical desorption study are shown in Figure 29.

Desorption studies then were conducted in a fixed-bed adsorber column containing 250 g of carbon and operated at a flow rate of 5.2 gpm/ft^2 (212 $L/m^2/min$). After the column reached exhaustion, high purity water was fed to the saturated adsorber for 30 days to evaluate desorption effects. A significant amount of CCl_4 was desorbed (Figure 30). To investigate the possibility of chromatographic phenomena, a 5-mg/L solution of humic acid was fed to a CCl_4-saturated column for 15 days. The effluent profile (Figure 31) strongly suggests displacement of CCl_4 by humic acid.

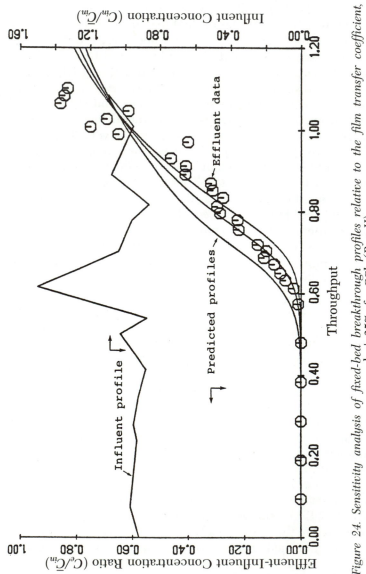

Figure 24. Sensitivity analysis of fixed-bed breakthrough profiles relative to the film transfer coefficient, $k_f \pm 25\%$, for CCl_4 (Run II).

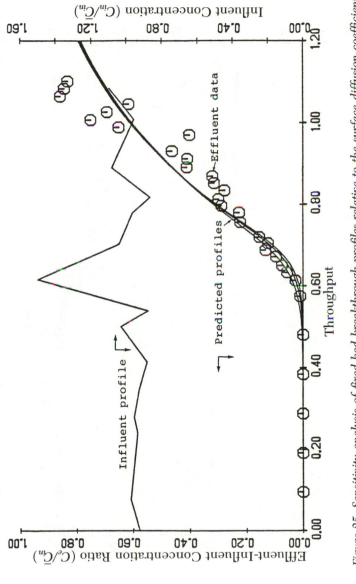

Figure 25. Sensitivity analysis of fixed-bed breakthrough profiles relative to the surface diffusion coefficient, $D_s \pm 25\%$, for CCl_4 (Run II).

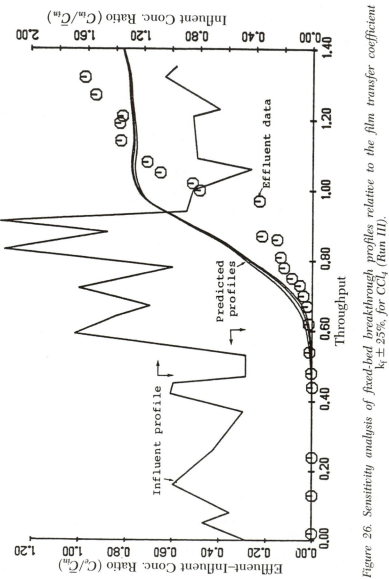

Figure 26. Sensitivity analysis of fixed-bed breakthrough profiles relative to the film transfer coefficient $k_f \pm 25\%$, for CCl_4 (Run III).

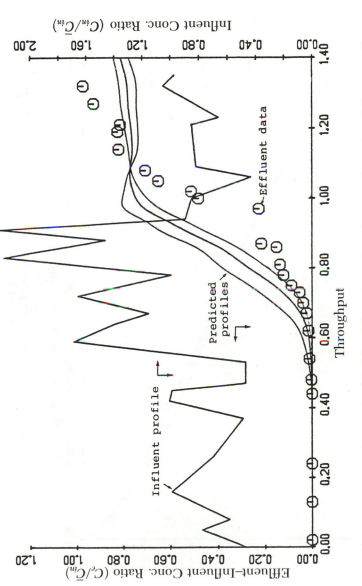

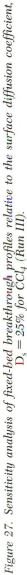

Figure 27. Sensitivity analysis of fixed-bed breakthrough profiles relative to the surface diffusion coefficient, $D_s = 25\%$ for CCl_4 (Run III).

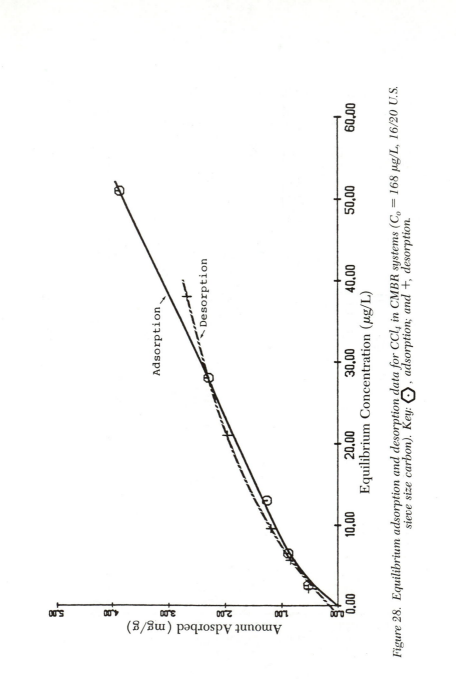

Figure 28. Equilibrium adsorption and desorption data for CCl_4 in CMBR systems ($C_o = 168$ μg/L, 16/20 U.S. sieve size carbon). Key: ⬡ *, adsorption; and +, desorption.*

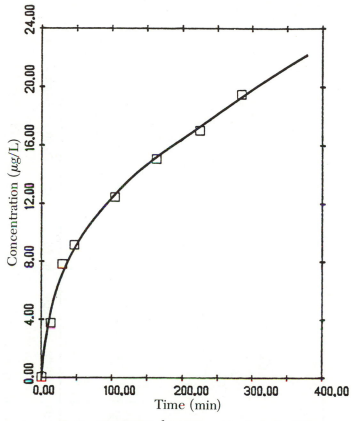

Figure 29. CMBR desorption rate study for CCl$_4$.

Competitive Adsorption. Fixed-bed adsorption studies with CCl$_4$ in the presence of 5 mg/L humic acid were conducted to investigate competitive adsorption phenomenon. These experiments were designed and executed in a manner similar to those described for the OFW background solutions and utilized a 3-cm glass adsorber containing 100 g of 16/20 carbon. Analogous conditions were used to facilitate comparison of model profile simulations in terms of estimated values for k_f and D_s.

All attempts to carry the fixed-bed adsorber operations to completion were hampered by extensive biological growth. Figure 32 presents typical influent and effluent data for CCl$_4$ in a 5-mg/L humic acid background solution for the initial part of the breakthrough profile. It is evident that the effects of the biological growth make it virtually impossible to reach any definite conclusions regarding the dynamic adsorption behavior of CCl$_4$ in the presence of humic acids.

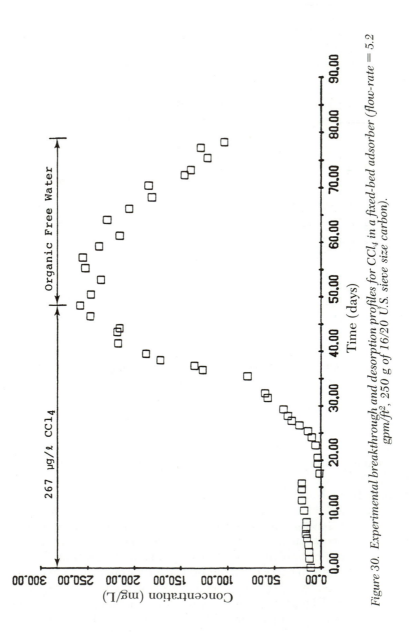

Figure 30. Experimental breakthrough and desorption profiles for CCl₄ in a fixed-bed adsorber (flow-rate = 5.2 gpm/ft², 250 g of 16/20 U.S. sieve size carbon).

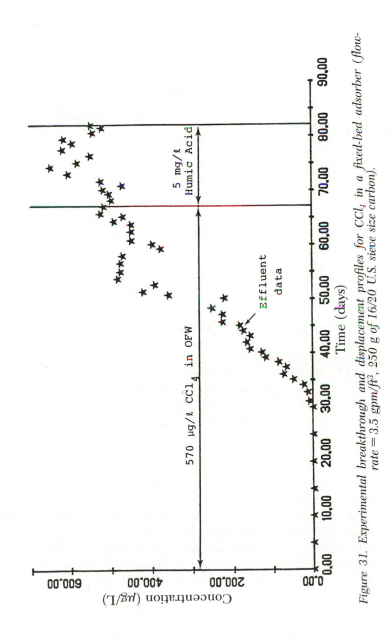

Figure 31. Experimental breakthrough and displacement profiles for CCl₄ in a fixed-bed adsorber (flow-rate = 3.5 gpm/ft³, 250 g of 16/20 U.S. sieve size carbon).

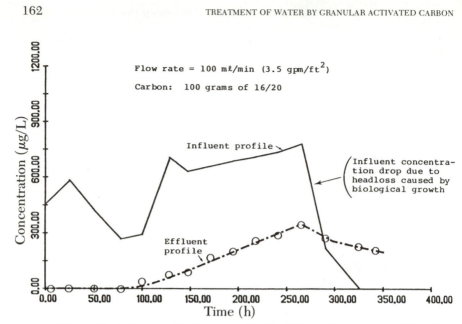

Figure 32. Influent and effluent profiles for CCl₄ in the presence of 5 mg/L humic acids in a fixed-bed adsorber [flow rate = 100 mL/min (3.5 gpm/ft²), 100 g of 16/20 U.S. sieve size carbon].

Efforts were then made to inhibit biological activity. The feed solution was dosed with 2 mg/L silver nitrate. This inhibitor was subsequently deemed unsatisfactory because it complexes with humic acid and changes the adsorption characteristics of the latter (*19, 22*).

Other alternatives were explored. Fixed-bed adsorbers saturated with CCl_4 were fed with a solution of 5 mg/L humic acid and CCl_4 for 10 days, beyond which it was impossible to continue the operation because of prolific biological growth. Figure 33 presents data from this experiment.

Two minicolumn adsorbers, one containing 2 g of carbon saturated with humic acid and the other 2 g of virgin carbon, were fed with a solution of CCl_4 in further experiments. The corresponding effluent breakthrough profiles for CCl_4 (Figure 34) indicated a substantial reduction in the capacity of the presaturated carbon for adsorption of CCl_4. Further analyses showed no humic acid residuals in the effluent from the presaturated column, indicating no humic acid displacement by CCl_4. Thus, competitive adsorption appears to play a significant role in the dynamics of CCl_4 removal in typical adsorber applications.

Summary and Conclusions

Adsorption–desorption equilibrium studies suggest that the uptake of CCl_4 by activated carbon is readily reversible. The release of substantial

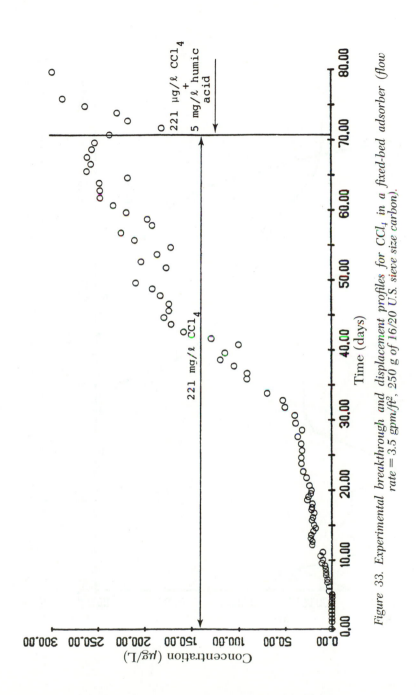

Figure 33. Experimental breakthrough and displacement profiles for CCl₄ in a fixed-bed adsorber (flow rate = 3.5 gpm/ft², 250 g of 16/20 U.S. sieve size carbon).

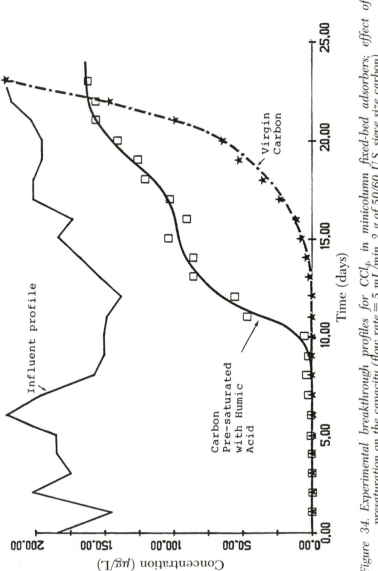

Figure 34. Experimental breakthrough profiles for CCl_4, in minicolumn fixed-bed adsorbers; effect of presaturation on the capacity (flow rate = 5 mL/min, 2 g of 50/60 U.S. sieve size carbon).

amounts of CCl_4 from fixed-bed adsorbers observed during dynamic desorption studies confirms the ready reversibility of the adsorption. These observations suggest that low energy van der Waals attraction (physical adsorption) predominates in the uptake of CCl_4 by carbon.

The MADAM adsorption model was generally able to simulate and predict the performance of fixed-bed adsorbers for CCl_4 removal from water. Predictions suffered, however, in cases where influent concentrations fluctuated drastically and/or where frequent air-scouring and backwashing were required to control headlosses caused by biological activity in the adsorbers.

The effectiveness of activated carbon for CCl_4 removal from water in practical applications is likely to be adversely affected by the competitive adsorption and displacement effects of background organic substances, such as humic acids.

Observations of apparently different rate controls in differently configured adsorber systems suggest that the mass transport (rate) limiting mechanism for CCl_4 adsorption by carbon in fixed-bed adsorbers depends in large measure on the physical characteristics of the system, including column type, hydraulic regime, and carbon particle size.

Acknowledgments

The research reported herein was sponsored in part by Research Grant No. R-804369 from the Water Supply Research Division, Municipal Environmental Research Laboratory, USEPA, Cincinnati, Ohio. The authors express their appreciation to Jim Long for his diligent and faithful technical assistance in the associated laboratory work.

Literature Cited

1. Grob, K.; Grob, G. *J. Chromatogr.* **1974**, *90*, 303–13.
2. Rook, J. J. *J. Water Treat. Examination* **1974**, *23*, 234–43.
3. U.S. Environmental Protection Agency, "National Organics Reconnaissance Survey for Halogenated Organics," Water Supply Research Division, Municipal Environmental Research Laboratory, Cincinnati, Ohio **1975**.
4. Ward, P. S. *J. Water Pollut. Control Fed.*, **1977**, *49*(4), 529–31.
5. Cairo, P. R.; Lee, R. G.; Aptowiz, B. S.; Blankenship, W. M. *J. Am. Water Works Assoc. 71*(8) 450–454, **1979**, *71*(8), 450–54.
6. National Cancer Institute (NCI), "Report on the Carcinogenesis Bioassay of Chloroform," Carcinogenesis Program, Division of Cancer Cause and Prevention, Bethesda, MD, March 1976.
7. Pirbazari, M.; Herbert, M.D.; Weber, W. J., Jr. In "Hydrocarbons and Halogenated Hydrocarbons in the Aquatic Environment"; Afghan, B. K.; Mackay, D., Eds.; Plenum: New York, 1980.
8. Bellar, T. A.; Lichtenberg, J. J. "The Determination of Volatile Organic Compounds at the $\mu g/L$ Level in Water by Gas Chromatography," USEPA, 670/4-74-009, Nov. 1974.
9 Morris, R. L.; Johnson, L. G. *J. Am. Water Works Assoc.*, *68*(9), 492–94 **1976**, *68*(9), 492–4.

10. Weber, W. J., Jr.; Pirbazari, M.; Herbert, M. D.; Thompson, R. In "Viruses and Trace Contaminants in Water and Wastewater"; Ann Arbor Science: Ann Arbor, Mich. 1977; Chapter 8.
11. Nicholson, A. A., Meresz, O. "Organics in Ontario Drinking Water: Part I. the Occurrence and Determination of Free and Total Potential Haloforms," Ontario Ministry of the Envi. Lab. Service Branch, presented at Pittsburgh Conf. of Anal. Chem. and Appl. Spect., 1976.
12. Henderson, J. E.; Peyton, G. R.; Glaze, W.II. "Identification and Analysis of Organic Pollutants in Water"; Ann Arbor Science: Ann Arbor, Mich., 1976; Chapter 7.
13. Mieure, J. P. *J. Am. Water Works Assoc.* **1977,** 69(1), 60–2.
14. Weber, W. J., Jr. "Physiocochemical Processes for Water Quality Control"; Wiley-Interscience: New York, 1972.
15. Weber, W. J., Jr. Crittenden, J. C. *J. Water Pollut. Control Fed.* **1975,** *46,* 5.
16. Crittenden, J. C.; Weber, W. J., Jr., *J. Environ. Eng. Div.,* ASCE **1978a,** *104*(EE2).
17. Crittenden, J. C.; Weber, W. J., Jr. *J. Environ. Eng. Div.,* ASCE, **1978c.**
18. Crittenden, J.C.; Weber, W.J. Jr. *J. Environ. Eng. Div.,* ASCE, **1978c,** *104*(EE6).
19. Pirbazari, M., Ph.D. Dissertation, The University of Michigan, Ann Arbor, Mich., 1980.
20. Pirbazari, M.; Weber, W. S., Jr. In "Chemistry in Water Reuse"; Cooper, W. S., Ed., Ann Arbor Science: Ann Arbor., 1981; Vol. 2.
21. Williamson, J. E.; Bazaire, K. E.; Geankopolis, C. J. *Ind. Eng. Chem. Fund* **1963,** *2*(2), 126.
22. Weber, W. J. Jr., Pirbazari, M.; Barton, D. In "Activated Carbon Adsorption of Organics from the Aqueous Phase"; Suffet I. H.; McGuire, I. M., Eds.; Ann Arbor Science: Ann Arbor, Mich., 1980 Vol 1. 317–319.
23. Hammerstrand, K. "Chloroform in Drinking Water," *Varian Instrument Applications,* **1976,** *10*(2).

RECEIVED for review August 3, 1981. ACCEPTED for publication April 1, 1982.

8

Recent Advances in the Calculation of Multicomponent Adsorption in Fixed Beds

CHI TIEN

Syracuse University, Department of Chemical Engineering and Materials Science, Syracuse, NY 13210

Recent advances in the calculation of multicomponent adsorption in fixed beds are presented. Starting with the knowledge of the single species isotherm data of the adsorbates, the dynamic behavior of multicomponent adsorption in fixed beds can be predicted using different models for intraparticle diffusion with the assumption that the multicomponent adsorption equilibrium can be estimated using the ideal adsorbed solution (IAS) theory. The complexities involved in these calculations depend largely on the intraparticle diffusion model employed.

THIS CHAPTER PROVIDES a systematic presentation of recent developments in the calculation of multicomponent adsorption in fixed beds. Particular emphasis is placed on the use of the thermodynamically derived equilibrium relationship (i.e., IAS isotherm) for the calculation of granular activated-carbon adsorption of trace organics from aqueous solutions. The prediction of the dynamic behavior of adsorption processes has been extensively studied during the last 30 years. Most of these studies, however, are concerned with the single species case; i.e., only one adsorbable species present in the liquid phase. For example, a thorough literature survey conducted by Beveridge in 1962 (1) on adsorption and other sorptive processes listed over 350 references and not one article dealt with multicomponent adsorption.

Recently, the application of activated carbon for waste and water treatment has been of interest. For such problems, the adsorbable species present in the aqueous phase number in the hundreds or even thousands. The need for a simple and yet sufficiently accurate algorithm capable of

0065-2393/83/0202-0167$9.25/0

handling an arbitrarily large number of species at dilute concentrations with a minimum amount of base information is obvious.

The dynamics of adsorption are defined by the stoichiometric relationship, the adsorption isotherm, and the rate expressions. Although there is no problem in formulating equations describing the relevent physical processes, several difficulties are encountered in multicomponent adsorption calculations. First, multicomponent mass transfer (i.e., multicomponent diffusion) is not well understood, and the rate parameters are not properly defined. Fortunately, at dilute concentrations, the effect due to interferences among species is negligible. Under the conditions prevailing in the application of carbon adsorption, mass transfer of the individual species can be described on the single species basis. The more difficult problem is the availability of the thermodynamic equilibrium data between the solution and the adsorbed phases. While this type of information, in principle, can be determined experimentally, the required experimental work and subsequent correlation of experimental data become impractical and even impossible as the number of adsorbates present in the solution increases. For practical considerations, it is necessary to employ a procedure that estimates the multicomponent equilibrium relationship from equilibrium data of single species only.

In general, calculations of sorptive processes in fixed beds can be classified according to two categories: those based on equilibrium theory and those based on nonequilibrium theory. The equilibrium calculation is made by ignoring the mass transfer effect and assumes the establishment of equilibrium conditions between the fluid and adsorbent throughout the bed. Thus, one may predict the column performance from stoichiometric considerations and equilibrium relationships without excessive calculation. The comprehensive work in this area is the h-transformation of Hellfferich and Klein (2). In a later publication, Tien et al. (3) extended the h-transformation to multicomponent adsorption calculation. The procedure applies to systems that obey the Langmuir adsorption isotherm.

The first attempt on the calculation of a multicomponent sorptive process was made by Dranoff and Lapidus (4). These authors considered the ion-exchange problems with two different exchangeable species present in the liquid phase. The rate expression was assumed to be second order and reversible. Later studies by Cooney and Strusi (5,6) suggested a procedure which, in effect, decomposes the bisolute adsorption into two single solute adsorption processes. It is not clear how this rather ingenious procedure can be extended to cases with more than two adsorbable species.

More recently, several studies on the calculation of carbon adsorption of organics from aqueous solutions have been carried out. A partial

Table I. Summary of Studies of Multicomponent
Adsorption Calculation

Investigator	Number of Adsorbable Species	Equilibrium Relationship	Rate Expression
Hsieh et al. (7)	arbitrary	Langmuir	combined liquid and particle phase mass transfer; linear driving force model
Crittenden and Weber (13)	two species	empirical	combined liquid and particle phase mass transfer; surface diffuse model
Wilde (8)	arbitrary	Langmuir	combined liquid and particle phase mass transfer; linear driving force model
Liapis et al. (14, 15)	two species	empirical	combined liquid and particle phase mass transfer; pore diffusion model
Liapis and Litchfield (16)	three species	empirical	combined liquid and particle phase mass transfer; prediffusion model

summary of these works is given in Table I. Generally speaking, most of these investigations deal with specific systems with empirically determined adsorption isotherm expressions. Only two studies (7,8) that used the Langmuir isotherm can be considered as general algorithms. The details of Wilde's (8) work, however, were not available. On the other hand, the work of Hsieh et al. (7) has one major disadvantage. The computational algorithms become less efficient when the mass transfer resistances of the solution and particle phases are comparable. Furthermore, the Langmuir isotherm is not sufficiently accurate to predict the adsorption equilibrium relationship for activated carbon.

It is apparent that a general algorithm cannot be developed using empirically determined isotherm expressions. The lack of multicomponent adsorption isotherm data and the experimental difficulty in obtaining them suggest that any general algorithm developed for multicomponent calculation must be based on an isotherm expression obtained from single component isotherm data of the individual species. A summary of the recent studies in this area carried out in the author's laboratory is presented.

Multicomponent Adsorption Isotherm

Two types of adsorption isotherm expressions are considered. Both expressions are structured so that the multicomponent adsorption equilibrium can be estimated from single component isotherm data of the individual species. The Langmuir isotherm expresses the concentration of the i-th species in the adsorbed phase, q_i, in equilibrium with a solution containing N adsorbates with concentrations $c_1, c_2, \ldots, c_N$ as follows:

$$q_i = \frac{a_i c_i}{1 + \sum_{j=1}^{N} b_j c_j} \tag{1}$$

where a_i and b_i are the Langmuir constants for the i-th adsorbate in its single species state. In other words, if the i-th is the only adsorbate in the solution, the isotherm expression becomes:

$$q_i = \frac{a_i c_i}{1 + b_i c_i} \tag{2}$$

The other general adsorption isotherm is based on the IAS (ideal adsorbed solution) theory. The IAS theory defines the adsorption equilibrium of the i-th species in the adsorbed phase, q_i, in equilibrium with a solution with concentrations $c_1, c_2, \ldots, c_N$; q_i is determined by the following set of equations:

$$q_i = q_T z_i \tag{3}$$

$$q_T = \left[\sum_{j=1}^{N} \frac{z_j}{q_j^o} \right]^{-1} \tag{4}$$

$$c_i = c_i^o(\pi_i, T) z_i \tag{5}$$

$$\pi_i = \frac{RT}{\lambda} \int_o^{c_i^o} \frac{q_i^o}{c_i^o} \, dc_i^o \tag{6}$$

where π_i is the spreading pressure defined as the difference between the interfacial tension of the pure solvent–adsorbent interface and that of the solution–adsorbent interface; λ is the area of the solution–adsorbent interface per unit of adsorbent; the superscript o denotes the single species state; q_i^o and c_i^o are related by the single species adsorption isotherm written as:

$$q_i^o = f(c_i^o \, \pi_i) \tag{7}$$

and z_i denotes the mole fraction of the i-th species in the adsorbed phase. By definition:

$$\sum_{i=1}^{N} z_i = 1 \qquad (8)$$

The procedure to find the adsorbed phase concentration in equilibrium with a solution of known concentration is shown in Table II. This trial and error procedure first assumes a spreading pressure at the equilibrium condition. The corresponding values of c_i^o can be found from Equation 6 and then used to calculate the mole fraction z_i. The satisfaction of Equation 8 provides the confirmation that the assumed value of π_i is correct. Once z_i and c_i^o are known, q_i^o and q_T can be found from Equations 7 and 4, respectively, which in turn makes it possible to determine q_i. The IAS theory yields a much better prediction of adsorption equilibrium data with carbon as the adsorbent than the Langmuir equation. A few examples of the relative accuracy of these two methods are given in Table III.

The implicit manner in which the equilibrium concentrations in the solution and adsorbed phases are related by the IAS theory may give the impression that its use in adsorption calculations is impractical. As shown in later sections, the complexities introduced by the IAS theory depend upon the model used in describing the intraparticle diffusion. Even in the worst case, the added computation effort is by no means prohibitive and can be adequately handled with available computation capability.

Computation Algorithms for Multicomponent Adsorption

In the following sections, a systematic presentation of the development of computation algorithms for multicomponent adsorption based on the Langmuir and IAS theory is given. A general rate of expression that takes into account the resistance in both the liquid and particle phases is used. The intraparticle diffusion is assumed to be described by the linear driving force model, the surface (or particle) diffusion model, or the pore diffusion model.

The macroscopic conservation equation describing multicomponent adsorption in fixed beds can be written as:

$$u\left(\frac{\partial c_i}{\partial z}\right) + \rho_b \frac{\partial q_i}{\partial \theta} = 0, \qquad i = 1, 2, \ldots, N \qquad (9)$$

with the meaning of the symbols given under *Nomenclature*. The rate expression of the adsorption process depends upon the model used and is discussed separately.

Table II. Equilibrium Concentrations Based on IAS Isotherm

Given Conditions

c_1
c_2 $\cdots$ c_N

$$\pi_i = \frac{RT}{\lambda} \int_o^{c_i^o} \frac{q_i^o}{c_i^o}\, dc_i^o$$

$$\sum_{i=1}^{N} z_i = 1$$

c_1^o
c_2^o $\cdots$ $c_N^o \rightarrow$

$z_i = \dfrac{c_i}{c_i^o}$

z_1
z_2 $\cdots$ z_N

$q_i^o = f(\pi_i c_i^o)$

q_1^o
q_2^o $\cdots$ q_N^o

$$q_T = \left[\sum_{i=1}^{N} \frac{z_i}{q_i^o} \right]^{-1}$$

q_T

$q_i = q_T z_i$

q_1
q_2 $\cdots$ q_N

Table III. Comparisons of Langmuir and IAS Isotherms for Certain Systems

Langmuir Isotherm

		q_1		q_2	
c_1	c_2	Experiment	Prediction	Experiment	Prediction
694	462	659	782	859	708
526	356	704	778	791	702
370	44	716	1112	614	190
342	37.5	666	1115	551	163
65	6.25	773	907	476	116
17	1.9	634	537	184	80

IAS Isotherm

		q_1		q_2	
c_1	c_2	Experiment	Prediction	Experiment	Prediction
694	492	659	722	859	714
526	356	704	842	791	460
370	44	716	892	614	461
342	37.5	666	713	551	432
65	6.25	773	825	476	410
17	1.9	634	796	184	241

Linear Driving Force Model. With the use of this model, the rate of intraparticle diffusion of the i-th species can be described by the difference between the surface concentration and the average concentration of the i-th species. The adsorption rate is given as:

$$\frac{\partial q_i}{\partial \theta} = \frac{3k_{l_i}}{a_p \rho_p}(c_i - c_{s_i}) = k_{s_i}(q_{s_i} - q_i) \tag{10}$$

or:

$$q_{s_i} - q_i = \frac{3k_{l_i}}{k_{s_i}a_p\rho_p}(c_i - c_{s_i}) \tag{11}$$

and:

$$q_{s_i} = f(c_{s_1}, c_{s_2}, \ldots, c_{s_N}) \tag{12}$$

The problem now is the numerical solution of Equations 9, 10, and 12 with the appropriate initial and boundary conditions. The dependent variables are $c_1, c_2, \ldots, c_N, c_{s_1}, c_{s_2}, \ldots, c_{s_N}, q_{s_1}, q_{s_2}, \ldots, q_{s_N}$; and $q_1, q_2, \ldots, q_N$, or a total of $4N$. The independent variables ae z and θ. The numerical method required for the solution is as follows. Referring to Figure 1, assuming the dependent variables are all known at point $(l-1, m)$, $(l-1, m-1)$, and $(l, m-1)$, the values of $c_1, c_2, \ldots, c_N$ and $q_1, q_2, \ldots, q_N$ can be found from the third-order algorithm developed earlier $(7,9)$.

A first estimation of c_i and q_i is:

$$\overset{(1)}{c_{i_{l,m}}} = c_{i_{l-1,m}} + c_{i_{l,m-1}} + c_{i_{l-1,m-1}} + \left(\frac{1}{u}\right) \frac{3}{a_p} \frac{\rho_b}{\rho_p} \Delta z$$

$$[-p_{i_{l-1,m}} - p_{i_{l-1,m-1}}] \tag{13}$$

$$\overset{(1)}{q_{i_{l,m}}} = q_{i_{l-1,m}} + q_{i_{l,m-1}} + q_{i_{l-1,m-1}} + \frac{3}{a_p \rho_p} \frac{c_{i_o}}{q_i^+} \Delta\theta$$

$$[p_{i_{l-1,m}} + p_{i_{l-1,m-1}}] \tag{14}$$

$$p_i = k_i(c_i - c_{s_i}) \tag{15}$$

and q_i^+ is a value of q_i in equilibrium with a solution of concentration c_1, $c_2, \ldots, c_N$.

An iteration calculation is carried out as follows:

$$\overset{(k)}{c_{i_{l,m}}} = \overset{(1)}{c_{i_{l,m}}} + \frac{3}{2} \frac{\rho_b}{u a_p \rho_p} [- \overset{(k-1)}{p_{i_{l,m}}} - p_{i_{l-1,m-1}} + p_{i_{l,m-1}} + p_{i_{l-1,m}}] \tag{16}$$

$$\overset{(k)}{q_{i_{l,m}}} = \overset{(1)}{q_{i_{l,m}}} + \frac{3}{2} \frac{c_{i_o}}{a_p \rho_p q_i^+} - [\overset{(k)}{p_{i_{l,m}}} + p_{i_{l-1,m-1}} - p_{i_{l,m-1}} + p_{i_{l-i,m}}] \tag{17}$$

where $p_i^{(k-1)}$ is the value of p_i evaluated with the $(k-1)$th iterated values of c_i and q_i at point (l,m). The iteration stops when the desired degree of convergence is achieved.

Once the values of $c_1, c_2, \ldots, c_N$ and $q_1, q_2, \ldots, q_N$ are known, the corresponding values of $c_{s_1}, c_{s_2}, \ldots, c_{s_N}$ and $q_{s_1}, q_{s_2}, \ldots, q_{s_N}$ can be found from Equations 11 and 12. The effect due to the functional form of the isotherm expression employed in describing the multicomponent adsorption equilibrium is manifested in this second step of calculation. If the Langmuir isotherm is used, the equilibrium relationship is explicitly given as:

$$q_{s_i} = \frac{a_i c_{s_i}}{1 + \sum\limits_{j=1}^{N} a_j c_{s_j}} = \frac{a_i c_{s_i}}{D} \tag{18}$$

where

$$D = 1 + \sum_{j=1}^{N} a_j c_{s_j} \tag{19}$$

Combining Equations 18 and 11, one has:

$$c_{s_i} \frac{a_i}{D} = \Phi_i(c_i - c_{s_i}) + q_i \tag{20}$$

where:

$$\Phi_i = \frac{3k_{l_i}}{a_p \rho_p k_{s_i}} \tag{21}$$

or:

$$c_{s_i} = \frac{c_i + q_i/\Phi_i}{1 + a_i/(\Phi_i D)} \tag{22}$$

Accordingly, the quantity D can be expressed as:

$$D = 1 + \sum_{j=1}^{N} b_j \frac{c_j + q_j/\Phi_j}{1 + a_j/(\Phi_j D)} \tag{23}$$

$$1 - D + \sum_{j=1}^{N} \frac{Db_j(c_j + q_j/\Phi_j)}{D + a_j/\Phi_j} = 0 \tag{24}$$

The values of interphase concentrations at point (l, m) can therefore be obtained by first solving for the value of D at point (l, m) from Equation 24. Once the value of D is known the values of c_{s_i} can be found from Equation 22 and the values of q_{s_i} can be found from Equation 18. The major computation effort involved is the finding of the root of an algebraic equation. This step can be accomplished using a standard algorithm such as Newton's iteration method.

A similar procedure can be developed when the IAS theory is used to estimate the multicomponent adsorption equilibrium. First, the single species isotherm data of all adsorbates are represented by the Freundlich expression:

$$q_i^o = A_i c_i^{o \frac{1}{n_i}} \tag{25}$$

The corresponding spreading pressure (note that all π_i's are assumed to be the same) can be found from Equation 6:

$$\Pi = \frac{\pi_i \lambda}{RT} = \int_o^{c_i^o} \frac{q_i^o}{c_i^o} \, dc_i^o = n_i A_i (c_i^o)^{\frac{1}{n_i}} \tag{26}$$

Since the interphase concentrations c_{s_i} and q_{s_i} are at equilibrium, the corresponding values in the single species state, $c_{s_i}^o$ and $q_{s_i}^o$ corresponding to the equilibrium spreading pressure, Π, can be written as:

$$c_{s_i}^{\ o} = \left(\frac{\Pi}{n_i A_i}\right)^{n_i} \tag{27}$$

$$q_{s_i}^{\ o} = A_i c_i^{o \frac{1}{n_i}} = \frac{\Pi}{n_i} \tag{28}$$

By combining Equations 27 and 28 with Equations 4 and 5, the interphase concentrations become:

$$q_{s_i} = \left[\sum_{j=1}^{N} \frac{n_j z_j}{\Pi_i}\right]^{-1} z_i = \Pi \left[\sum_{j=1}^{N} n_j z_j\right]^{-1} z_i \tag{29}$$

and:

$$c_{s_i} = \left(\frac{\Pi}{n_i A_i}\right)^{n_i} z_i \tag{30}$$

Substituting Equations 29 and 30 into Equation 20 yields:

$$\left[\sum_{j=1}^{N} \frac{n_j z_j}{\Pi}\right]^{-1} z_i = q_i + \Phi_i c_i - \Phi_i \left[\frac{\Pi}{n_i A_i}\right]^{n_i} z_i \tag{31}$$

Solving for z_i gives:

$$z_i = \frac{q_i + \Phi_i c_i}{\Phi_1 \left[\dfrac{\Pi}{n_i A_i}\right]^{n_i} + \left[\displaystyle\sum_{j=1}^{N} \dfrac{n_j z_j}{\Pi}\right]^{-1}}$$

$$z_i = \frac{q_i + \Phi_i c_i}{\Phi_i \left[\dfrac{\Pi}{n_i A_i}\right]^{n_i} + (\Pi/S)} \tag{32}$$

where:

$$S = \sum_{j=1}^{N} n_j z_j \tag{33}$$

Combining Equations 32 and 33 gives:

$$S = \sum_{i=1}^{N} n_i \frac{(q_i + \Phi_i c_i)}{\Phi_i \left[\dfrac{\Pi}{n_i A_i} \right]^{n_i} + (\Pi/S)} \tag{34}$$

Also be definition, $\sum_{i=1}^{N} z_i = 1$. From Equation 32, the following expression can be obtained:

$$\sum_{i=1}^{N} z_i = \sum_{i=1}^{N} \frac{q_i + \Phi_i c_i}{\Phi_i \left[\dfrac{\Pi}{n_i A_i} \right]^{n_i} + (\Pi/S)} = 1 \tag{35}$$

Equations 34 and 35 can be written as:

$$f(\Pi, S) = \sum_{i=1}^{N} \frac{q_i + \Phi_i c_i}{\Phi_i \left[\dfrac{\Pi}{n_i A_i} \right]^{n_i} + (\Pi/S)} - 1 \tag{36}$$

$$g(\Pi, S) = \sum_{i=1}^{N} \frac{n_i (q_i + \Phi_i c_i)}{\Phi_i \left[\dfrac{\Pi}{n_i A_i} \right]^{n_i} + (\Pi/S)} - S \tag{37}$$

Accordingly, with $c_1, c_2, \ldots, c_N$ and $q_1, q_2, \ldots, q_N$ known at point (l, m), the interphase concentration can be found by solving Equations 36 and 37. Once the values of Π and S are found, the mole fractions of the adsorbed phase, z_i, can be obtained from Equation 32. With z_i known, the interphase concentrations can be found from Equations 29 and 30 directly.

Surface (or Particle) Diffusion Model. The rate expression is given as:

$$\frac{\partial q_i}{\partial \theta} = \frac{6 D_i}{a_p^2} \int_o^\theta \left(\frac{\partial q_{s_i}}{\partial \lambda} \right) \left[\sum_{n=1}^{\infty} e^{-D_i} \left(\frac{n\pi}{a_p} \right)^{2(\theta - \lambda)} \right] d\lambda = \frac{3 k_{l_i}}{a_p \rho_p} (c_i - c_{s_i}) \tag{38}$$

and Equation 12.

The numerical method for the solution of Equations 9, 12, and 38, with appropriate initial and boundary conditions, can be stated as follows. Again refer to Figure 1. By assuming that all dependent variables at point $(l-1, m)$, $(l-1, m-1)$, and $(l, m-1)$ are known, the values of c_1, $c_2, \ldots, c_N$ and $q_1, q_2, \ldots, q_N$ at point (l, m) can be found in the same way as described in Equations 13–17. Once the solution and adsorbed phase concentrations are known, the interphase concentrations can be found from Equations 12 and 38. The integral of Equation 38 at $z = l\Delta z$ and $\theta = m\Delta\theta$ [i.e., point (l, m)] can be obtained by evaluating the integration along the vertical axis $z = l\Delta z$. The values of q_{s_i} are known at all grid points along this axis with the exception of point (l, m). The integral can be written as:

$$\int_o^{m\Delta\theta} \left(\frac{\partial q_{s_i}}{\partial \lambda}\right) \sum_{n=1}^{\infty} e^{-D_i \left(\frac{n\pi}{a_p}\right)^2 (m\Delta\theta - \lambda)} d\lambda$$

$$= \int_{(m-1)\Delta\theta}^{m\Delta\theta} \left(\frac{\partial q_{s_i}}{\partial \lambda}\right) \sum_{n=1}^{\infty} e^{-D_i \left(\frac{n\pi}{a_p}\right)^2 (m\Delta\theta - \lambda)} d\lambda$$

$$+ \int_o^{(m-1)\Delta\theta} \left(\frac{\partial q_s}{\partial \lambda}\right) \sum_{n=1}^{\infty} e^{-D_i \left(\frac{n\pi}{a_p}\right)^2 (m\Delta\theta - \lambda)} d\lambda$$

$$\simeq (q_{s_{i_{l,m}}} - q_{s_{i_{l,m-1}}}) g_i + F_i \tag{39}$$

where:

$$g_i = \sum_{n=1}^{\infty} e^{-D_i \left(\frac{n\pi}{a_p}\right)^2} \frac{\Delta\theta}{2} \tag{40}$$

$$F_i = \int_o^{(m-1)\Delta\theta} \left(\frac{\partial q_{s_i}}{\partial \lambda}\right) \sum_{n=1}^{\infty} e^{-D_i \left(\frac{n\pi}{a_p}\right)^2 (m\Delta\theta - \lambda)} d\lambda \tag{41}$$

The term F_i can be evaluated with a standard quadrature formula since values of q_{s_i} are known at point $(l, 0)$, $(l, 1), \ldots, (l, m-1)$. Substituting Equation 39 into 38 after rearrangement gives:

$$q_{s_i} = \psi_i(c_i - c_{s_i}) + \left(q_{s_{i_{l,m-1}}} + \frac{F_i}{g_i}\right) \tag{42}$$

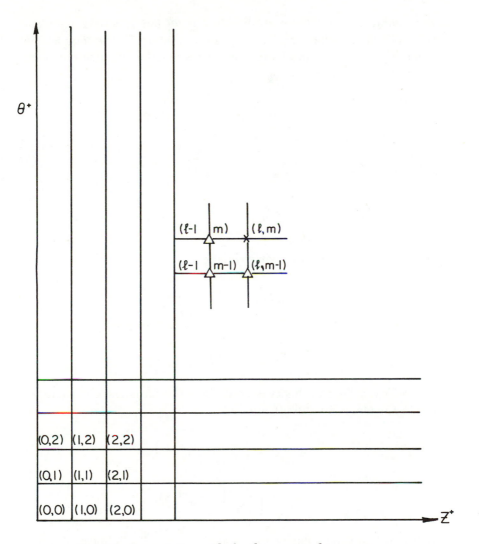

Figure 1. Computation grids for the numerical integration.

when:

$$\psi_i = \frac{k_{l_i}a_p}{2\rho_p D_i} \cdot \frac{1}{g_i} \tag{43}$$

All dependent variables (q_{s_i}, c_i, and c_{s_i}) refer to the values at point (l, m) unless specified otherwise. Equation 42 is similar to Equation 20 with ψ_i and $(q_{s_{i_{l,m-1}}} + F_i/g_i)$ corresponding to Φ_i and q_i, respectively. The results obtained before can be adapted readily. If the equilibrium relationship is given by the Langmuir expression, the interphase concentrations can be found from:

$$c_{s_i} = \frac{c_i + \left(q_{s_{i_{l,m-1}}} + \dfrac{F_i}{g_i}\right)/\psi_i}{1 + \dfrac{a_i}{\psi_i D}} \tag{44}$$

$$q_{s_i} = \frac{a_i c_{s_i}}{D} \tag{45}$$

and D is the root of the following equation:

$$1 + \sum_{i=1}^{N} b_i \frac{c_i + \left(q_{s_{i_{l,m-1}}} + \dfrac{F_i}{g_i}\right)/\psi_i}{1 + \dfrac{a_i}{\psi_i D}} - D = 0 \tag{46}$$

Similarly, if the IAS theory is used, the interphase concentrations are given in Equations 29 and 30 and:

$$z_i = \frac{\psi_i c_i + \left(q_{s_{i_{l,m-1}}} + \dfrac{F_i}{g_i}\right)}{\psi_i \left[\dfrac{\Pi}{n_i A_i}\right]^{n_i} + (\Pi/S)} \tag{47}$$

and Π and S are the roots of the following pairs of nonlinear algebraic equations:

$$\sum_{i=1}^{N} \frac{\psi_i c_i + \left[q_{s_{i_{l,m-1}}} + \dfrac{F_i}{g_i}\right]}{\psi_i \left[\dfrac{\Pi}{n_i A_i}\right]^{n_i} + \dfrac{\Pi}{S}} - 1 = 0 \tag{48}$$

$$\sum_{i=1}^{N} \frac{n_i\left[\psi_i c_i + \left(q_{s_{il,m-1}} + \dfrac{F_i}{g_i}\right)\right]}{\psi_i\left[\dfrac{\Pi}{n_i A_i}\right]^{n_i} + \dfrac{\Pi}{S}} - S = 0 \qquad (49)$$

Pore Diffusion Model.   If the pore diffusion model is assumed to be descriptive of the adsorption process and if the adsorbent particle is assumed to be spherical:

$$\frac{\partial q_i}{\partial \theta} = \frac{3k_{l_i}}{a_p \rho_p}(c_i - c_{s_i}) = \frac{3}{a_p \rho_p} D_{\text{pore}_i}\left(\frac{\partial \alpha_i}{\partial r}\right)_{r=a_p} \qquad (50)$$

where α_i is the concentration of the i-th species present in the pore fluid. For a given position with the bed (i.e., a fixed z value), α_i is a function of θ and the radial distance measured from the center of the adsorbent and $\alpha_i|_{r=a_p} = c_{s_i}$. Similar to the development of the previous two cases, and assuming that all dependent variables are known at points $(l-1, m)$, $(l-1, m-1)$, and $(l, m-1)$, the values of c_i can be found from Equations 13–17. However, to find the values of c_{s_i} at point (l, m), the appropriate pore diffusion equations have to be solved. The governing equations can be written as:

$$\rho_p \frac{\partial \beta_i}{\partial \theta} = \frac{1}{r^2} D_{\text{pore}_i} \frac{\partial}{\partial r}\left[r^2 \frac{\partial \alpha_i}{\partial r}\right] \qquad 0 < r < a_p \qquad (51)$$

$$\frac{\partial \alpha_i}{\partial r} = \frac{\partial \beta_i}{\partial \theta} = 0 \qquad \text{at} \qquad r = 0 \qquad (52)$$

$$k_{l_i}(c_i - c_{s_i}) = D_{\text{pore}_i}\left(\frac{\partial \alpha_i}{\partial r}\right) \qquad \text{at} \qquad r = a_p \qquad (53)$$

$$\alpha_i = c_{s_i} \qquad \text{at} \qquad r = a_p \qquad (54)$$

and:

$$\beta_i = f(\alpha_1, \alpha_2, \dots, \alpha_N) \qquad (55)$$

where α_i is the concentration of the i-th species present in the pore liquid and β_i is the concentration of the i-th species in the adsorbed phase. Customarily, α_i and β_i are assumed to be in equilibrium as shown in Equation 55.

A number of algorithms have been developed for the solution of the pore diffusion equation. Regardless of the specific algorithm used, Equation 51 enables the prediction of β_i at $r = 0$, $\Delta r, \ldots$, $(n-1)\Delta r$ $[n\Delta r = a_p]$ and $\theta = m\Delta\theta$ from the values of α_i at the preceding time interval [i.e., $\theta = (m-1)\Delta\theta$]. Once the values of β_i are known, the corresponding equilibrium values of α_i have to be found from the equilibrium relationship (Equation 55). If the Langmuir expression is used, one has:

$$\alpha_i = \frac{D\beta_i}{a_i} \tag{56}$$

and:

$$D = 1 + \sum_{j=1}^{N} b_j \alpha_j$$

$$D = 1 + \sum_{j=1}^{N} \frac{\beta_j}{a_j} b_j = 1 + D \sum_{j=1}^{N} \frac{\beta_j}{a_j} b_j \tag{57}$$

or:

$$D = \frac{1}{1 - \sum_{j=1}^{N} \frac{b_j}{a_j} \beta_j} \tag{58}$$

namely, Equations 56 and 58 can be used to calculate the values of α_i at the interior points of the adsorbent particles. Similarly, for the IAS theory case, one has:

$$\Pi = n_i A_i \alpha_i^{(0)\left(\frac{1}{n_i}\right)} \tag{59}$$

$$\alpha_i^{(0)} = \left(\frac{\Pi}{n_i A_i}\right)^{n_i} \tag{60}$$

$$\beta_i^{(0)} = A_i \alpha_i^{(0)\left(\frac{1}{n_i}\right)} = \frac{\Pi}{n_i} \tag{61}$$

$$\alpha_i = \alpha_i^{(0)} z_i = \left(\frac{\Pi}{n_i A_i}\right)^{n_i} z_i \tag{62}$$

$$\beta_i = \left[\sum_{j=1}^{N} \frac{z_i}{\beta_i^{(0)}}\right]^{-1} z_i \tag{63}$$

From Equation 63 and the definition $\sum_{j=1}^{N} z_j = 1$, one has:

$$\left[\sum_{j=1}^{N} \frac{-z_j}{\beta_j^{(0)}}\right] = \frac{1}{\sum_{j=1}^{N} \beta_j} \tag{64}$$

and:

$$z_i = \frac{\beta_i}{\sum_{j=1}^{N} \beta_j} \tag{65}$$

By combining Equations 61 and 64:

$$\sum_{j=1}^{N} \frac{z_j}{\beta_i^{(0)}} = \frac{1}{\Pi} \sum_{j=1}^{N} n_j z_n = \frac{1}{\sum_{j=1}^{N} \beta_j}$$

or:

$$\Pi = \left(\sum_{j=1}^{N} \beta_j\right)\left(\sum_{j=1}^{N} n_j z_j\right) \tag{66}$$

Thus, with β_i known, the mole fraction of the adsorbed phase, z_i, and Π can be found from Equations 65 and 66 based on which one can calculate α_i from Equation 62.

Further Refinement of Computation

The development concerning the use of the IAS isotherm was based on the assumption that the single species data can be represented by the Freundlich isotherm expression (i.e., Equation 20). This assumption is not true for all cases. A general treatment can be made by representing the available adsorption isotherm data piecewise, or:

$$q_i^o = (A_i)_k (c_i^o)^{\frac{1}{n_{ik}}} \qquad \text{for} \qquad (c_i^o)_{k-1} < c_i^o < (c_k^o)_k \tag{67}$$

Following the same procedure as before, for the linear driving force model, the calculation of the interphase concentrations at point (l,m) from the knowledge of the bulk phase concentrations can be made using:

$$q_{s_i} = \left[\sum_{j=1}^{N} \frac{(n_j)_k z_j}{\Pi + (\delta_j)_k} \right]^{-1} z_i \tag{68}$$

$$c_{s_i} = \left[\frac{\Pi + (\delta_i)_k}{(n_i)_k (A_i)_k} \right]^{(n_i)_k} z_i \tag{69}$$

$$z_i = \frac{q_i + \Phi c_i}{\Phi_i \left[\dfrac{\Pi + (\delta_i)_k}{(n_i)_k (A_i)_k} \right]^{(n_i)_k} + \left(\dfrac{\Pi}{S} \right)} \tag{70}$$

where:

$$(\delta_i)_k = (n_i)_k (A_i)_k (c_i^o)_{k-1}{}^{(\frac{1}{n_i})_k}$$

$$- \sum_{m=1}^{k-1} (n_i)_m (A_i)_m \left\{ [(c_i^o)_m]^{(\frac{1}{n_i})_m} - [(c_i^o)_{m-1}]^{\frac{1}{(n_i)_{m-1}}} \right\} \tag{71}$$

The quantities Π and S are the roots of the following equations:

$$f(\Pi,S) = \sum_{i=1}^{N} \frac{q_i + \Phi_i c_i}{\Phi_i \left[\dfrac{\Pi + (\delta_i)_k}{n_i(A_i)_k} \right]^{(n_i)_k} + \dfrac{\Pi}{S}} - 1 \tag{72}$$

$$g(\Pi,S) = \sum_{i=1}^{N} \frac{(n_i)_k (q_i + \Phi_i c_i)}{[\Pi + (\delta_i)_k] \left\{ \Phi_i \left[\dfrac{\Pi + (\delta_i)_k}{(n_i)_k(A_i)_k} \right]^{(n_i)_k} + \dfrac{\Pi}{S} \right\}} - \frac{S}{\Pi} \tag{73}$$

Equations 72 and 73 reduce to Equations 36 and 37 when the Freudlich isotherm is applicable over the entire concentration range. Similar expressions can be obtained for the other two models.

Sample Calculation Results and Discussion

The presented algorithms clearly indicate the degree of complication in the numerical calculation of multicomponent adsorption with the use of two types of adsorption isotherms and three kinds of rate models. The effect due to the isotherm functional form is limited to the calculation of the interphase concentration. For the pore diffusion model, there is little difference in terms of computational complexity between the use of the IAS theory and the Langmuir expression. For the other two cases, the surface diffusion model and the linear driving force model, calculation of the interphase concentrations requires the solution of two nonlinear algebraic equations when the IAS theory is used, in contrast to the solution of one nonlinear algebraic equation when the Langmuir expression is used. The computation effort required for the calculation of the bulk phase concentrations is much less than that for the calculation of the interphase concentration when the rate expression is given by the linear driving force model. Accordingly, the greatest impact in terms of the increase in computation effort due to the use of the IAS theory occurs in the linear driving force case.

Another important factor in selecting the isotherm expression is the accuracy of the expression in representing the equilibrium relationship. The advantage of the IAS theory is obvious. Thus, for the case of the linear driving force model and the surface diffusion model, there is a trade-off between the increase in computational complexity against the improved accuracy in representing equilibrium data.

Sample calculations of breakthrough curves of specific cases were made using both the Langmuir and IAS isotherm expressions and the linear driving force rate expression. The specific cases considered pertain to fixed bed adsorption of aqueous solutions of p-nitrophenol (PNP), p-chlorophenol (PCP), and propionic acid (PA) (in various combinations) with granular activated carbon. These cases were chosen because of the availability of relevant experimental data (*10*).

The conditions used in the sample calculation are listed in Table IV. The seven cases listed include three cases of single species adsorption, three cases of binary adsorption, and one case in which all three species were present in the aqueous phase. The liquid phase mass transfer coefficient was estimated using the empirical correlation of Schlunder (*11*) and the k_{s_i} values were calculated from the expression:

$$k_{s_i} = \frac{15 \, D_i}{a_p{}^2} \tag{74}$$

The transfer coefficient values used are given in Table V.

The single species adsorption data for the three organics (PNP, PCP, and PA), with Lurgi Type B 10 II carbon were determined by Merk (*12*). Merk fitted the data by the Freundlich expression piecewise, and his values were used in the sample calculations. In addition, these data also

Table IV. Conditions and CPU Time of the Sample Calculations

CPU Time (s) (IBM 370/155)

Case	Substrate	Method I: Langmuir Isotherm	Method II: IAS Isotherm	Method III: Simplified IAS Isotherm	Inlet Conc., c_o (mmol/L)	Superficial Velocity u(cm/s)	Column Length z	Column Length Δz	Time (h) θ	Time (h) $\Delta\theta$
1	PNP	11	28	26	1.02	6.43×10^{-2}	10	0.25	80	0.5
2	PCP	12	28	26	1.0	6.12×10^{-2}	10	0.25	80	0.5
3	PA PNP	11	26	25	10.5 1.01	6.12×10^{-2}	18	0.45	8	0.05
4	PCP PNP	32	95	85	0.99 1.02	6.12×10^{-2}	10	0.25	80	0.5
5	PA PCP	34	88	77	100.0 1.0	6.12×10^{-2}	10	0.25^a	80	0.5^a
6	PA PNP	34	90	77	100.0 1.0	6.12×10^{-2}	10	0.25^a	80	0.5^a
7	PCP PA	48	128	111	0.98 100.0	$6.12 \times 10^{-}$	10	0.25^a	80	0.5^a

[a] $\Delta z = 0.5$ cm and $\Delta\theta = 0.25$ h were used in Method I calculation.

Table V. Mass Transfer Parameters

Substance	D_l (cm^2/s)	D_s (cm^2/s)	k_l $(cm/s)^a$	k_s (s^{-1})
PNP	9×10^{-6}	2×10^{-8}	1.7×10^{-3}	6.89×10^{-5}
PCP	9×10^{-6}	2×10^{-8}	1.7×10^{-3}	6.89×10^{-5}
PA	9×10^{-6}	1.5×10^{-7}	1.7×10^{-3}	5.16×10^{-4}

Note: $\rho_p = 0.59$ g/cm^3, $\rho_b = 0.39$ g/cm^3, and $a_p = 6.6 \times 10^{-2}$ cm.
[a]Corresponding values estimated from the correlation suggested by Vermeulen et al. (*17*) are found to be approximately within 10% of the tabulated values:

$$\frac{3k_l}{a_p} = \frac{\dfrac{2.62}{1 - 0.4}(D_l u)^{0.5}}{(2a_p)^{1.5}}$$

were approximated by both the Langmuir and Freundlich expressions over their entire concentration range.

For each sample calculation, three predictions were made. For the three single species adsorption cases, breakthrough curves were predicted with the use of the Langmuir isotherm (Method I), the Freundlich isotherm piecewise (Method II), and the Freundlich isotherm (Method III). For the other four cases with more than one species, breakthrough curves were predicted with the Langmuir expression (Method I), the Freundlich expression piecewise (Method II), and the Freundlich expression (Method III).

The comparison between the predicted breakthrough curves and experiments for the single species cases are shown in Figures 2–4. The predictions from Methods II and III were essentially the same, and their agreement with data is somewhat better than Method I. Considering the very low concentrations involved and the inherent difficulty of determining concentrations accurately at such low levels, all three methods can be considered to give reasonably good prediction.

The two species adsorption calculations (Figures 5–7), however, revealed different results. Method II gave consistently better results than either Method I or III. The prediction of the overshoot of the breakthrough curve with the Langmuir isotherm (Method I) differs significantly from experiments. This result is also true for the three species case (Figure 8).

The computer processing unit (CPU) time required for these calculations is given in Table IV. The difference in computation among the three methods resides in the calculation of interphase concentrations. When the Langmuir isotherm is used, the problem is the solution of one nonlinear algebraic equation. When the IAS theory is used, a pair of equations must be considered. The use of a single Freundlich expression

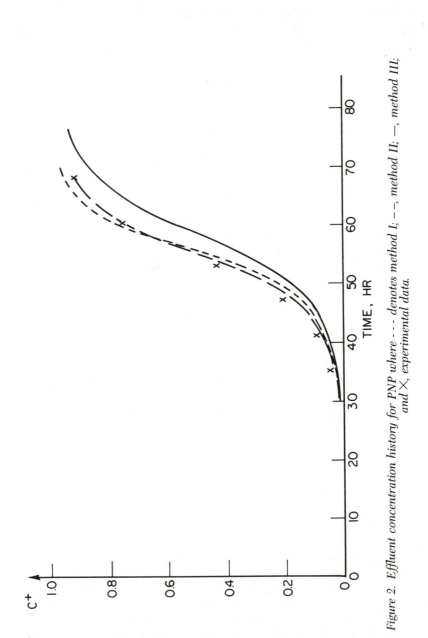

Figure 2. Effluent concentration history for PNP where - - - denotes method I; - -, method II; —, method III; and ✕, experimental data.

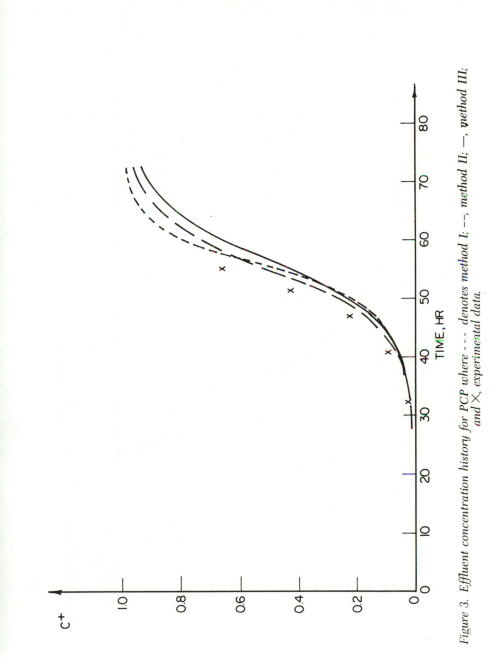

Figure 3. Effluent concentration history for PCP where - - - denotes method I; – –, method II; —, method III; and ✕, experimental data.

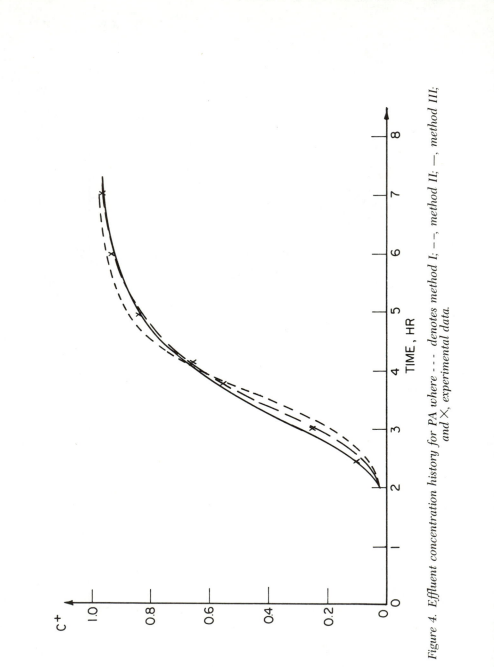

Figure 4. *Effluent concentration history for PA where - - - denotes method I; - -, method II; —, method III; and ×, experimental data.*

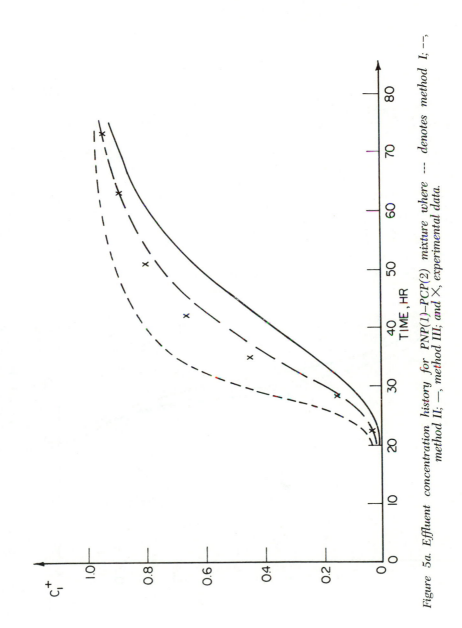

Figure 5a. Effluent concentration history for PNP(1)–PCP(2) mixture where --- denotes method I; --, method II; —, method III; and × experimental data.

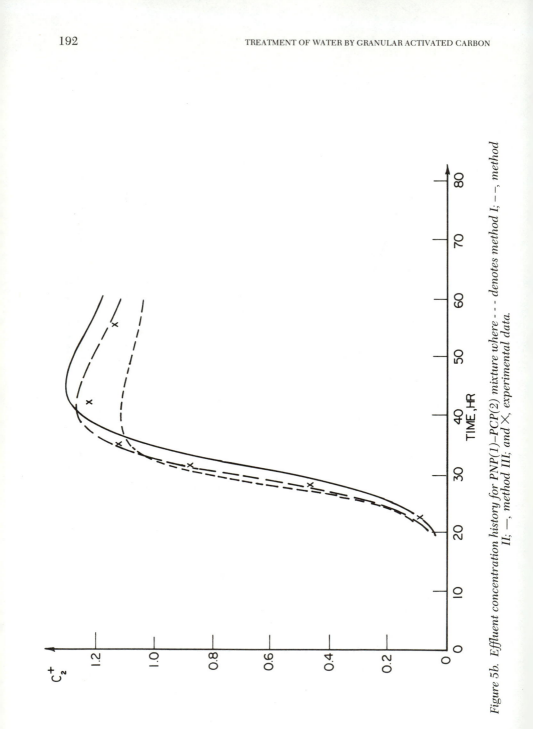

Figure 5b. Effluent concentration history for PNP(1)–PCP(2) mixture where - - - denotes method I; – –, method II; —, method III; and $\times$, experimental data.

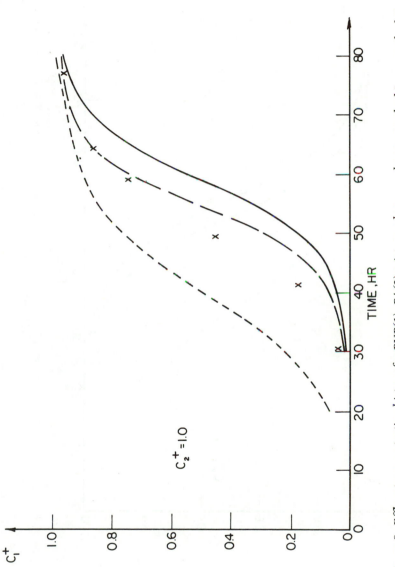

Figure 6. Effluent concentration history for PNP(1)–PA(2) mixture where - - - denotes method I; – –, method II; —, method III; and × experimental data.

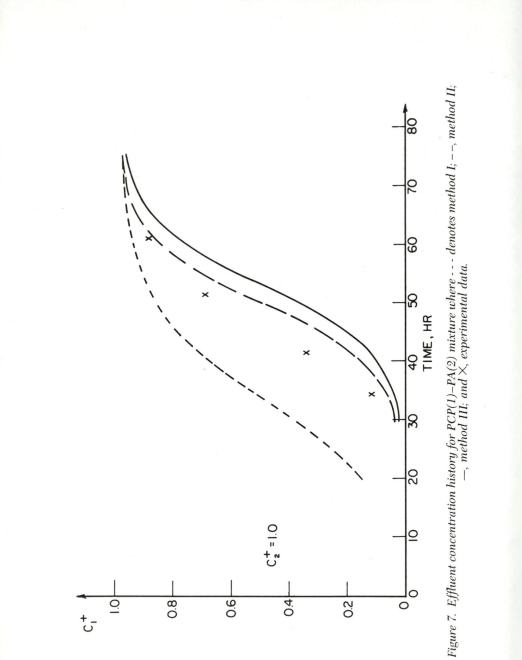

Figure 7. Effluent concentration history for PCP(1)–PA(2) mixture where - - - denotes method I; – –, method II; —, method III; and $\times$, experimental data.

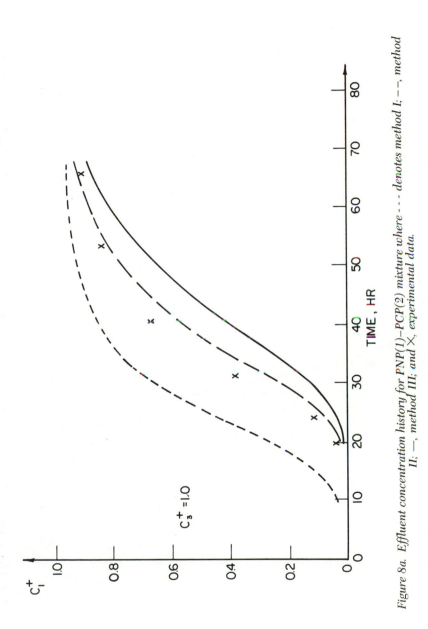

Figure 8a. Effluent concentration history for PNP(1)–PCP(2) mixture where - - - denotes method I; – –, method II; —, method III; and ×, experimental data.

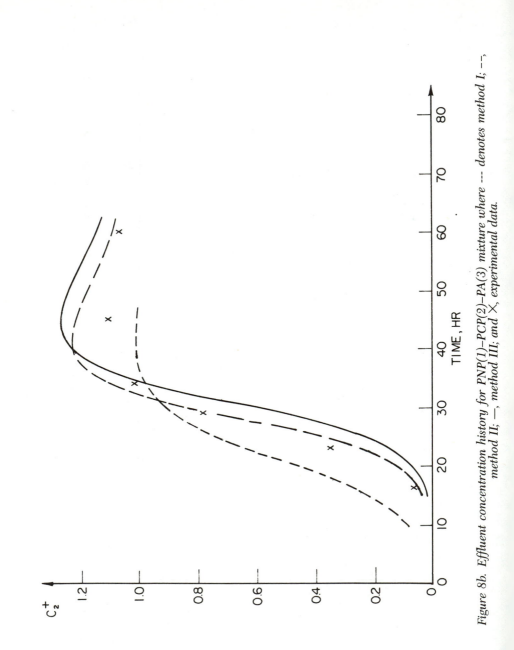

Figure 8b. Effluent concentration history for PNP(1)–PCP(2)–PA(3) mixture where -- denotes method I; – –, method II; —, method III; and ✕, experimental data.

(Method III) as compared with its use in a piecewise manner (Method II) for the single species isotherm data, affects the CPU time only marginally, although Method II gives much better accuracy.

Intuitively, from the number of equations required for solution, one may conclude that the time requirement with the use of the IAS theory should be twice as large as the Langmuir isotherm case. The approximate validity of this argument is substantiated by the sample calculation results. The results are shown in Table IV. The ratios of the CPU time of Methods I and II range from 2.2 to 2.7. It is reasonable to believe that this behavior is generally true.

A number of conclusions can be made from this work. First, the IAS theory, in spite of its apparent complexity, does not cause prohibitive computing requirements. Even in the most severe cases, the increase in CPU time is approximately 150%. The more important consideration in CPU time is not the functional form of the adsorption isotherm; the model of the rate expression is far more important.

Second, the accuracy of the prediction of breakthrough curve depends upon the accuracy of the isotherm expression as demonstrated in Figures 1–6. Method III did not yield significantly better results than Method I since the simple Freundlich expression does not represent data accurately. On the other hand, the piecewise fitting of the single species adsorption data with the Freundlich expression is a general and effective way to apply the IAS isotherm in multicomponent calculations.

The good agreement, as indicated by Figures 4–6, suggests that the linear driving force model is adequate in characterizing fixed bed adsorption. Accordingly, there does not appear to be compelling need to consider the use of other models such as the pore diffusion and the surface diffusion models.

Nomenclature

A_i	Freundlich constant of the i-th species
$(A_i)_k$	Freundlich constant of the i-th species over a concentration range of $(c_i)_{k-1} < c_i < (c_i)_k$
a_i, b_i	Langmuir constants of the i-th species
a_p	particle radius
c_i	concentration of the i-th species of the solution phase
c_i^o	concentration of the i-th species of the solution phase in its single species state
c_{s_i}	concentration of the i-th species in the solution phase and the solution–absorbent interphase
$c_{s_i}^o$	value of c_{s_i} with the i-th species in its single species state
D	quantity defined by Equation 19 or 57

D_i surface diffusion coefficient of the i-th species

D_{pore_i} pore diffusion coefficient of the i-th species

F_i quantity defined by Equation 41

g_i quantity defined by Equation 40

k_{l_i}, k_{s_i} liquid phase and particle phase transfer coefficients

l, m indices

n_i Freundlich exponent of the i-th species

$(n_i)_k$ Freundlich exponent of the i-th species in the concentration range $(c_i)_k < c_i < (c_i)_k$

P_i quantity defined by Equation 15

q_i concentration of the i-th species in the adsorbed phase

q_i^o concentration of the i-th species of the solution phase in its single species state

q_{s_i} concentration of the i-th species of the adsorbed phase at the solution–adsorbent interphase

q_T quantity defined by Equation 4

R ideal gas law constant

S quantity defined by Equation 34

T absolute temperature

u superficial velocity of fluid through bed

z axial distance of bed

z_i mole fraction of the i-th species in the adsorbed phase

Greek Letters

α_i concentration of the i-th species in the pore liquid of an adsorbed particle

β_i concentration of the i-th species in the adsorbed phase of an adsorbent particle

$(\delta_i)_k$ quantity defined by Equation 71

$\Delta\theta$ increment of corrected time

Δz increment of axial distance

Φ_i quantity defined by Equation 21

ψ_i quantity defined by Equation 43

ρ_p particle density

θ corrected time, defined as $t - z\varepsilon/u$ with t as real time and ε as porosity of bed

λ area of solution–adsorbent interface per unit of adsorbent

π_i spreading pressure of i-th species

Π quantity defined by Equation 26

Acknowledgment

This study was performed under Grant No. CPE 75-08893, National Science Foundation.

Literature Cited

1. Beveridge, G. S. G. "A Survey of Interphase Reaction and Exchange in Fixed and Moving Beds," Univ. of Minnesota, 1962.
2. Helfferich, F.; Klein, G. "Multicomponent Chromatography—Theory in Interference," Marcel Dekker, New York, 1970.
3. Tien, C.; Hsieh, S. C.; Turian, R. M. *AIChE J*, **1976**, *22*, 498.
4. Dranoff, J. S.; Lightfoot, E. N. *Ind. Eng. Chem. Fundamentals,* **1958**, *50*, 1648.
5. Cooney, D. C.; Strusi, R. P. *Ind. Eng. Chem. Fundamentals,* **1966**, *5*, 25.
6. Cooney, D. C.; Strusi, R. P. *Ind. Eng. Chem. Fundamentals,* **1972**, *11*, 123.
7. Hsieh, S. C.; Turian, R. M.; Tien, C. *AIChE J.,* **1977**, *23*, 213.
8. Wilde, K. A. In "Activated Carbon Adsorption of Organics from the Aqueous Phase," Suffet I. H.; McGuire, M. J., Eds., Ann Arbor Science, Ann Arbor, Mich., 1980, Volume I, Chapter 13.
9. Vanier, C. R. Ph.D. Dissertation, Syracuse Univ., Syracuse, N.Y., 1970.
10. Tien, C.; Thodos, *AIChE J.,* **1960**, *6*, 364.
11. Schlunder, E. U., "Einfuhrung in die Warme-und Stoffubertragung," Uni-Test, Vieweg Verleg Braunschweig, 2, Aufl., 1975.
12. Merk, W., "Konkurrierende Adsorption Verschiedoner Organischer Wasserinhaltsstoffe in Aktivkohlefiltern," Dissertation, Karlsruhe, 1978.
13. Crittenden, J. C.; Weber, W. J., Jr. *J. Environ. Eng. Div.,* ASCE, **1978**, *104* (EE 6), 1175.
14. Balzli, M. W.; Liapas, A. I.; Rippin, D. W. T. *Inst. Chem. Eng.,* **1978**, *56*, 145.
15. Liapis, A. I.; Rippin, D. W. T. *Chem. Eng. Sci.,* **1978**, *33*, 593.
16. Liapis, A. I.; Litchfield, R. J. *Chem. Eng. Sci.,* **1980**, *35*, 2366.
17. Vermeulen, T.; Klein, G.; Hiester, N. K. In "Chemical Engineers Handbook," 5th ed., Perry R. H.; Chilton, C. H. Eds., McGraw-Hill, New York, N.Y., 1973.

RECEIVED for review August 3, 1981. ACCEPTED for publication March 2, 1982.

Controlling Mechanisms for Granular Activated-Carbon Adsorption Columns in the Liquid Phase

M. R. ROSENE

Calgon Corporation, Pittsburgh, PA 15230

Nonadsorptive phenomena contributed to granular activated-carbon column performance as measured by the breakthrough curve. A model was developed that examines the impact of three possible mechanisms that control the development of the breakthrough curve and allows the use of the mass transfer zone in the interpretation of breakthrough curves. The developed concepts help provide a sound basis for the interpretation of granular activated-carbon column adsorptive dynamics.

THE USE OF GRANULAR ACTIVATED-CARBON (GAC) columns for the removal of undesirable organics in both industrial and municipal applications has been widely discussed (*1–12*). Mechanistic studies of column operation have mainly centered on the adsorption process as the rate-limiting step. While this approach is valid for many cases, nonadsorptive phenomena also contribute to column performance as measured by the breakthrough curve. In particular, the impact of the axial or longitudinal dispersion phenomenon has been largely ignored as an important aspect of column operation. Consequently, the need exists for an approach that delineates the conditions under which this axial dispersion phenomenon cannot be neglected and that describes the physical phenomena underlying both the mass transfer limited and axial dispersion mechanisms of GAC column performance through the interpretation of breakthrough data.

0065-2393/83/0202-0201$06.00/0

Theory

The GAC kinetic adsorption process in the liquid phase can be considered to consist of several steps. Each step can be argued to have several substeps and even the exact number of steps may be debated, however; the approach here considers only the following:

1. *Bulk Diffusion*—This step includes the transport of the adsorbate in the bulk liquid, either by mixing or molecular diffusion.

2. *External Mass Transfer* (Film Diffusion)—This step involves the transport of the adsorbate across the boundary layer at the external surface of the particle.

3. *Intraparticle Mass Transfer* (Pore Diffusion)—This step concerns the rate at which the adsorbate is transferred from the surface to the particle interior.

4. *Micropore Adsorption*—This rate involves the adsorption step itself and is generally considered to be very rapid.

Of these steps, intraparticle mass transfer is generally recognized as the most common rate-limiting step in adsorption (1–12). A useful parameter for describing the mass transfer within the adsorbent particle is the effective intraparticle diffusivity (D).

The basic mathematics of diffusion into a sphere are given by Crank (13) and further modified by Dedrick and Beckmann (2). If a homogeneous diffusion rate is assumed within the particle, then from Dedrick and Beckmann:

$$E = \begin{cases} \dfrac{6}{\sqrt{\pi}}\sqrt{\dfrac{Dt}{a^2}} - \dfrac{3\,Dt}{a^2}, & \sqrt{\dfrac{Dt}{a^2}} \leq 0.4 \\[2ex] 1 - \dfrac{6}{\pi^2}\exp\left[\dfrac{\pi^2\,Dt}{a^2}\right], & \sqrt{\dfrac{Dt}{a^2}} \geq 0.4 \end{cases} \qquad (1)$$

where D is the effective diffusivity, t is the time, a is the effective particle radius, and E is the fraction of the equilibrium capacity obtained in time t. Rearranging Equation 1 for $\sqrt{Dt/a^2} \leq 0.4$ gives:

$$\sqrt{D} = \frac{a}{\sqrt{t}}\left(\frac{1}{\sqrt{\pi}} \pm \sqrt{\frac{1}{\pi} - \frac{E}{3}}\right) \qquad (2)$$

The effective diffusivity is used to predict the breakthrough curve of a single component from a GAC column when intraparticle mass transfer is the rate-controlling phenomenon. The procedure involves the use of a standard compound (in this case p-nitrophenol) and the concept of the mass transfer zone (MTZ) as approximated here:

$$\text{MTZ} = \left(\frac{t_s - t_b}{t_s} \right) L \tag{3}$$

where t_s is the time to saturation in a column, t_b is the time to initial breakthrough, and L is the column length.

This approximation is valid when the MTZ is small compared to the total column length. The experimentally determined effective intra-particle diffusivity, calculated from Equation 2 using an experimental procedure described later, for the standard compound is correlated with the width of the MTZ experimentally determined for a set of column conditions where intraparticle mass transfer is the controlling factor. This ratio then, can be used to calculate the MTZ width for other compounds under intraparticle controlled conditions from their determined effective intraparticle diffusivities. Once the effective diffusivity is obtained, the procedure for predicting the breakthrough curve by computer is relatively straightforward.

The other step that controls adsorption is external mass transfer. This step involves transport of the adsorbate across the boundary layer or film at the external surface of the particle. This phenomenon is a function of the film diffusion rates for the adsorbate of interest as well as factors such as liquid shear forces which affect the thickness of the boundary layer. Liu (*14*) isolated the contribution of external mass transfer to the kinetics of GAC column operation using short columns and determined that the external mass transfer rate only controls in the early stages of his column runs. This result is not unexpected as the external surface of the adsorbent particle is saturated rapidly relative to the internal surface and intraparticle mass transfer quickly becomes the rate-limiting adsorption step.

The prediction of breakthrough curves when the controlling mechanism is external mass transfer could be accomplished in much the same manner as for intraparticle diffusion. Here, however, the mass transfer rate across the boundary layer must be estimated for the conditions to be used in the calculations. This step involves more experimental work since the film boundary response to a change in flow rate would have to be determined for the adsorbent of interest.

In addition to the impact of the adsorption process on GAC column operation, the potential contributions of fluid flow effects in the packed bed must be considered.

For example, an aspect of fluid flow requiring careful consideration is that of axial or longitudinal dispersion in a packed GAC bed. This effect is attributable to backmixing and turbulence within the bed and in some cases results in significant spreading of the mass transfer zone. Ebach (15) investigated the phenomenon of axial dispersion in water flowing through beds of packed solids and established that, for a range of interstitial linear velocities, an empirical correlation for calculating the effective axial dispersion (D_a) (along the length of the column) was given by:

$$D_a = Kd_pU^b \tag{4}$$

where K and b are constants, d_p is the effective particle diameter, and U is the effective interstitial linear velocity.

This correlation is valid for the range of linear velocities normally encountered in GAC column operations. This expression does not account for any adsorption that may occur on the bed of packed soils. To treat this effect, Lapidus and Amundson (16) developed the mathematics for the adsorptive retention of the axial dispersion wavefront. From their derivation we have:

$$\left(\frac{\partial C}{\partial R}\right)_{R=\gamma} = \sqrt{\frac{ZU}{4\pi R^2 D_a}} \tag{5}$$

where C is the concentration in the effluent, Z is the column length, U is the interstitial velocity, R is a reduced time variable equal to (Ut/Z), and the quantity $(\partial C/\partial R)$ $R = \gamma$ is the slope of the breakthrough curve when C is equal to 50% of the influent level.

Combining Equation 5 and D_a calculated from Equation 4, it is possible to calculate the slope of the breakthrough curve. A major assumption in the derivation is that adsorption equilibrium is obtained locally within the column. Therefore, this calculation is only valid when the adsorption step itself is not the rate-limiting mechanism in the development of the mass transfer zone.

If these conditions hold, it is possible to predict the breakthrough curve for a single component when axial dispersion is the controlling mechanism. From Equation 4, the necessary data to calculate D_a are the interstitial linear velocity, the effective particle diameter, and the constants K and b which are given by Ebach. Then, by using Equation 5, the slope of the breakthrough curve at the 50% breakthrough level can be calculated by inputting the calculated value for D_a and the column length. When this procedure is computerized and combined with an isotherm capacity, predicted breakthrough curves for the axial dispersion controlled column runs are easily calculated.

We now consider how the MTZ, as expressed in Equation 3, can be used to distinguish between the three potential controlling phenomena discussed (i.e., intraparticle mass transfer, external mass transfer, and axial dispersion). First, consider an experiment in which a GAC column of fixed bed height B_1 is first exhausted on an influent stream which is applied at a hydraulic loading rate F_1. The loading rate is then increased to F_2, and the experiment is repeated. Under these conditions, the MTZ will exhibit a different behavior depending on the controlling phenomena. For example, in the case of intraparticle mass transfer, if we assume that the wavefront is contained, then the MTZ increases roughly in proportion to the increase in flow from F_1 to F_2. In other words, the rate of penetration of the particle is independent of the external flow conditions, and a doubling of the flow doubles the volume between initial breakthrough and saturation. The time required to reduce the influent concentration within the column to zero does not change since that is controlled by the rate in which the adsorbent particles become exhausted. This situation results in the MTZ, as expressed in Equation 3, increasing proportionally to the increase in flow.

In the case of external mass transfer, the described experimental conditions result in a different response in the MTZ. Here, as the flow is increased from F_1 to F_2, velocity shear forces affect the hydrodynamic boundary layer, decreasing the resistance to adsorbate mass transfer. Again, if we assume the wavefront is contained, the time to initial breakthrough decreases or increases with the magnitude depending on the relative impact of the increase in flow versus the increased rate of mass transfer across the boundary layer. The time to saturation decreases in response to the increase in mass transfer rate. The overall effect of these responses on the value of the MTZ is indeterminate with either a net decrease or increase possible.

In the case of axial dispersion controlled breakthrough, the adsorption step is not limiting, and the breakthrough curve behaves according to Equations 4 and 5. Under the conditions of this experiment, this situation results in an increase in the magnitude of the axial dispersion D_a in direct proportion to the increase in flow. It is apparent from Equation 5 that this increase has little effect on the slope of the breakthrough curve since the increase in D_a is offset by a like increase in U. Thus, with the slope constant and the time to 50% breakthrough reduced, the MTZ increases.

This discussion reveals a pitfall into which the unwary can easily stumble. If, for example, a column experiment is run at a fixed bed depth and two different flows, an increase in the MTZ less than proportional to the increase in flow cannot be used as unequivocal evidence for either external mass transfer phenomena or axial dispersion. A second set of experimental conditions is necessary to resolve this question. In this experiment, the flow at F_1 is held constant and the bed depth is increased from B_1 to B_2. Under these conditions, intraparticle mass transfer exhibits no change in the MTZ since the interval between t_b and t_s is determined by the effective intraparticle diffusivity, and increases in t_s are exactly offset by an increase in L.

The case of external mass transfer is nearly identical. The mass transfer rate across the boundary layer is constant since the flow is constant and the interval between t_s and t_b is again a constant. Therefore, the MTZ does not change.

In the case of axial dispersion, however, the impact of the column length is described by Equation 5. By using this expression, the slope of the breakthrough curve is found to be proportional to the square root of the increase in bed depth. This finding means that the MTZ shows an increase directly related to the increase in bed depth from B_1 to B_2.

Table I summarizes the effect of the controlling mechanism on the MTZ for the two experimental test conditions discussed.

Results and Discussion

Consider how the model just described can be used to interpret the results of experimental column runs. Figure 1 presents two column experiments for phenol removal. Both columns were run with the same empty bed contact time (EBCT) but at surface loadings which differed by sixteenfold (0.3 gpm/ft^2 vs. 4.8 gpm/ft^2). When the controlling mechanism is intraparticle mass transfer, these two curves plotted as percent breakthrough versus time would be expected to coincide. However, as can be seen, the curves do not overlay. Past experience showed that phenol adsorption at a surface loading of 4.8 gpm/ft^2 is intraparticle mass transfer controlled (12); thus, the marked decrease in the MTZ with the increase in flow from 0.3 gpm/ft^2 is an indication that a change of controlling mechanism occurred. A question nevertheless remains as to which al-

Table I. Effect of Controlling Mechanism on MTZ

Experimental Conditions	Intraparticle (Pore)	External (Film)	Axial
Increase flow/ bed depth constant	Increase MTZ	Decrease or increase MTZ	Increase MTZ
Constant flow/ increase bed depth	No change in MTZ	No change in MTZ	Increase MTZ

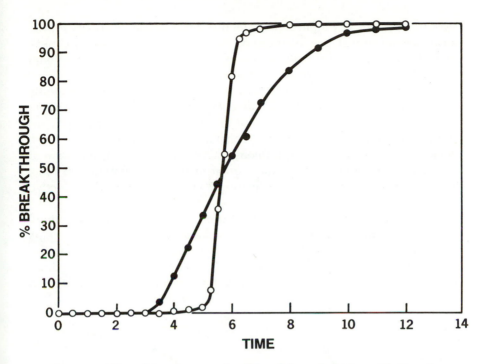

Figure 1. Effect of flow rate on phenol breakthrough (100 mg/L). Key: O, 4.8 gpm/ft², and ●, 0.3 gpm/ft².

ternate mechanism does control under the low flow conditions. Based only on the data of Figure 1, an argument could be made for either external mass transfer or axial dispersion. The definitive experiment would be to conduct another run at the same surface loading and a different bed depth. If the controlling mechanism is external mass transfer, no change should be expected in the MTZ. By contrast, if axial dispersion is the controlling factor, then a decrease in the bed depth would result in a reduction in the MTZ.

Figure 2 presents the results of just such an experiment. The original curve has been replotted for comparison, and the second curve shows the breakthrough for a bed depth 0.37 times that for the first curve. Calculation of the MTZ for the second curve reveals a decrease in direct relation to the reduction in bed depth. Therefore, this case can be ascribed to a shift of mechanistic control from intraparticle mass transfer to axial dispersion as the flow rate is reduced.

As discussed earlier, computerized predictions of single component breakthrough curves can be made for any of the three mechanisms presented, assuming the proper input data are available. For intraparticle mass transfer, the key input is the effective intraparticle diffusivity for the adsorbate–adsorbent pair of interest. The data can be obtained experimentally by using a modification of the high-pressure minicolumn (HPMC) technique described by Rosene et al. (*17*).

The experiment is conducted by running the column at sufficiently high surface loadings to obtain immediate breakthrough of the adsorbate concentration in the column effluent of 90% of the influent value. This value allows approximation of the concentration throughout the entire column as the average of the influent and effluent values. Integration of the breakthrough curve gives the carbon loading as a function of time. These data, when entered into Equation 2 yield a value for the effective diffusivity.

Table II presents effective intraparticle diffusivity data determined by this technique for *p*-nitrophenol in combination with three adsorbents at two concentration levels. The first three runs on carbon A show the technique gives reproducibility within 10%. The average of these three runs is a value of 5.9×10^{-9} cm^2/s. Reducing the concentration to 50 ppm

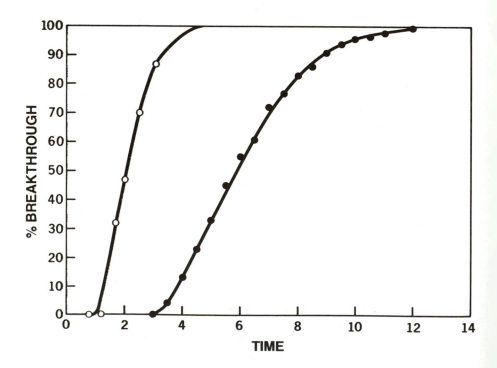

Figure 2. Phenol breakthrough (100 mg/L, 0.3 gpm /ft^2) vs. bed depth. Key: O, 0.37 original bed depth; and ●, original bed depth.

Table II. Effective Intraparticle Diffusivities

Carbon	p-Nitrophenol Concentration (mg/L)		E	D (cm^2/s)
	Inlet	Average		
A	100	94	0.597	6.2×10^{-9}
A	100	97	0.589	6.0×10^{-9}
A	100	96	0.565	5.4×10^{-9}
A	50	46.6	0.584	5.9×10^{-9}
B	50	46.5	0.122	2.0×10^{-10}
C	50	47.8	0.087	1.0×10^{-10}

showed no significant change in the diffusivity, as expected from the homogeneous diffusion model. By contrast, the effect of different adsorbents under the same conditions showed marked differences. Both carbon B and carbon C (experimental carbons) had much lower effective diffusivities, 2.0×10^{-10} cm^2/s and 1.0×10^{-10} cm^2/s, respectively.

An example of how the computer prediction method can be used is illustrated in Figures 3 and 4. Here, two column experiments for the removal of p-nitrophenol by carbon A are presented. The column bed depths are identical in each case, but the surface loading in Figure 3 is 8.7 gpm/ft^2 while in Figure 4 it is 2.9 gpm/ft^2. The actual data are represented by the individual points, and the dashed and solid lines represent the intraparticle diffusion and axial dispersion model predictions, respectively. At the higher flow, intraparticle diffusion is seen to be the controlling factor; however, when the flow is reduced to the lower rate, the curve does not adopt the sharp S-shape predicted by the intraparticle diffusion model but is much broader as expected from axial dispersion control. Agreement between the intraparticle model prediction and the experimental data in Figure 3 is excellent. The fit to the axial model is fairly good and is quite gratifying since the prediction is based on only an isotherm capacity and Ebach's correlation.

Conclusions

The model presented examines the impact of each of three possible mechanisms that control the development of a GAC column effluent breakthrough curve. The model allows the use of the MTZ in the interpretation of breakthrough curves as a means of determining the dominant mechanism. In a number of cases, the model has provided logical explanations for what otherwise would have been very puzzling results. These concepts should help to provide a sound basis for the interpretation of GAC column adsorptive dynamics.

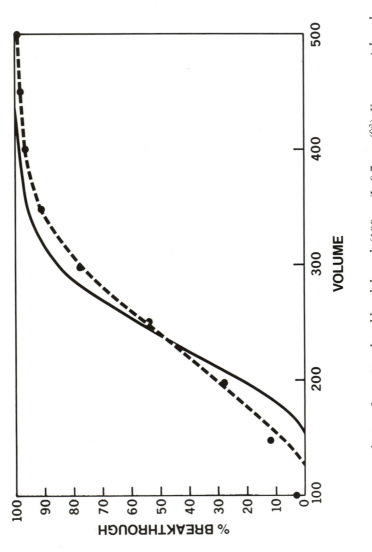

Figure 3. Computer predictions for p-nitrophenol breakthrough (100 mg/L, 8.7 gpm/ft²). Key: —, axial; and - - ●- -, intraparticle.

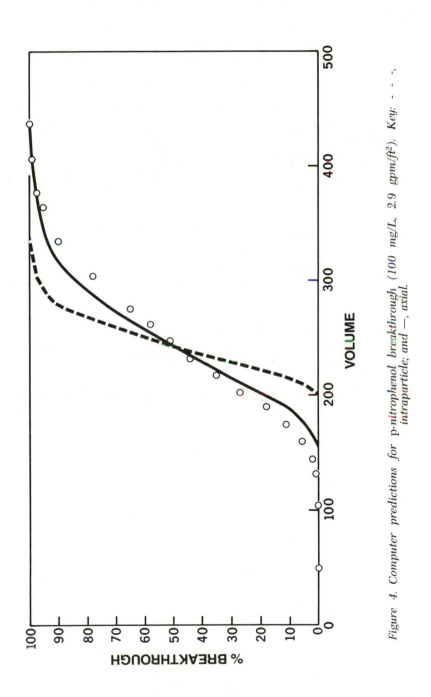

Figure 4. Computer predictions for p-nitrophenol breakthrough (100 mg/L, 2.9 gpm/ft²). Key: - - -, intraparticle; and —, axial.

Literature Cited

1. Weber, W. J., Jr.; Morris, J. C. *J. Sanitary Eng. Div.* **1963**, *89*, (542), 31–59.
2. Dedrick, L.; Beckmann, R. B. *Chem. Eng. Progr., Symp. Ser.* **1967**, *63* (74), 68–78.
3. Knoblauch, K.; Jüntgen, H.; Peters, W. *Chem. Ing. Tech.* **1969**, *41*, 798–805.
4. Knoblauch, K.; Jüntgen, H. *Chem. Ing. Tech.* **1970**, 42(2), 77–81.
5. Mattson J. S.; Kennedy, F. W. *J. Water Pollution Control Fed.* *43*, pp. 2210–2217 **1971**, *43*, 2210—17.
6. Spiridakis N. J.; Brown, L. F. presented at the AIChE 67th Annual Meeting, Washington, D.C., Dec 1974.
7. Suzuki M.; Kawazoe, K. *J. Chem. Eng. Jpn.* **1975**, *8*, 379–82.
8. Suzuki, M.; Kawazoe, K. *J. Chem. Eng. Jpn.* **1974**, *7*, 346–50.
9. Suzuki, M.; Kawai, T.; Kawazoe, K. *J. Chem. Eng. Jpn.* **1976**, *9*, 203–8.
10. Hashimoto, K.; Miura, K.; Nagata, S. *J. Chem. Eng. Jpn.* **1975**, *8*, 368–73.
11. Nagy, L. G.; Fóti, G.; Kuty, H.; Schay, G. "Equilibrium and Kinetic Studies of Liquid Adsorption on Porous Activated Carbon," Proceedings of the International Conference on Colloid and Surface Science, Wolfram, E.; Ed. Akad. Kiado: Budapest, Hungary 1975, pp. 107–115.
12. Zogorski, J. S.; Faust, S. D. AIChE Symposium Series Water-I: Physical, Chemical Wastewater Treatment, 1976, pp. 54–65.
13. Crank, J. "Mathematics of Diffusion"; Claredon Press, Oxford, England, 1956.
14. Liu, K. Ph. D. Dissertation, University of Michigan, Ann Arbor, Mich., 1980.
15. E. A. Ebach, Ph.D. Dissertation, University of Michigan, Ann Arbor, Mich., 1957.
16. Lapidus L.; Amundson, N. R. *J. Phys. Chem.* **1952**, *56*, 984–88.
17. Rosene, M. R.; Deithorn, R. T.; Lutchko, J. R.; Wagner, N. J. In "Activated Carbon Adsorption of Organics from the Aqueous Phase", Suffet, I. H.; McGuire, M. J. Eds; Ann Arbor Science: Ann Arbor, Mich., 1980; Vol. 1, Chapter 15.

RECEIVED for review August 3, 1981. ACCEPTED for publication March 18, 1982.

Evaluation of Activated Carbon by the Dynamic Minicolumn Adsorption Technique

L. J. BILELLO and B. A. BEAUDET

Environmental Science and Engineering, Inc., P.O. Box ESE, Gainesville, FL 32602

Refinements in a previously published carbon evaluation method are described, and data are presented comparing data obtained from the dynamic minicolumn adsorption technique with pilot study data on five industrial process wastewaters. Also presented are data applying this method to screening granular activated carbon for removal of trihalomethane precursors and trihalomethanes from a potable water supply. The dynamic minicolumn adsorption technique is shown to be an effective, rapid method for determining representative carbon adsorption capacities and usage rates for specific organic components of actual wastewaters and potable waters. These data can be used to determine if a carbon treatment for a given water is feasible and to provide a rough estimate for system economics.

T HE EVALUATION OF GRANULAR ACTIVATED CARBON (GAC) for removal of toxic organics from industrial wastewaters or the removal of trihalomethanes (THM) or THM precursors generally requires lengthy, comprehensive, and costly studies. Without the resources or knowledge of how to apply the theoretical predictive models, design engineers are limited to the equilibrium isotherm for screening the use of carbon in a given application or in choosing the most efficient carbon. Often pilot column studies requiring 1 or more months of testing at considerable cost are begun based on isotherm results, and the researcher can be confident only that such pilot data are valid for the conditions and period of time the test was run.

This paper addresses work performed to refine an earlier published carbon evaluation method using a packed column of about 2 mm in diameter and a high-pressure pump. The refined method and procedures are described, and data obtained from the dynamic minicolumn adsorption technique (DMCAT) are compared with pilot study data on five industrial process wastewaters. Data applying the DMCAT method to screen GAC for removal of THM precursors and THM from a potable water supply also are presented, as are suggestions regarding the use of the DMCAT method in the field of carbon adsorption application.

Background

Between 1977 and 1980, IERL-Ci[1] performed pilot-scale column tests on site at various organic chemical plants. The purpose of this program was to generate carbon performance data regarding the removal of specific organic contaminants from organic chemical wastewaters. The pilot tests were conducted by using a mobile pilot plant equipped with 10-cm (4-in.) internal diameter, 1.8-m (6-ft) high columns, and a mobile analytical laboratory equipped with a gas chromatograph and a total organic carbon (TOC) analyzer. Each test was conducted using a reactivated bituminous coal-based carbon from an industrial source with a minimum of three carbon columns in tandem. Each column provided 10–20 min empty-bed contact time (EBCT). The pilot-scale tests generated excellent data on carbon performance, system design criteria, and carbon usage rates for adsorption of specific organic contaminants from real-world wastewaters. The testing costs at each plant, however, were considerable.

EPA recognized the need for a rapid evaluation technique which would provide reliable information regarding dynamic GAC performance beyond that obtained from an isotherm. We were requested to evaluate a published method of rapid carbon evaluation (1, 2).

The technique described by Rosene et' al. (1) was based on the principle of high-pressure liquid chromatography (HPLC). The system consisted of a high-pressure liquid pump and a small stainless steel column about 2 mm in diameter containing pulverized activated carbon. Influent containing a known concentration of a single component in pure water was pumped through the column at an unspecified pressure. The effluent concentration was monitored for breakthrough of the compound, and the adsorptive capacity of the carbon was calculated from the known mass of carbon in the column and the volume of liquid processed. The small column technique was able to provide reliable estimates of carbon loading for these single components, even at low influent concentrations where isotherm interpretation is difficult.

[1]U.S. Environmental Protection Agency Industrial Environmental Research Laboratory.

Later (2), a similar system, called the high-pressure minicolumn (HPMC) system, was used to investigate the adsorptive capacities of several activated carbons for single organic components of interest in potable water containing different background TOC concentrations. Chloroform and benzene were the specific organic compounds studied.

In these studies, an unspecified pressure was used to rapidly establish a mass transfer zone within the column and to obtain data used for evaluation of the organic removal efficiency in relation to the length of the mass transfer zone. Flow rates were in the range of 2–3 cc/min, using a fine mesh activated carbon packed to a depth of 20–25 mm. Particle size was affected with the finer mesh carbon having a shorter mass transfer zone and a longer service life before breakthrough. The particle sizes investigated were not identified, but the bulk of the work was reported as using carbon in the 200 × 325-mesh range.

Due to the combined effects of particle size, flow distribution, and wall effects, direct scale-up of minicolumn breakthrough profiles to full-scale systems is difficult. However, relative comparison of carbons by HPMC methods are informative for both adsorptive capacities and kinetics of an adsorbent under dynamic test conditions.

This minicolumn method first was evaluated to determine its feasibility for use on complex high-strength, multicomponent systems of organic chemicals wastewater and to determine the reliability of carbon loading and/or design data obtained using the method. Later evaluations were conducted on potable water to determine the relative cost of GAC for removal of THM and precursors as compared to other removal alternatives.

Experimental Design

System Description. The system used to perform DMCAT consists of a high-pressure precision flow metering pump, a closed sample reservoir, and a stainless steel minicolumn (2.25 mm internal diameter × 70 mm long). Figure 1 presents a schematic diagram of this system. To successfully apply the DMCAT system to measure the adsorption characteristics of volatile organic compounds, two major design requirements existed. First, the system had to be constructed of inert materials to avoid contamination of the sample. Second, the flow and concentration of the influent had to be kept constant throughout each run. All materials contacting the sample were composed of Teflon or stainless steel. The influent sample was held in a sealed, collapsible air sampling bag. By using this bag, the influent end of the system was totally closed, thereby preventing volatile losses or pickup of organics from the laboratory atmosphere.

A solvent pump (Altex 110A) commonly used in analytical HPLC and capable

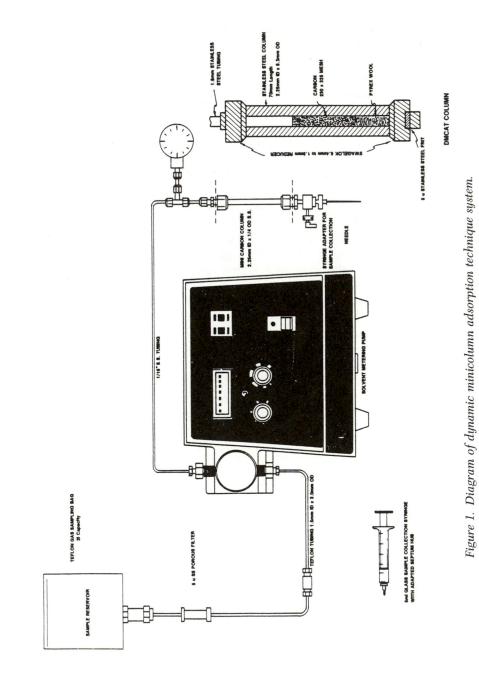

Figure 1. Diagram of dynamic minicolumn adsorption technique system.

of maintaining constant flow ($\pm$ 0.03 mL/min) at pressures up to 6,000 pounds per square inch (psi) was chosen. The measured effluent was collected in a graduated cylinder. A pressure gauge rated at $\pm$ 1% full-scale accuracy from 0 to 3,000 psi was positioned just ahead of the minicolumn for measurement of actual column operating pressure.

A 5-μm stainless steel frit was used as an inline filter when necessary to remove suspended solids and oils and to protect the pump. As glass wool plug at the bottom of the column was used to prevent carbon from plugging the discharge tube.

Experimental Considerations.   Although prior researchers had used the high-pressure minicolumn technique to develop carbon loadings for single component systems, limited information on the actual procedures and conditions for the study was presented. Before evaluating the DMCAT method for comparison with isotherms and pilot data, experiments replicating the previous studies using similar equipment and operating conditions as described in the literature were run. Conditions for the procedure not specified included particle size of the carbon, carbon preparation and packing techniques, and operating pressure.

Effects of particle size showing that a finer mesh carbon processed a greater volume prior to a breakthrough had been previously demonstrated. Initial studies used a minus 200-mesh virgin carbon prepared by grinding and wet sieving. Chloroform in organic-free water at a concentration of about 11.5 mg/L was chosen as the initial test contaminant. Pressures exceeded the recommended maximum pressure of 4,000 psi for the pump. These runs were terminated before a breakthrough occurred.

To reduce pressure build-up, carbon was sieved to a particle size range of 200 $\times$ 325-mesh. Runs were completed producing a chloroform breakthrough curve as shown in Figure 2, which occurred earlier than similar previous work. Since adsorption is particle size dependent, the particle size was narrowed to a 230 $\times$ 325-mesh range by grinding and dry sieving. Pressure in subsequent runs could be maintained below 100 psi. Figure 3 is the breakthrough curve for chloroform at 14.8 mg/L and a 230 $\times$ 325-mesh carbon. The breakthrough occurs at about 160 mL as compared with 40 mL for the coarser carbon, consistent with the earlier work.

Reproducible bed length for a given carbon weight is necessary for a meaningful evaluation. Maximum packing is desirable to assure intimate contact of the fluid with the carbon and to eliminate channeling and wall effects. The tube diameter, 2.25 mm, made it difficult to load the carbon without some being held up on the walls.

The best method evaluated used a vibrator tool adjusted to a low setting. Figure 4 shows the consistency of this method for reproducibility of measured bed lengths versus carbon weights for both virgin and reactivated carbon. As expected, the denser reactivated carbon packed to a shorter bed length than the virgin carbon.

Measurements of bed length were repeated after the run during which the carbon was wetted and packed more tightly, if possible. A slight drop of less than 5% in bed length was observed.

The prior work had been labeled as a high-pressure technique. After resolving particle size and packing procedures, experiments were performed under an induced pressure to determine the effects of pressure on adsorption. Again, chloroform and virgin bituminous-based carbon (230 $\times$ 325-mesh) were used at pressures of less than 100, 600, 1,000, and 2,200 psi. Figure 5 shows the

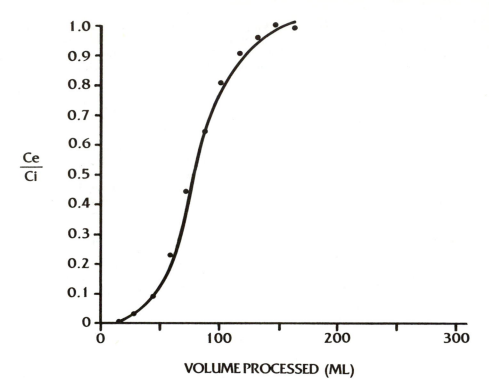

Figure 2. DMCAT breakthrough curve for chloroform (11.5 mg/L), coarser carbon (50 mg) grind, wet screened and tamped. Conditions: 200 × 325 mesh; flow = 2.5 cc/min; pressure = 0–3500 psi; L = 20 mm; and Zm = 18.0 mm.

breakthrough curve for each of these runs, showing no apparent dependency of the initial breakthrough or of the point of saturation $(CE/CI = 1.0)$ on test column pressure. Subsequent runs were performed at system pressure below 100 psi.

Experimental Procedures.   SYSTEM OPERATION.   The minicolumn system is assembled as shown in Figure 1. The sample reservoir was filled with the solution to be tested and all excess air was expelled. The reservoir was connected to the pump inlet line and the system was allowed to purge. The minicolumn was connected, and the pump was set at the desired flow rate. Sample effluent was collected immediately in a graduated cylinder.

At intervals of 5–15 mL, depending on influent characteristics, effluent samples were collected for analysis. The sample collection technique consisted of a syringe-to-column connection which allowed the sample to be pumped directly into the 5-mL graduated syringe. The 5-mL portion of sample was injected into a 60-mL bottle with Teflon-lined septum and subsequently analyzed for organics by headspace gas chromatography. The 5-mL sample volumes removed for analyses were accounted for in the total sample volume processed. A column pressure reading was taken with each sample. Column influent samples were collected at intervals of approximately 30 mL.

ISOTHERM PROCEDURE.   Carbon was weighed into a 60-mL bottle fitted with

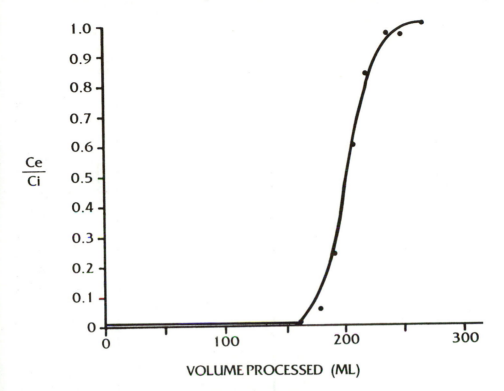

Figure 3. DMCAT breakthrough curve for chloroform (14.8 mg/L), finer carbon grind, dry screened and vibrated (50.2 mg). Conditions: 230 × 325 mesh; L = 29.0 mm; Zm = 11.3 mm; and < 100 psi.

screw caps and Teflon-lined septa. Then 25 mL of wastewater was transferred to each bottle, including a bottle not containing carbon. The bottles were attached to a wrist-action shaker and shaken for 2 h. Following shaking, the bottles were allowed to stand for 15 min. Then 1 mL of the headspace was removed with a gas-tight syringe and analyzed by gas chromatography. All standards were prepared using identical techniques with the exception of the addition of carbon. In all carbon preparation, the carbon was not allowed to come into contact with the laboratory atmosphere in order to prevent organic pickup, and washings were performed with organic-free water.

PRECURSOR EVALUATION. Formation of THM from a water supply is dependent upon the presence, nature, and concentration of the so-called precursors and of the chlorine concentration and the time allowed for formation. Evaluations of carbon for removal of precursors must therefore be conducted by taking in-plant water samples at the likely point for carbon application contacting with carbon, chlorinating, and ceasing formation at a time representative of the plant's treatment and distribution system. Analysis of water samples for THM should then be conducted.

A procedure was developed by which influent and effluent samples from the minicolumn are each contacted with chlorine directly in the septum vial and held

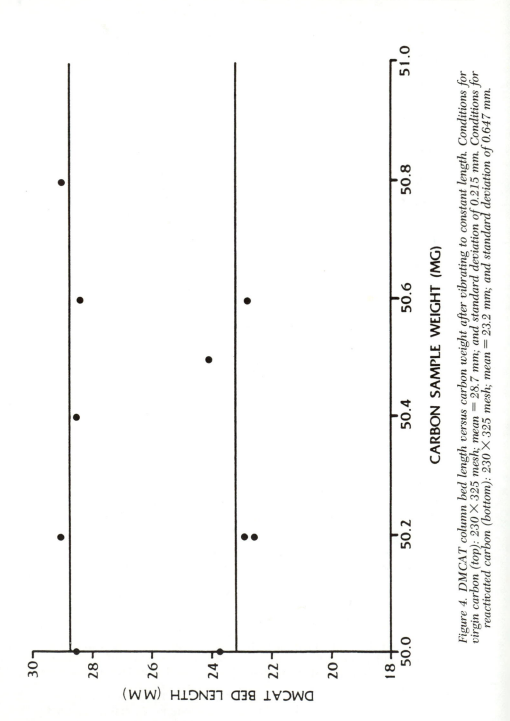

Figure 4. DMCAT column bed length versus carbon weight after vibrating to constant length. Conditions for virgin carbon (top): 230 × 325 mesh; mean = 28.7 mm; and standard deviation of 0.215 mm. Conditions for reactivated carbon (bottom): 230 × 325 mesh; mean = 23.2 mm; and standard deviation of 0.647 mm.

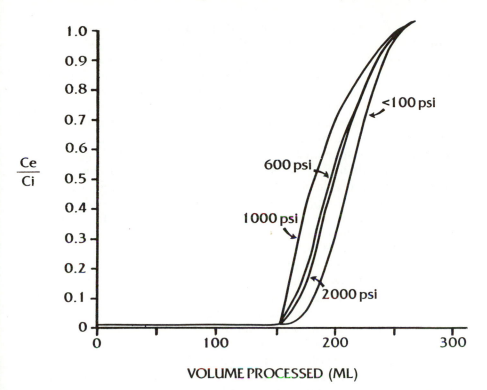

Figure 5. Evaluation of pressure on DMCAT. Conditions: chloroform (14.8 mg/L), and virgin carbon (230 × 325 mesh).

for the appropriate formation time. Formation is then ceased by introducing sodium thiosulfate, after which THM analysis is performed.

Results

Once a reliable experimental method for DMCAT was established, runs were performed for a series of synthetic multicomponent and actual wastewaters. The results of these runs were compared to equilibrium isotherms and/or dynamic pilot column test results available for the same or similar wastewaters.

Pure Component Benzene–Toluene Solution.   A pure multicomponent solution of benzene–toluene in organic-free water was selected to test the adsorptive characteristics of a virgin and a regenerated carbon in the DMCAT system. This test was designed primarily to determine the applicability of the DMCAT for multicomponent systems, since prior investigators had reported results solely for single component systems.

The concentrations of benzene and toluene in the test solution were chosen to be comparable to the concentrations in an actual light hydro-carbon cracking (LHC) waste stream for which considerable background data are available. Figures 6 and 7 represent DMCAT breakthrough curves based on benzene removal for the virgin and regenerated bitum-inous coal-based activated carbon. These curves are representative of classical performance expected from activated carbon in that the virgin carbon processed a greater volume before breakthrough than the reactivated carbon for a small molecule such as benzene. The length of the experimental carbon bed (L) is shown on these as well as subsequent breakthrough curves. The mass transfer zone (Zm) for the virgin carbon is shorter and more sharply defined, and the breakthrough of toluene (shown only for the reactivated carbon) occurs as expected after benzene has reached saturation on the carbon (effluent concentration in equili-brium with the influent). The results of this test indicate that the DMCAT system does indeed represent expected carbon adsorption performance.

Figure 8 is a benzene isotherm for a similar concentration of benzene and toluene using the same reactivated carbon as for the DMCAT test. Table I shows that the carbon loading calculated from the DMCAT run at benzene saturation compares favorably with that predicted from the isotherm at initial wastewater concentration.

Evaluation on Actual Wastewater. The next DMCAT run was performed on an actual wastewater stream for which isotherm and dynamic pilot column data were available for comparison. This waste-water, which contains benzene and toluene as the major specific con-taminants, is a quench water from an LHC unit. During June 1979, when the on-site pilot studies and isotherms were being conducted, the average concentrations of benzene and toluene were 53 and 8.3 mg/L, respec-tively. The wastewater also contained an average of 16 mg/L oil and grease and 200 mg/L TOC. When samples of the wastewater were collected in January 1980 from the same stream for the DMCAT studies, process changes had reduced the average concentrations of benzene and toluene to 22 and 3.6 mg/L, respectively. Other wastewater characteristics remained relatively unchanged. Table II lists the organic and conventional pollutants present in this wastewater.

Since previous work with the DMCAT system involved pure com-ponent solutions, it was not known whether the system would be capable of physically processing the actual wastewater. On-line filtration using a stainless steel frit was required to prevent plugging of the minicolumns.

Following filtration, reproducible DMCAT runs with well-defined breakthrough were obtained, as shown in Figures 9 and 10. Figure 11 presents a reactivated carbon isotherm for benzene in the LHC waste-water, and Figure 12 presents a breakthrough curve for benzene obtained

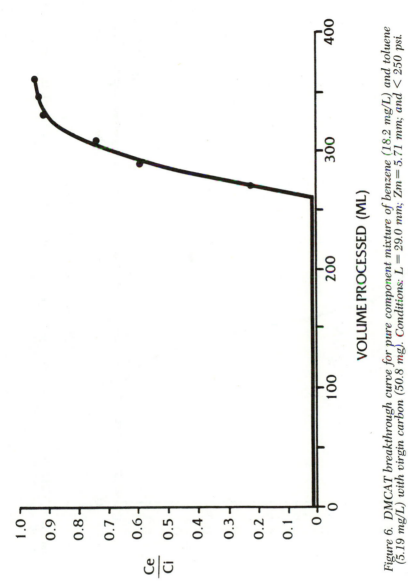

VOLUME PROCESSED (ML)

$$\frac{Ce}{Ci}$$

Figure 6. DMCAT breakthrough curve for pure component mixture of benzene (18.2 mg/L) and toluene (5.19 mg/L) with virgin carbon (50.8 mg). Conditions: L = 29.0 mm; Zm = 5.71 mm; and < 250 psi.

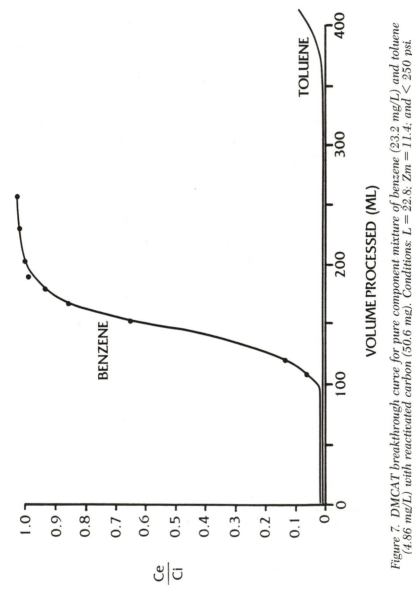

Figure 7. DMCAT breakthrough curve for pure component mixture of benzene (23.2 mg/L) and toluene (4.86 mg/L) with reactivated carbon (50.6 mg). Conditions: $L = 22.8$; $Zm = 11.4$; and < 250 psi.

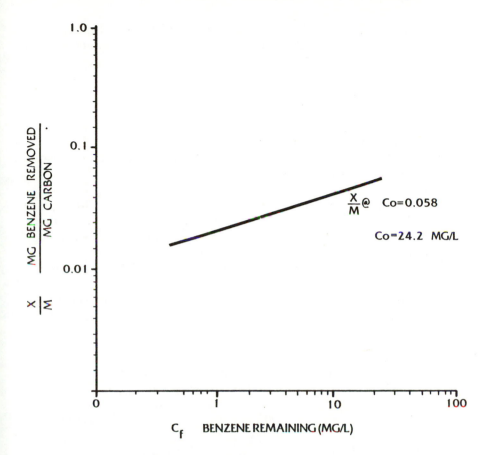

Figure 8. Pure component benzene; isotherm for benzene with reactivated carbon.

during dynamic pilot column testing performed for the same wastewater using the same carbon.

These figures also show a column effluent concentration for benzene greater than that of the influent concentration ($CE/CI > 1$). This is an indication that benzene was desorbed from the carbon as a result of competition for adsorption sites by the more adsorbable toluene molecule. Prediction of this behavior by the DMCAT is valuable in determining the order of adsorption as well as potential effluent quality from a spent carbon column.

One objective of an activated carbon feasibility study is to obtain data on carbon loading and carbon usage rates, which are necessary to design the system and to obtain a preliminary estimate of operating costs. Typically, this information is derived from breakthrough curves obtained

Table I. Comparison of Loadings for DMCAT with Isotherm for Benzene on Pure Component Mixture for Reactivated Carbon

Loading (mg/mg) at Initial Concentration

Isotherm	0.058 (24.2 mg/L benzene)
DMCAT	0.066 (23.2 mg/L benzene)

during a pilot study. Table III is a comparison of carbon loading for benzene predicted from the DMCAT runs on LHC wastewater compared to those estimated from the LHC isotherm and pilot study.

The loadings at breakthrough, defined as 10 μg/L benzene concentration, predicted from the DMCAT runs are 0.185 and 0.190 mg/mg, respectively. Carbon loading predicted from the pilot system at 10-min EBCT, for the same 10 μg/L breakthrough, is 0.182 mg/mg. Prediction of a carbon loading from an isotherm at this breakthrough level is difficult due to the extrapolation of the isotherm near the limits of detection for benzene of 10 μg/L.

Table IV is a comparison of the carbon usage rates in pounds per 1,000 gallons predicted from the DMCAT and pilot system runs. These

Table II. LHC Wastewater Characteristics

Conventionals

Parameter	Minimum	Average	Maximum
Temperature (°C)	14	22	33
pH	4.0	5.0–6.0	11
TSS (mg/L)	3.0	12	130
TOC (mg/L)	110	184	396
Oil and Grease (mg/L)	< 10	16	230

Organics

Parameter	Minimum	Average	Maximum
Benzene (mg/L)	21	53	71
Toluene (mg/L)	5.4	8.3	12
PNA's			
Naphthalene (mg/L)	6.7	5.5	18
Phenanthrene (mg/L)	0.1	1.3	4.7
Anthracene (mg/L)	0.038	0.600	2.20
Pyrene (mg/L)	0.064	0.370	1.60
Acenaphthalene (mg/L)	< 0.001	0.200	0.800
Fluorene (mg/L)	< 0.001	< 0.001	< 0.001

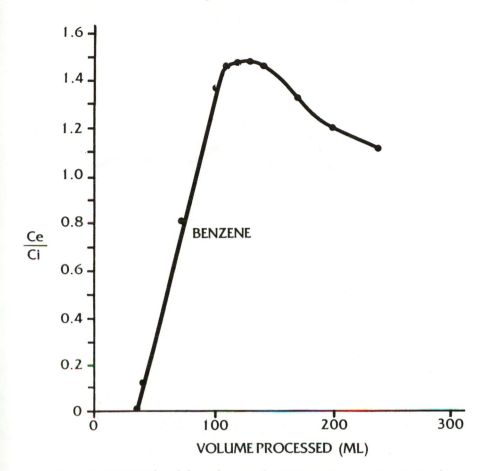

Figure 9. DMCAT breakthrough curve for LHC wastewater. Run. 1. Conditions: benzene (24.5 mg/L); toluene (4.1 mg/L); reactivated carbon (50.2 mg); L = 22.9 mm; and Zm = 13.7 mm.

usage rates, which represent the carbon consumption necessary to maintain the desired 10-μg/L effluent quality, are required for system design. Carbon usage rates are a function of the adsorber contact time and approach a minimum with increasing contact time. Note that the DMCAT predicted usage rate is well within the range of the 10-min and 60-min contact times evaluated in the pilot study.

Use of DMCAT as a Predictive Technique.   Knowledge of probable duration of a carbon pilot study as well as the order in which compounds will break through is useful for a meaningful evaluation and in substantially reducing analytical testing costs. The DMCAT method was used to predict the duration of a pilot study using a synthetic benzene and chlorinated

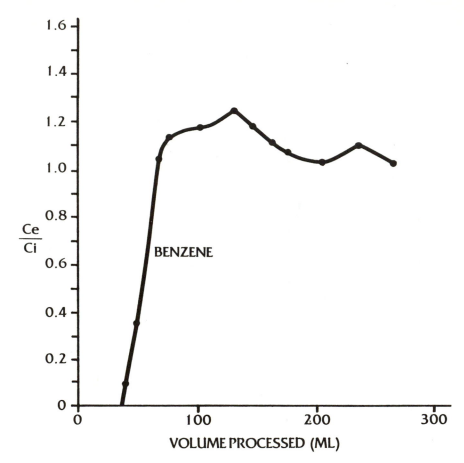

Figure 10. DMCAT breakthrough curve for LHC wastewater. Run 2. Conditions: benzene (19.7 mg/L, average); toluene (3.12 mg/L, average); reactivated carbon (50.0 mg); L = 23.7 mm; and Zm = 10.83 mm.

benzene wastewater. Because of the volatility of these components, a relatively constant concentration of pre-prepared column feed could not be maintained. Continuous feed preparation was felt appropriate, but an indication of the duration of the study to staff and maintain feed properly was essential. DMCAT runs were performed on a synthetic mix, with the average characteristics shown in Table V. Benzene was the first component to break through even though monochlorobenzene was at higher concentrations, as shown in Figure 13, for a DMCAT run at 2.5 cc/min. This rate represented a carbon usage of 14 lb/1,000 gallons. For the typical pilot run using 10 lb of carbon in each of three 10-cm (4-in.) diameter columns and contact times of 20, 40, and 60 min, an estimated run duration of 14 days was predicted. Actual run time was 21 days.

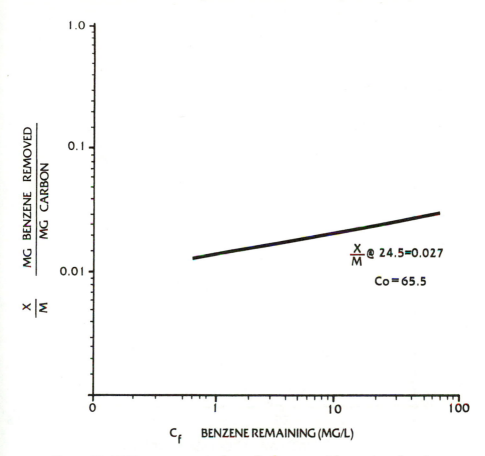

Figure 11. LHC wastewater isotherm for benzene with reactivated carbon.

Knowledge of the first compound, benzene, to break through also was useful in reducing analytical costs. If benzene was not detected in a column effluent, analysis for other compounds of interest was not continued.

Figure 13 also shows that a peak in the benzene effluent concentration about two and one-half times the influent value occurs at approximately the same volume processed as when monochlorobenzene begins to break through, indicating stripping of benzene. This peak is also shown for the pilot column run (Figure 15) to occur at approximately the same point in the run. Upon further processing, the effluent concentration approaches the same concentration as the influent. Influent concentrations for the pilot column are shown in Table VI.

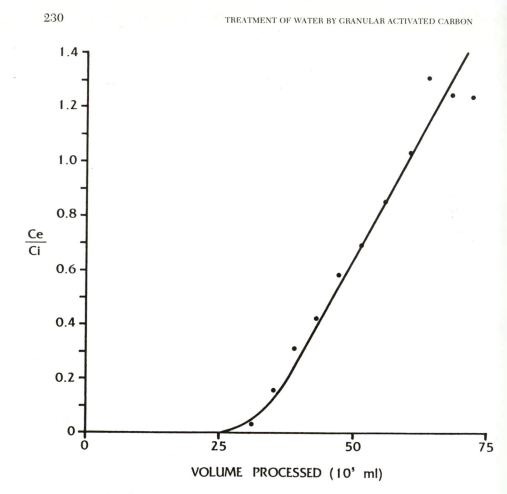

Figure 12. Pilot-scale run of LHC wastewater on reactivated carbon. Conditions: 3946 g; L = 1016 mm; average benzene concentration = 54 mg/L.

Table III. Comparison of Loadings for DMCAT with Isotherm and Pilot Study on LHC Wastewater

		At Breakthrough (mg/mg)	At Saturation (mg/mg)
DMCAT	Run 1	0.0185	0.030
	Run 2	0.0190	0.0296
Isotherm		—	0.027
Pilot Study		0.0182	0.037

**Table IV. Comparison of Predicted Carbon Usage Rates for DMCAT
with Pilot Study for LHC Wastewater**

	Contact Time (min)	Usage (lbs/1000 gallons)
DMCAT	0.027	11.04
Pilot Study	10	13.1
	60	8.25

Figure 14 represents breakthrough curves for a DMCAT run at 1.0 cc/min for the benzene–chlorinated benzene synthetic wastewater. This run, at a lower flow rate than the previous run, was performed to obtain more data prior to breakthrough in order to better define breakthrough curves for this system. Breakthrough and saturation of both compounds occur at exactly the same volumes processed, as for the 2.5-cc/min run, indicating that carbon loading data predicted from DMCAT breakthrough curves appear not to be a function of flow rate.

Figure 15 presents breakthrough results for both benzene and monochlorobenzene for the pilot-scale column test (EBCT = 20 min), again using the same reactivated carbon used for the previous DMCAT and pilot-scale runs. As in the DMCAT runs, breakthrough of monochlorobenzene occurs coincident with a peak in effluent benzene concentration.

Table VII shows a comparison of loadings predicted from both DMCAT and pilot-scale tests. The loadings at both breakthrough and saturation for the two DMCAT runs are nearly identical. These DMCAT-predicted loadings at breakthrough are 24% higher than those obtained from the pilot-scale test, while at saturation they are nearly identical. Usage rates predicted from DMCAT are 14.06 lb/1,000 gallons and 18.34 lb/1,000 gallons for the pilot run. Note that the 20-min EBCT in the pilot-scale test does not represent the optimum contact time.

Use of DMCAT for Evaluating Carbon at Varying Influent Concentration.   An on-site pilot study was performed at a vinyl chloride monomer plant. The process wastewater from this plant contained ethylene dichloride (EDC) at an average concentration of 1,260 mg/L and

**Table V. Synthetic Benzene–Chlorinated Benzene Wastewater
for DMCAT Runs Prepared in Unchlorinated Well Water**

Compound	Concentration (mg/L)
Benzene	53
Monochlorobenzene	203
o-Dichlorobenzene	23.1
p-Dichlorobenzene	24.5

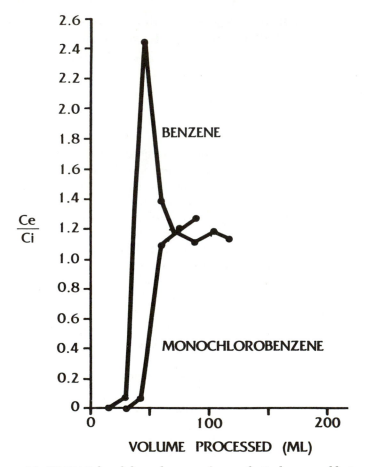

Figure 13. DMCAT breakthrough curve for synthetic benzene-chlorinated benzenes with reactivated carbon at 2.5 cc/min. Conditions: carbon (50.7 mg); and L = 24 mm.

chloroform at an average concentration of 25.0 mg/L. A steam stripper was treating this wastewater to average concentrations of 53.0 mg/L EDC and 0.04 mg/L chloroform. Pilot carbon column tests were performed on both the influent and effluent of the stripper to compare the treatment efficiency and cost effectiveness of GAC with the steam stripper as well as to determine the cost of carbon treatment for further reduction in organic concentration.

The pilot column data, including carbon usage rates, are presented in Table VIII and the corresponding breakthrough curves are shown in Figures 16 and 17. The carbon usage rate for the stripper influent stream at a nominal 60-min EBCT was 60 lb/1,000 gallons, based on an EDC

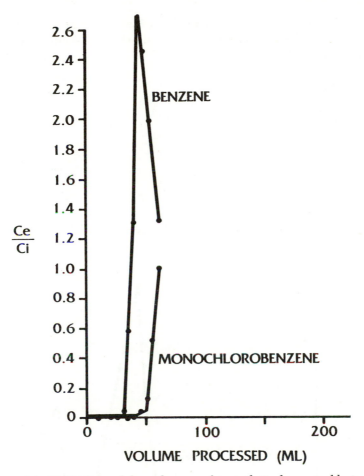

Figure 14. DMCAT breakthrough curve for synthetic benzene-chlorinated benzenes with reactivated cabon at 1 cc/min. Conditions: carbon (50.6 mg); and L = 24 mm.

breakthrough of 10 μg/L. The usage rate based on chloroform break-through of 10 μg/L at 60-min EBCT was 47.0 lb/1,000 gallons. Carbon consumption for treatment of the stripper influent would be significant due to the high concentration of the organics and the relatively low carbon adsorptive capacity for both EDC and chloroform.

The carbon usage rates determined by the pilot test performed on the stripper effluent stream were 10.5 lb/1,000 gallons for EDC removal at 22-min EBCT and 16.0 lb/1,000 gallons for chloroform removal at 12-min EBCT, with breakthrough of both compounds defined at 10 μg/L.

Samples for minicolumn tests were obtained 10 months after completion of the pilot study. During the intervening time, the characteristics

Table VI. Synthetic Benzene–Chlorinated Benzene Wastewater for Pilot Run Prepared in Unchlorinated Well Water

Compound	Concentration (mg/L)
Benzene	54.1
Monochlorobenzene	253
o-Dichlorobenzene	32.6
p-Dichlorobenzene	32.7

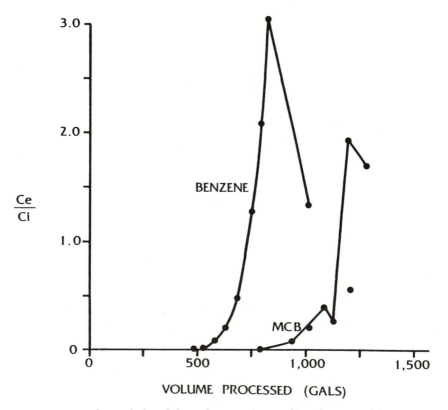

Figure 15. Pilot-scale breakthrough curve for synthetic benzene-chlorinated benzene wastewater. Conditions: reactivated carbon (3955 g;) L = 1125 mm; flow rate = 430 mL/min; and empty bed contact time = 20 min.

Table VII. Comparison of Loading for DMCAT with Isotherm and Pilot Study on Synthetic Benzene–Chlorinated Benzene Wastewater

	At Breakthrough (mg/mg)		At Saturation (mg/mg)	
DMCAT	1 cc/min	0.030	1 cc/min	0.033
	2.5 cc/min	0.030	2.5 cc/min	0.034
Pilot Study		0.0243		0.0331

of both the process wastewater (stripper influent) and stripper effluent changed considerably. DMCAT tests were performed on the new process wastewater containing 135 mg/L EDC and 2.03 mg/L chloroform. The DMCAT tests used the same regenerated bituminous coal carbon as the pilot tests. The flow rate for the DMCAT tests was 1 cc/min.

The carbon usage rates obtained from the DMCAT breakthrough curves shown in Figure 18 were 43.8 lb/1,000 gallons for EDC and 26.6 lb/1,000 gallons for chloroform at the same $10\text{-}\mu g/L$ breakthrough.

The differences in influent contaminant concentration between the pilot-tested wastewater and the minicolumn tested wastewater preclude direct comparison of minicolumn predicted carbon usage rates with pilot test predicted values. It is possible, however, to predict a usage rate at the new wastewater concentrations by extrapolation from the experimentally determined pilot test conditions and to compare this extrapolation with the minicolumn results.

Adsorptive capacity is dependent on the concentration of the adsorbate. Bohart and Adams (3) showed that the adsorptive capacity (carbon usage rate) of a given system (carbon and adsorbate) is directly proportional to the logarithm of the ratio between different feed concentrations. Although there are limitations to this principle, it is useful for estimating the carbon usage rate of a system at various feed concentrations. Two or more experimental points will establish the slope of the carbon usage rate versus feed concentration line which can then be used for predicting usage rates at other concentrations. Figures 19 and 20 depict the relationship between carbon usage rate and influent concentration using pilot test data for EDC and chloroform, respectively. These figures also demonstrate that the usage rates predicted by the minicolumn at an intermediate influent concentration are in reasonable agreement with extrapolated values.

Use of DMCAT as a Screening Tool. GAC is considered a leading alternative for the control of THM in potable water supplies by either direct removal of THM or by reduction of THM precursors. In addition to evaluating the point of application within the treatment scheme for optimizing carbon efficiency, a rapid evaluation of carbon to compare its cost performance with other alternatives is useful to reduce or eliminate costly pilot studies.

Table VIII. Comparison of Pilot-Scale and Minicolumn Performance for EDC/VCM Wastewater

Compound	Influent Conc. (mg/L)		Usage Rates (lb/1,000 gallons)		Carbon Loading (mg/mg)		
	Pilot, Avg	Minicolumn	Pilot	DMCAT	At Breakthrough		At Saturation
					Pilot	Mini-column	DMCAT
1,2-Dichloroethane							
Stripper Influent[a]	1260	135	Col 1 77.11	43.8	0.164	0.026	0.039
			Col 2 63.4	—	—	—	—
			Col 3 60.2	—	—	—	—
Stripper Effluent[b]	53.25		Col 1 26.2	—	—	—	—
			Col 2 10.5	—	—	—	—
Chloroform							
Stripper Influent	25.0	2.03	Col 1 51.5	26.6	0.004	0.0003	—
			Col 2 52.8	—	—	—	—
			Col 3 47.5	—	—	—	—
Stripper Effluent	0.038		Col 1 16.0	—	—	—	—

[a]Stripper Influent: Col 1 = 16-min EBCT; Col 2 = 36-min EBCT; and Col 3 = 58-min EBCT
[b]Stripper Effluent: Col 1 = 12-min EBCT; Col 2 = 22-min EBCT; and Col 3 = 27-min EBCT
Source: Environmental Science and Engineering, Inc., 1980.

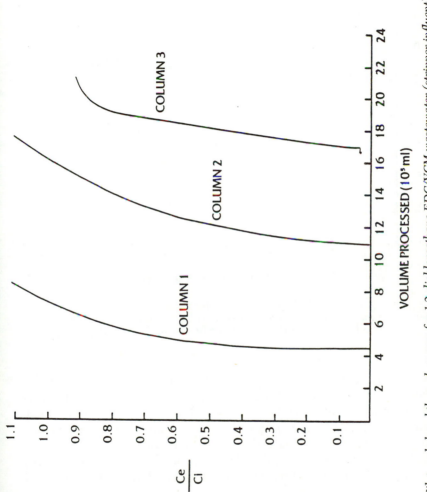

Figure 16. Pilot-scale breakthrough curves for 1,2-dichloroethane EDC/VCM wastewater (stripper influent) on reactivated carbon. Conditions for columns 1 through 3: 16 min EBCT, 36 min EBCT, and 59 min EBCT.

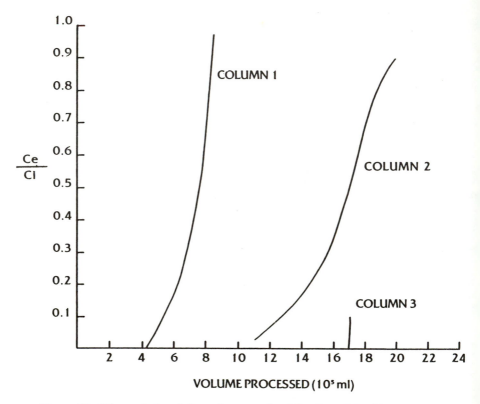

Figure 17. Pilot-scale breakthrough curves for chloroform EDC/VCM waste-water (stripper influent) on reactivated carbon. Conditions for columns 1 through 3: 16 min EBCT, 36 min EBCT, and 59 min EBCT.

More than a dozen treatment alternatives and combinations for the control of THM were studied recently to meet the proposed 0.10-mg/L THM maximum contaminant level (MCL) at a major Florida water treatment plant. GAC was considered along with chlorine dioxide, ozone, air stripping, and chloramines as the technologies most likely to meet the MCL. Evaluations of GAC in pilot plant typically last 4–6 weeks and are costly due to the number of required THM and THM formation potential analyses.

To evaluate GAC for the control of THM by the removal of pre-cursors, a sample was obtained of the raw water before any chlorination. Raw water was first run through a column containing 50.8 mg of carbon. Samples of the minicolumn effluent were collected at 20-mL intervals, and influent samples were collected at approximately 100-mL intervals. The samples were then contacted with at least 50 mg/L chlorine by microliter injection, resealed, and held for 48 h, the average detention

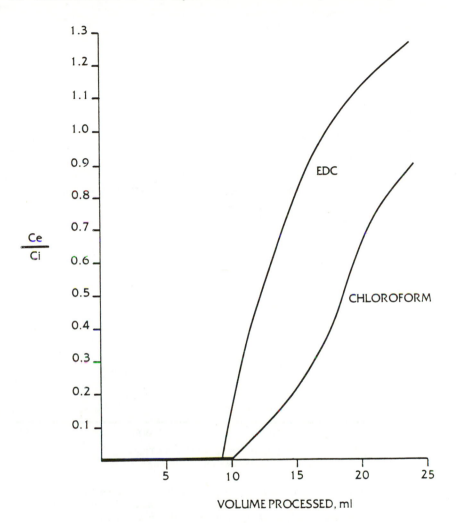

Figure 18. DMCAT breakthrough curve for stripper influent wastewater on reactivated carbon (50.4 mg); L = 23.8 mm, flow rate = 0.5 cc/min.

time in distribution system, before quenching with sodium thiosulfate. Because of the small sample volume, 50 mg/L was the minimum practical chlorine dose. Finally, the samples were analyzed for THM using gas chromatographic headspace analysis.

Figure 21 is the breakthrough curve of total trihalomethane (TTHM) concentration in the effluent versus the volume processed through the column. Using a breakthrough criterion of 0.10 mg/L TTHM, the carbon was spent after processing 70 mL of water, i.e., column effluent exceeded 0.10 mg/L TTHM. The carbon usage rate based on the volume processed

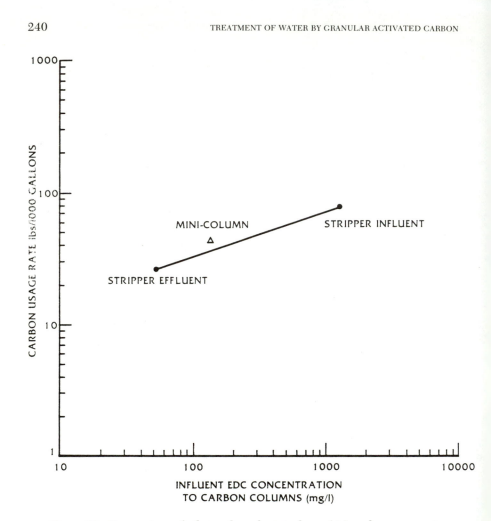

Figure 19. Comparison of pilot-scale and minicolumn (△) carbon usage rates for 1,2-dichloroethane EDC/VCM wastewater. Conditions: stripper effluent (12 min EBCT), and stripper influent (16 min EBCT).

and the weight of carbon in the column was 6.2 lb/1,000 gallons processed. The resultant treatment cost was representative of the potential for removal of THM precursors by carbon, which could be compared with other treatment alternatives.

To evaluate GAC for the direct removal of THM to compare with other direct removal technologies such as air stripping, samples of water settled before filtration were obtained from the plant. Since they had been chlorinated in the plant during the treatment process, the samples were quenched directly after collection to avoid further THM formation.

Two runs were performed on settled water. THM concentration for

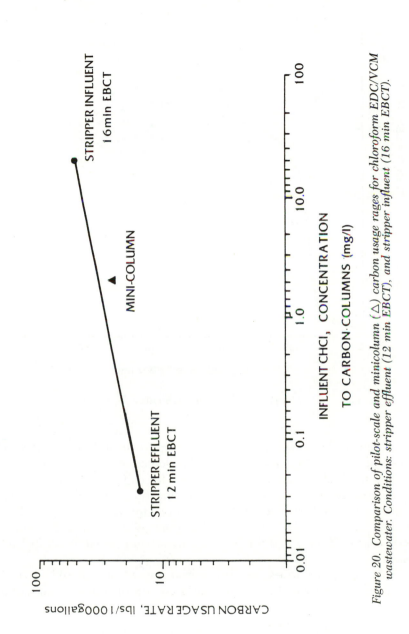

Figure 20. Comparison of pilot-scale and minicolumn (△) carbon usage rages for chloroform EDC/VCM wastewater. Conditions: stripper effluent (12 min EBCT), and stripper influent (16 min EBCT).

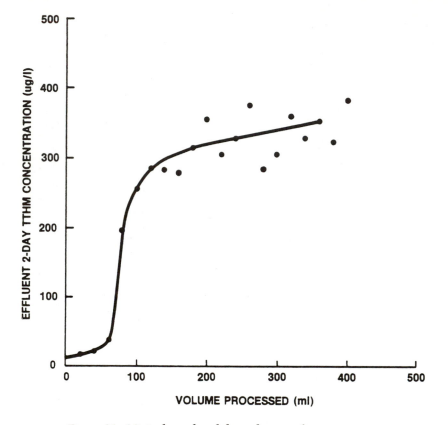

Figure 21. Minicolumn breakthrough curve for raw water.

the column influent was determined for approximately every 100 mL processed, while effluent samples were collected at 20-mL intervals and analyzed for THM by gas chromatographic headspace analysis.

Figures 22 and 23 show corresponding breakthrough curves. The volumes processed prior to a 0.10 mg/L breakthrough were 490 mL for Run 1 and 450 mL for Run 2. The carbon usage rate, based on the average of two runs (470 mL) and the 50.1 mg of carbon present, is 0.9 lb/1,000 gallons.

Table IX summarizes the daily carbon usage rates based on a 60-million-gallons/day plant flow and yearly pounds of replacement carbon required, assuming a 10%/cycle loss in reactivation and handling.

Additional pilot studies to determine required contact time were eliminated from the scope of this project because of the high estimated treatment cost with carbon, $0.16/1,000 gallon minimum, versus that for

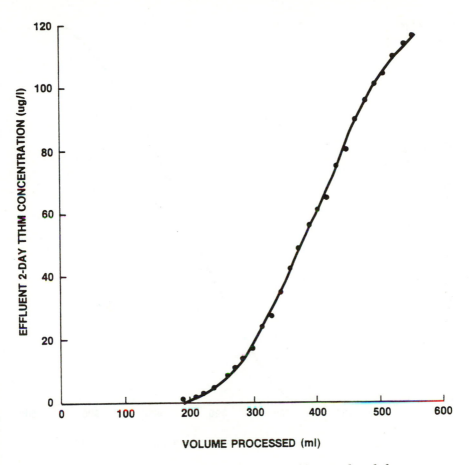

Figure 22. Minicolumn breakthrough curve for chlorinated settled water, Run 1.

other technologies such as air stripping with an estimated treatment cost of about $0.06/1,000 gallons.

Summary

The use of a DMCAT as discussed in this report is an effective, rapid method for determining representative carbon adsorptive capacities and usage rates for specific organic components in real-world wastewaters and potable waters. These data can be used to determine if a carbon treatment for a given water is feasible and to provide a rough estimate for system economics. The technique is capable of generating reproducible data in a rapid fashion using simple equipment.

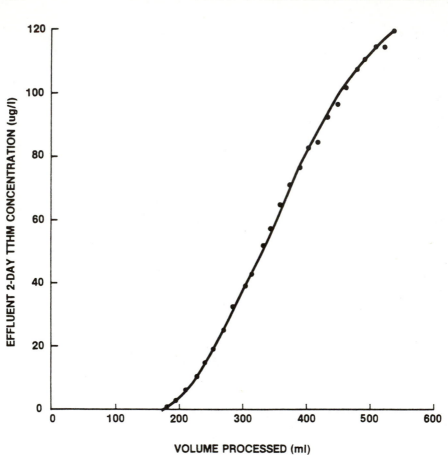

Figure 23. Minicolumn breakthrough curve for chlorinated settled water,
Run 2.

The DMCAT technique can be applied successfully within a wide range of operating pressures and flow rates. Carbon particle size should, however, be kept within a narrow range such as 230 × 325 mesh to avoid rapid pressure build-up and to eliminate the variation of adsorptive efficiency as a function of particle size.

Based on the data from a limited number of waste streams, DMCAT-predicted loadings and usage rates compare favorably with those obtained from pilot-scale tests. The DMCAT technique appears to be more direct than an equilibrium isotherm in predicting carbon loadings, in that a sharp, well-defined breakthrough curve is obtained with DMCAT. Because DMCAT is easily conducted in a closed system, loss of volatile

Table IX. Summary of Carbon Usage Rates for Control of THM

Treatment Location	Minicolumn Usage Rate (lb/1,000 gallons)	Projected Daily Usage Rate at 60 MGD[a] (lb)	Approximate Yearly Carbon Consumption, 10% Reactivation Loss, 60 MGD
Raw Water	6.2	372,000	12,483,000
Settled Water	0.9	54,000	1,971,000

[a]MGD = million gallons per day.
Source: Environmental Science and Engineering, Inc., 1980.

components normally experienced with standard isotherm techniques does not occur.

Results from the DMCAT technique are consistent with existing data on the relative performance of various carbons as observed by the breakthroughs obtained for a virgin and a reactivated carbon on a similar pure component and with exhibiting the dependency of carbon adsorption on concentration.

Although DMCAT provides representative information on carbon usage rates, data to scale-up for adsorber system design (contact time) cannot be obtained. Insufficient information is available to correlate mass transfer zones or minicolumn contact times with those of pilot or actual systems. Additional work on several carbon systems, either pilot or full scale, is indicated to determine if a correlation can be established.

DMCAT is extremely useful for screening the use of GAC for various applications and to provide approximate cost estimates to compare with other treatment alternatives. DMCAT also can be used to compare the performance of available carbons to select the most cost-effective carbon for the application. Loss of performance or shift in adsorbate removal can be monitored by frequent DMCAT runs with reactivated carbon on the actual waste stream to provide a more meaningful monitoring parameter than a nonspecific parameter such as iodine number. DMCAT also can be useful in determining the effect of pretreatment on carbon performance such as optimizing pH. With some modification in the DMCAT test apparatus, DMCAT can be used to evaluate solvent and steam regeneration techniques.

Literature Cited

1. Rosene, M. R.; Ozcan, M.; Manes, M. *J. Phys. Chem.* **1976**, *23*, 80.
2. Rosene, M. R.; Deithun, R. T.; Lutchko, J. R.; Wayner, W. J. "High Pressure Technique for Rapid Screening on Activated Carbons for Use in Potable Water", Calgon Corp., 1979.
3. Bohart, G. S.; Adams, E. Q. *J. Am. Chem. Soc.*, **1920**, *42*, 523–44.
4. Beaudet, B. A.; Bilello, L. J.; Kellar, E. M.; Allan, J. M. "Removal of Specific Organics from Wastewater by Activated Carbon Adorption, Evaluation of a Rapid Method for Determining Carbon Usage Rates", presented at the 35th Annual Industrial Waste Conference, Purdue University, 1980.
5. Beaudet, B. A.; Kellar, E. M.; Bilello, L. J.; Turner, R. J. *Proc. Water Wastewater Equipment Manuf. Assoc.*, in press.

RECEIVED for review August 3, 1981. ACCEPTED for publication June 29, 1982.

Adsorption Equilibria in Multisolute Mixtures of Known and Unknown Composition

BERND R. FRICK and HEINRICH SONTHEIMER

Bereich Wasserchemie am Engler-Bunte-Institut der Universität Karlsruhe, Richard-Willstätter-Allee 5, D-7500 Karlsruhe 1, Federal Republic of Germany

The influence of the composition of multisolute mixtures on the adsorptive uptake of individual compounds and on the adsorption of total organics is demonstrated. Based on these dependencies, adsorption isotherms can be interpreted relative to mixture composition. A method is proposed that characterizes multisolute mixtures as a solution of three compounds with defined adsorption properties, and computation results based on this procedure are presented. The prediction of adsorption equilibria can be improved greatly as can the prediction of granular activated-carbon filter performance when multisolute mixtures are adsorbed. Experimental data for humic acid solutions and groundwater are used to confirm the applicability of the proposed method.

K NOWLEDGE ON THE PROCESS of activated-carbon adsorption from the aqueous phase has been enlarged during the past 15 years. The major results of intensive research are, from our point of view, several mathematical models that describe adsorption equilibria and adsorption kinetics for single solute systems as well as for mixtures containing two or three organic solutes (*1–5*). These models provide a basis for the design of granular activated-carbon (GAC) treatment plants and cost estimates, and they can be used to evaluate pilot plant studies.

As most theories and models need single solute adsorption data for compounds present in water, the most important provision in applying such models is to know the components of the mixture and their concentrations. However, these requirements restrict, or even prevent, the application of this design tool when the prediction of GAC performance in water treatment is desired. The reason for this is the variety of

0065-2393/83/0202-0247$0.700/0

compounds in the water and the lack of analytical methods to detect them.

To overcome this difficulty, the total of the organics present frequently was treated mathematically as a single compound by using equilibria and kinetic data obtained from experiments where the organics concentration was measured as a sum by determining dissolved organic carbon (DOC) [total organic carbon (TOC)] by UV absorbance or fluorescence. With this method, the interactions between the different compounds are neglected, and the prediction of chromatographic effects or of the decreased adsorption capacity of the carbon, which influences GAC performance drastically, is impossible (3,6).

It was the aim of this study to find a method by which the models developed for predicting adsorption behavior of defined multisolute mixtures with a limited number of components could be applied to the calculation of adsorption equilibria in multisolute mixtures of unknown composition.

To overcome the drawback of missing information on the composition and adsorption properties of the mixture constituents, a procedure grouping compounds with similar adsorption behavior, i.e., reducing the number of components, seemed to be a way to consider the heterogeneity of those mixtures.

By means of this hypothetical mixture and the models referred to previously, the actual adsorption behavior can be simulated as a response of competitive adsorption. Before one can define such a reduced mixture, the number of fictive components, and their adsorption properties, it is necessary to know more about the influence of competitive adsorption on the uptake of individual compounds and the adsorption behavior of total organics. This knowledge is essential as the selection of such a hypothetical mixture should be based on experimental equilibrium data determined with the unknown mixture.

Adsorption Equilibria of Individual Solutes in Mixtures

Data on adsorption equilibria in mixtures containing two solutes have been reported primarily as adsorption isotherms for the individual species (1). To illustrate the influence of the other solute present, the equilibrium concentration or the initial concentration of this compound is used as a parameter according to the general isotherm equations:

$$q_i = f(c_i, c_j)c_j = \text{constant} \quad \text{or} \quad q_i = f(c_i, c_{0,j})c_{0,j} = \text{constant} \quad (1)$$

where the subscript i denotes the component for the adsorption isotherm of interest and subscript j denotes the other solute present.

In general, the batch determination of the isotherms for mixture adsorption is carried out by varying the composition of the initial solution and by holding constant the ratio of the solution volume, L, to the amount of carbon, m.

The experimental procedure to examine adsorption equilibria for mixtures of unknown composition is to add different amounts of carbon to constant volumes of the solution which is identical in initial composition for all points of the resulting isotherm.

To interpret results of batch equilibrium experiments with complex mixtures of undefined composition, adsorption isotherms with defined mixtures containing two or three organic solutes were determined using the method of variable liquid volume—carbon amount ratio (L/m) and constant initial solution composition

The isotherms so obtained show a different shape from those presented in the literature. An explanation of this effect is given in Reference 7. Based on the characteristic shape of the isotherms, a qualitative interpretation can be made, which evaluates the adsorbability of the other mixture components relative to that of the individual species.

An example of adsorption equilibria of individual compounds in a ternary mixture is shown in Figure 1. Six solutions, each containing different initial concentrations of p-nitrophenol, p-hydroxybenzoic acid, and 4-phenolsulfonic acid were agitated with four different amounts of powdered activated carbon (Norit ROW 0.8 S, Table I). The liquid phase concentrations of the three individual compounds at equilibrium were determined by UV photometric analysis. Figure 1 illustrates the individual isotherms of these compounds when adsorbed simultaneously. The numbers attached to the isotherms refer to the six different initial solutions.

The adsorbability of the compounds as a single solute decreases from p-nitrophenol to sulfonic acid as shown in Table II where the Freundlich isotherm parameters are listed.

In a log–log scale, all six individual isotherms for p-nitrophenol are straight lines, while the isotherms for the more weakly adsorbed hydroxybenzoic acid and phenolsulfonic acid are curved.

The curved isotherms of individual species adsorbed from a mixture are typical if one or more of the other mixture constituents have a better adsorbability as a single solute than that of the species of concern. The most strongly adsorbed component in a mixture always shows a straight line in a log–log isotherm plot, i.e., the Freundlich isotherm equation can be applied. However, the Freundlich constant K decreases as the adsorption capacity is decreased by the other solutes and the isotherm exponent n increases in comparison to that of the single solute adsorption isotherm.

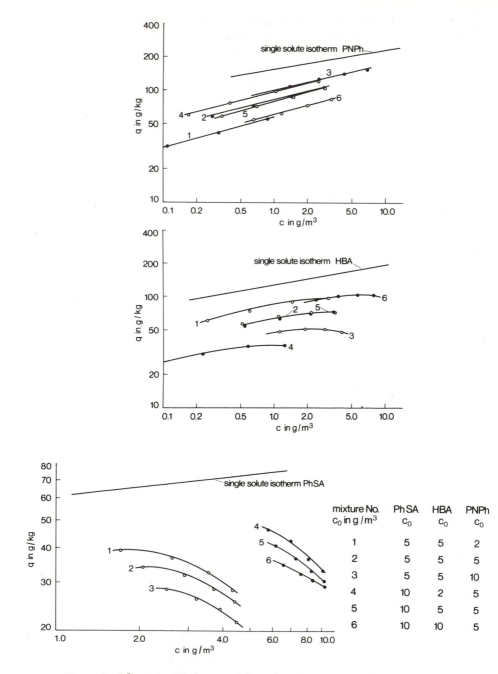

Figure 1. Adsorption isotherms of the individual components when adsorbed
simultaneously from a ternary mixture; initial composition of the mixture is
identical for the isotherms with the same numbers attached.

Table I. Activated-Carbon Properties

Property	ROW 0.8 S[a], Cylindrical Shape		F 300[b], Irregular Shape	
Particle density	0.64	g/cm^3	0.85	g/cm^3
Apparent density	0.40	g/cm^3	0.54	g/cm^3
Pore volume	1.03	cm^3/g	0.66	cm^3/g
Mean diameter (based on a sphere of identical volume)	1.24	mm	1.40	mm

[a]Norit, The Netherlands.
[b]Chemviron, Belgium

These dependencies were confirmed by equilibrium data in mixtures of up to six compounds (8). An extreme example of the influence of competitive adsorption on isotherms is given in Figure 2. The two solutes, *p*-nitrophenol and tetrachloroethane, have very different single solute adsorbabilities as indicated by the two single solute adsorption isotherms. If adsorbed simultaneously, the weakly adsorbed tetrachloroethane is strongly influenced by *p*-nitrophenol, resulting in an extremely curved isotherm.

These results indicate that it is possible to get an estimate of the mixture composition in view of the adsorbability of the components, if one can determine the adsorption isotherm of an individual and known compound already present in the mixture or specially added for this purpose.

Table II. Freundlich Isotherm Parameters and Properties of the Single Solutes (Norit ROW 0.8 S)

Substance		K (g/kg)	n	mw	pK
Phenolsulfonic acid (PhSA)	OH / SO$_3$H$^+$	60.8	0.12	174.2	<1/8.7
p-Hydroxybenzoic acid (HBA)	OH / COOH	120.0	0.20	138.1	4.0
p-Nitrophenol (PNPh)	OH / NO$_3$	141.0	0.20	139.1	7.2

Note: Freundlich isotherm: $q = Kc^n$.

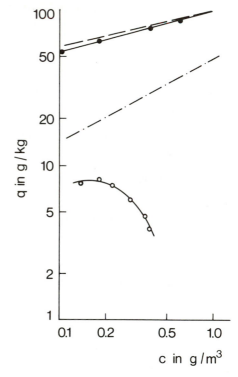

Figure 2. Adsorption isotherms of p-nitrophenol and tetrachloroethane as single solutes and for a binary mixture of constant initial composition. Conditions: C_0 PNPh = 1.0 g/m³, C_0 tetrachloroethane = 0.42 g/m³, and AC: P630. Key: – –, PNPh single solute; —●—, PNPh in mixture; –·–, tetrachloroethane single solute, and —○—, tetrachloroethane in mixture.

Influence of Mixture Composition on Adsorption Isotherm for Total Organics

By using a summary concentration parameter that covers all organics, the adsorption isotherm for total organics can be determined. Dissolved organic carbon (DOC) was used to detect the organics concentration. Subsequently, the isotherm, which relates the total residual concentration, c_T, to the solid phase loading, q_T, is referred to as the overall isotherm.

As the overall isotherm results from a superposition of the individual adsorption isotherms of the mixture components, it can be expected that it is also influenced by the mixture composition and that there are some features which can be used to interpret experimental results.

The nonadsorbable fraction of a mixture can be determined from the overall isotherm. Even with the addition of large amounts of carbon, there is no reduction in the total solute concentration, c_T; thus, the carbon loading decreases rapidly at a constant equilibrium concentration (9).

By extrapolating the isotherms shown in Figure 3, the initial concentration for the nonadsorbable compounds is indicated by the intersection with the axis, representing the residual concentration, c_T. The

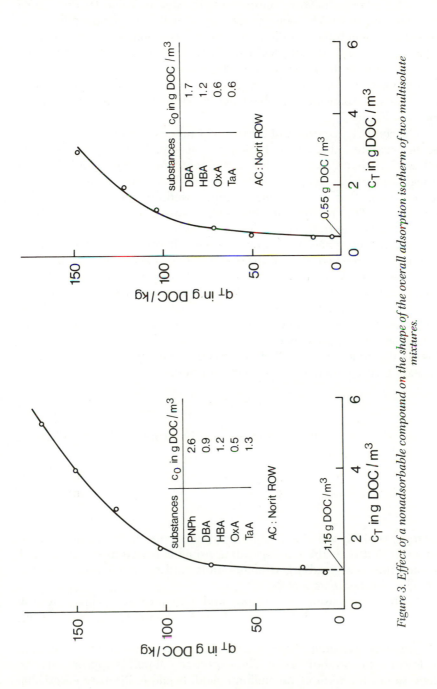

Figure 3. Effect of a nonadsorbable compound on the shape of the overall adsorption isotherm of two multisolute mixtures.

adsorbability of the mixture components is listed in Table III on a DOC concentration basis. The comparison between these concentrations and the initial concentration of the nonadsorbing component, represented by tartaric acid (TaA), shows good agreement within the tolerance of the DOC measurement. To cover this region of the overall isotherm, the carbon amounts have to be large enough to remove all adsorbable compounds. Carbon amounts above 500 mg/L were used for the mixtures shown in Figure 3.

A second characteristic of the overall isotherm can be observed in the region near the initial concentration, i.e., at equilibrium points obtained with small carbon dosages. If the solution contains poorly adsorbable and strongly adsorbable compounds, the slope of the overall isotherm is much steeper than that of the single solute isotherms of the individual compounds. The reason for this is the strong competition at high equilibrium concentrations, where the strongly adsorbed components are removed in preference. The extent of the isotherm slope depends on the distribution of the two characteristic compounds. This feature is illustrated in Figure 4. Phenolsulfonic acid represents the weakly adsorbed substance, and p-nitrophenol and hydroxybenzoic acid represent the substances with good adsorbability. For the mixture corresponding to the upper isotherm, the ratio of the initial concentrations between weakly and strongly adsorbing substances is 7.55; the ratio for the other isotherm is 2.94. If the first three equilibrium points were fitted by a Freundlich equation to get a pseudo single solute isotherm:

$$q_T = K_T c^{n_T} \tag{2}$$

then the slope, n_T, of the straight line can be expressed by the following empirical correlation:

$$n_T = n_2 + 0.6 n_2 \frac{c_{01}}{c_{02}} \frac{K_2 - K_1}{K_1} \tag{3}$$

where n_i and K_i are the Freundlich parameters of the single solute isotherms and c_{0i} is the corresponding initial concentration. Subscript 1 denotes the weakly adsorbed compounds, and subscript 2 denotes the strongly adsorbed compounds.

With an increasing ratio c_{01}/c_{02} and with increasing differences in adsorbability, i.e., large differences in $K_2 - K_1$, the total isotherm slope, n_T, also increases.

A common characteristic for both DOC isotherms can be observed in the lower concentration range. The experimental points approximate the single solute isotherm of the sulfonic acid, because all other adsorbable

Table III. Freundlich Isotherm Parameters for Single Solutes (Norit ROW) (Values Related to DOC Concentration)

Substance	K (g DOC/kg)	n
p-Nitrophenol (PNPh)	83.2	0.20
p-Hydroxybenzoic acid (HBA)	80.6	0.20
p-Diazobenzenesulfonic acid (DBA)	57.6	0.21
Oxalic acid (OxA)	26.6	0.27
Phenolsulfonic acid (PhSA)	27.9	0.12
Tartaric acid (TaA)	—	—

compounds have been removed from the fluid phase; thus, the DOC isotherm represents the single solute adsorption behavior of the weak adsorber.

With the aid of the characteristic shape of the isotherm as a consequence of specific adsorptive interactions, it is possible to interpret experimental overall adsorption data with regard to mixture compounds of defined adsorption behavior.

By using an adsorption isotherm of humic acid, which has been

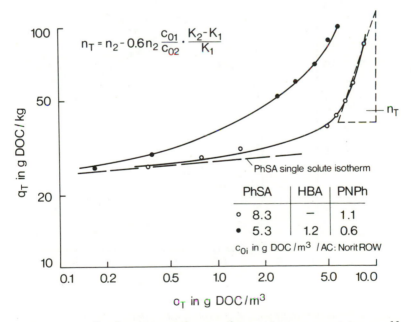

Figure 4. Overall adsorption isotherms of two mixtures containing weakly and strongly adsorbable compounds.

isolated from the regenerate of a macroreticular ion-exchange resin plant (10), it was proved that the dependencies found in defined mixtures also can be observed in complex mixtures of practical interest (Figure 5). The steep part of the isotherm near the initial concentration indicates that there are weakly as well as strongly adsorbing substances. The flat part represents the single solute behavior of the weak adsorber, and the descending part stands for the nonadsorbable fraction of the humic acid.

Simultaneous Adsorption Equilibria with Adsorption Models

Adsorption experiments with mixtures of known composition offer the opportunity to test adsorption models for the prediction of simultaneous adsorption equilibria. Because adsorption in mixtures strongly depends on the adsorption behavior of each compound, only those models that contain single solute adsorption isotherms of the different components were tested as input parameters.

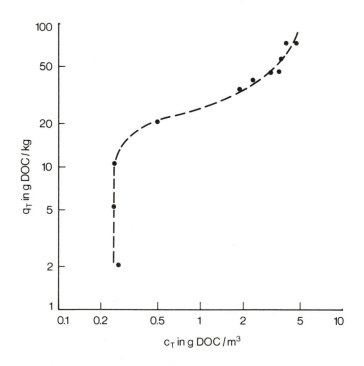

Figure 5. DOC isotherm of a multisolute mixture (humic acid from a groundwater with a high organic loading). Conditions: humic acid/Norit ROW, and $C_{OT} = 5.5$ g DOC/m^3.

Two models that meet this requirement are the ideal adsorbed solution (IAS) theory, developed by Myers and Prausnitz (*11*) and Radke and Prausnitz (*12*), and the simplified competitive adsorption model (SCAM) of Baldauf et al. (*13*), which is an extension of the method proposed by Kidnay and Myers (*14*). In the IAS theory, single solute adsorption is described by a Freundlich isotherm; at low concentrations Henry's law is valid (*2, 14*). In SCAM, single solute adsorption also is expressed by a Freundlich isotherm. Input parameters for both models include initial concentrations of the components and isotherm parameters of the single solutes. The experimental conditions give the liquid–solid ratio (L/S).

The IAS model was used for mixtures with up to three components, and SCAM was used for mixtures of up to four components. If the single solute behavior of the components does not differ too greatly, both models yield similar results. The predictive results will be better by using the IAS theory in those cases where there are appreciable differences in the single solute adsorbability (*5*). One reason seems to be the restriction that the single solute isotherms of the Freundlich type used in SCAM are assumed to be valid over the entire concentration range; when applying the IAS model, the Freundlich isotherm equation, enlarged by Singer et al. (*2*), was used.

As both models describe the relations found in the experiments qualitatively and, in most cases, also quantitatively, they can be applied for such purposes when dealing with unknown mixtures that have to be reduced to defined systems.

Prediction of Adsorption Equilibria in Multisolute Mixtures of Unknown Composition

Displacement effects and other phenomena due to competitive adsorption can be predicted only when the unknown mixtures are treated as a multicomponent mixture. To achieve this, the single solute adsorption data (e.g., Freundlich or Langmuir parameters) and the initial concentrations have to be known. Because analytical methods fail in determining the mixture composition, a procedure has to be developed that allows the reduction of multisolute systems to a defined mixture. This method, referred to here as sorption analysis, is based on adsorption experiments with the unknown mixture.

It was assumed that the organic background can be divided into three groups with specific single solute behavior. This selection is based on the experimental observations in mixtures of defined composition as shown previously in chapter 3. By means of these three components, the adsorption equilibrium should be predicted for the sum of the organics and their influence on the adsorption of selected compounds in water.

The definition of the single solute adsorbability of the components is as follows: a nonadsorbable fraction, a weakly adsorbable fraction, and a strongly adsorbable fraction. Figure 6 lists the experimental work necessary to get the input information for the sorption analysis. Information on the nonadsorbable fraction (*see* Point A), as well as on the single solute behavior of the weakly adsorbed substances (Point B) and on the distribution ratio between strongly and weakly adsorbed substances (Point C), can be derived from the shape of the overall isotherm as outlined in chapter 3.

The characteristic mixture isotherm of the tracer substance, i.e., a known compound added to the unknown mixture and detected by single substance analysis, helps in deciding whether the mixture contains compounds which adsorb better than the tracer substance or whether all of them adsorb worse.

These qualitative or semiquantitative estimates, obtained by computer analysis, are input parameters of a nonlinear curve fitting computer routine, which varies the single solute isotherm constants and the initial concentrations of the fictive mixture components.

The parameters are varied systematically so that the deviations between the measured isotherm points and those calculated with a

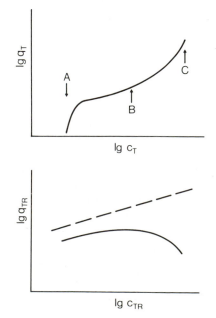

Figure 6. Adsorption experiments as basis for the sorption analysis of multisolute mixtures of unknown composition. The plot on the top represents an overall isotherm with equal volume of the unknown mixture and varying carbon dosage (~10–1000 mg carbon/L solution); determination of the total equilibrium concentration by DOC measurement. The plot on the bottom represents tracer substance isotherm with equal volume of the unknown mixture + amount of the tracer substance + varying carbon dosage (~ 10–50 mg carbon/L solution); determination of the tracer equilibrium concentration by UV or GC measurements.

competitive adsorption model are minimized. Figure 7 is a flow chart of this procedure.

Depending on the experimental input, e.g., overall isotherm or tracer compound isotherm, the two- or three-component versions of IAS or SCAM have to be used. The nonadsorbable fraction is assumed to be a inert compound and thus does not have to be considered in the competitive adsorption model. As a result of the sorption analysis, one gets a mixture reduced to three components, which shows an adsorption behavior equivalent to that of the real mixture. The isotherm constants and the initial concentrations of the fictive compounds can be used for the calculation of breakthrough profiles or for further prediction of adsorption equilibria.

The core of the parameter determination is a competitive adsorption model. IAS (*1, 2*) and SCAM (*5*) were used when the total isotherm was taken as the experimental input. For reasons of computation speed, only SCAM was used when the tracer isotherm had to be fitted.

Before applying the described procedure to unknown mixtures, it was tested with multisolute mixtures of known composition. Table IV presents the results obtained for a mixture of the poorly adsorbed phenolsulfonic acid and the strongly adsorbed p-nitrophenol. The overall isotherm and the hydroxybenzoic acid isotherm, added to the mixture as a tracer substance, provided the experimental background. The parameters for the two fictive mixture compounds obtained from the sorption analysis show a suitable agreement with the data of the actual mixture. In most cases, the sorption analysis, in connection with the IAS model, yields the more exact parameter values for the initial concentrations as well as for the single solute isotherm constants K and n.

An example of the application of sorption analysis to multisolute mixtures of unknown composition is shown in Figure 8. It can be seen that the measured DOC isotherm of the humic acid solution can be calculated as a response to the simultaneous adsorption in a three-component mixture. The Freundlich isotherm constants, K and n, and the initial concentrations, c_0, listed in the figure were the input for the prediction with SCAM. Beside the description of the overall adsorption, it is the aim of such a treatment to predict the competition between the organic background and selected single compounds which have to be removed from water.

A comparison between the measured p-nitrophenol (PNPh) isotherm in the mixture with humic acid (HA), the overall isotherm of which is illustrated in Figure 8, and the calculated isotherm is given in Figure 9. The mixture of PNPh and HA was mathematically treated as a four-component system, with the three fictive components of the humic acid (Figure 8) and the single solute isotherm parameters (Freundlich

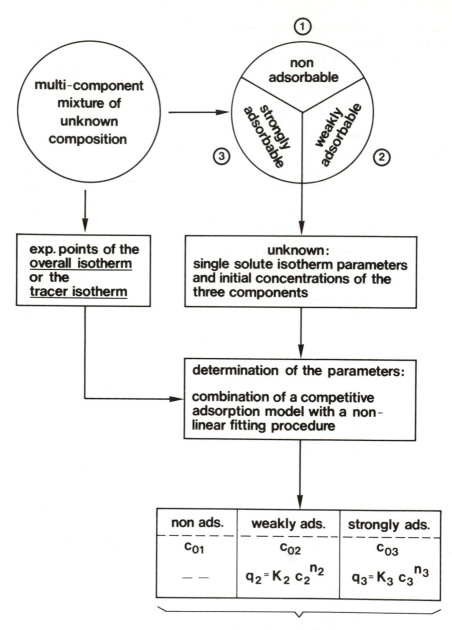

Figure 7. Flow chart of the sorption analysis procedure.

Table IV. Single Solute Isotherm Parameters and Initial Concentration Determined by the Sorption Analysis as Test for the Significance of the Proposed Method

Actual Mixture Parameters	Equivalent Mixture	
	SCAM	IAS
PhSA		
$c_0 = 8.3$ g DOC/m^3	7.7	8.3
$K = 27.9$	35.4	24.9
$n = 0.12$	0.19	0.17
PNPh		
$c_0 = 1.1$ g DOC/m^3	1.7	1.1
$k = 89.0$	87.4	95.6
$n = 0.20$	0.14	0.12

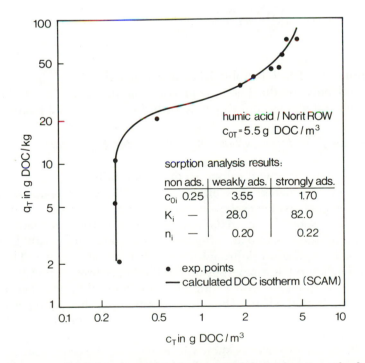

Figure 8. Result of the sorption analysis applied to the humic acid solution referred to in Figure 5.

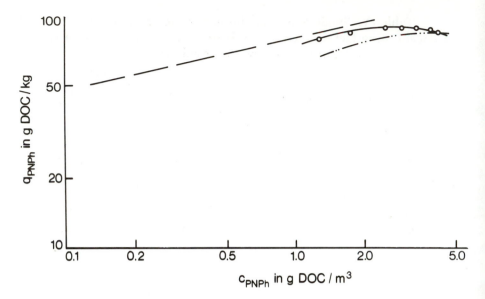

Figure 9. p-Nitrophenol adsorption isotherm in mixture with the humic acid referred to in Figures 5 and 8. Conditions: $C_0PNPh = 5.2$ g DOC/m^3, and $C_0HA = 5.5$ g DOC/m^3. Key: $--$, PNPh single solute isotherm; $-O-$, PNPh in mixture with humic acid; and $-\cdot\cdot-$ PNPh isotherm calculated with SCAM.

isotherm) of the PNPh in Table III. The calculated isotherm indicates the strong influence of the humic acid components on the *p*-nitrophenol adsorption and agrees approximately with the experimental points.

Competitive Adsorption Influencing GAC Filter Performance

A number of investigators studied the influence of adsorption equilibria on chromatographic effects in GAC filters (*4,6,14*). Apart from this phenomenon, which especially occurs if traces of organic compounds have to be removed, e.g., haloform peak concentrations in the filter effluent (*15*), there is also an influence of the compounds with different adsorbability on the breakthrough curve of the total organics. The total breakthrough curve is a superposition of the breakthrough profiles of the individual components which differ in their time of appearance in the effluent and in their shape. Thus, concentration profiles can be observed which are quite different from those when the adsorption of single solutes was investigated.

Such a curious curve is shown in Figure 10. A humic acid solution with *p*-nitrophenol (c_0 PNPh = 2.1 g DOC/m^3) was fed to an activated-carbon filter of 3 cm I.D. and 10-cm bed height. In a relatively short time,

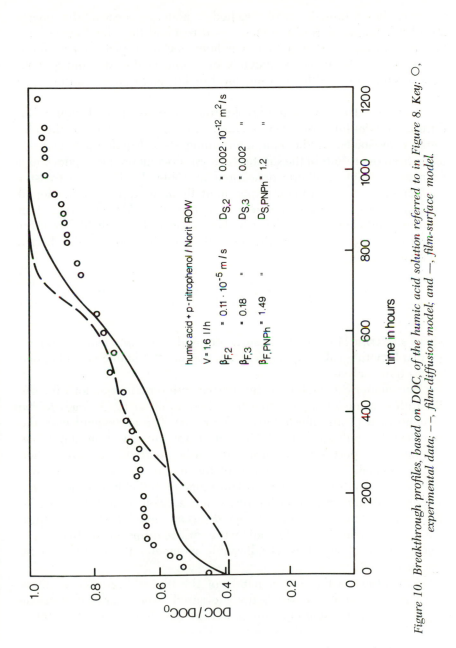

Figure 10. Breakthrough profiles, based on DOC, of the humic acid solution referred to in Figure 8. Key: ○, experimental data; – –, film-diffusion model; and —, film-surface model.

the total breakthrough curve reached a plateau, because the poorly adsorbable compounds of the humic acid reached their feed concentration or showed a concentration overshoot due to displacement effects. The strongly adsorbable compounds of humic acid and p-nitrophenol appear later in the effluent; thus, the breakthrough curve continues to ascend.

For the calculation of the breakthrough profile (dashed line) a film diffusion model for three components was used which contains the film transfer resistance as the rate-determining step. Equilibria data were taken from the results of the sorption analysis documented in Figure 7 and from single solute isotherm data listed in Table III. The mass transfer coefficients were determined experimentally in an apparatus developed by Hölzel (16). The batch kinetic experiment yields an overall film transfer coefficient for humic acid. The measured overall film transfer coefficient, $\beta_{F,M}$, relates to the individual coefficients of the fictive compounds as follows:

$$\beta_{F,M} = \frac{c_{01}}{c_{0T}} \beta_{F,1} + \frac{c_{02}}{c_{0T}} \beta_{F,2} \tag{4}$$

where subscript M denotes measurement and subscripts 1 and 2 denote the two hypothetical compounds characterizing the humic acid obtained by sorption analysis.

The numerical solution of the partial differential equations for the film diffusion model applied to a fixed-bed adsorber (3, 5) leads to the concentration profile illustrated in Figure 10. The agreement with the experiment is only qualitative. The reason for the deviation may be an additional transport resistance inside the carbon particle. This assumption seems reasonable because the size of the humic acid compounds was analyzed to be in the molecular weight range of 500–2000. A model that contains the film mass transfer resistance as well as the intraparticle transport resistance is the film surface diffusion model (3, 4). The diffusivity in the particle is characterized by the surface diffusion coefficient, D_S, which can be obtained as an adjustable parameter from batch kinetic experiments (3).

In contrast to the determination of the film diffusion coefficients, there is no relation by which the measured overall surface diffusion coefficient can be split up into individual coefficients for the fictive compounds obtained by the sorption analysis.

To show the influence of intraparticle transfer on the shape of the breakthrough curve, a computation was made taking D_S for the two humic compounds to be 0.002×10^{-12} m^2/s and that for p-nitrophenol to be 1.2×10^{-12} m^2/s. The assumption of D_S for the humic acid is based on kinetic measurements with bank filtrated river water (*17*). The surface diffusivity of PNPh was determined experimentally with GAC ROW 0.8 S.

As can be seen from the calculated curve (solid line) in Figure 10, it is useful to consider intraparticle mass transfer and to try to extend the proposed sorption analysis in view of adsorption kinetics.

The application of the sorption analysis to a case of practical interest is shown as a last example. Figure 11 presents the overall isotherm of a groundwater strongly contaminated with natural organics, mostly humic

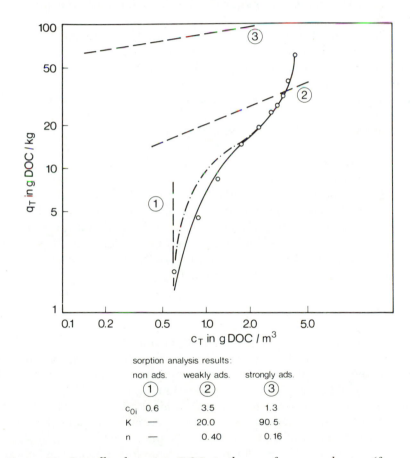

sorption analysis results:

	non ads.	weakly ads.	strongly ads.
	①	②	③
c_{0i}	0.6	3.5	1.3
K	—	20.0	90.5
n	—	0.40	0.16

Figure 11. Overall adsorption DOC isotherm of a groundwater (from Hanover) pretreated by flocculation and ozonation. The bottom of the figure documents the results of the sorption analysis which characterizes the multisolute mixture as a three-component system. Conditions: $C_{0T} = 5.4$ g DOC/m³, and AC: F 300. Key: ○, measured isotherm; —·—, calculated isotherm; and —, single solute isotherm of the three components.

acids. The three straight lines correspond to the single solute isotherm of the fictive mixture components which result from the sorption analysis. The initial concentrations are characterizing mixture composition (bottom of Figure 11).

As a first step, the groundwater is flocculated, and then it is ozonated and fed to a GAC filter. Concentration profiles based on DOC were measured in bed heights of 2.5 m and 5.0 m over 1 year. Figure 12 illustrates the measured and calculated breakthrough curves, assuming that film diffusion is the rate-detemining step. Equilibria parameters and feed concentrations were taken from the table in Figure 11; the overall mass transfer coefficients, $\beta_{F,M}$, were determined in a batch experiment to be 2.2×10^{-6} m/s.

During the first months of the filter run, there was good agreement between the measured and calculated curves. Deviations after a longer operation period can be explained by the influence of pore or surface diffusion (intraparticle mass transfer resistance) and by biodegradation of the organics due to ozone pretreatment (0.5 mg O_3/mg DOC). The calculated concentration profiles become constant after 5 months and balance out at a DOC/DOC_0 ratio of about 84%. The strongly adsorbable fraction, i.e., Component 3 in Figure 11, which is 24% of the total initial concentration, is totally adsorbed during the first 12 months and does not appear in the filter effluent.

This result shows that the plateau in the breakthrough curve need not necessarily be a consequence of biological activity but also can be caused by multicomponent adsorption. On the other hand, if there is biodegradation, the amount removed biologically may be smaller than that calculated on the assumption that the adsorption capacity has been exhausted when the effluent concentration profile reaches steady state.

Summary

Adsorption equilibria in multisolute mixtures are strongly affected by competitive adsorption. Thus, the interpretation of adsorption data obtained by adsorption experiments in mixtures of unknown composition is only possible if one considers multisolute equilibria. The main dependencies concerning total as well as individual adsorption were found by investigating adsorption equilibria in defined multicomponent systems.

A combination of experimental work and modeling, namely sorption analysis, was developed which reduces multicomponent mixtures of unknown composition to well-defined systems containing three compounds. By this grouping method, competitive adsorption models can be applied to the prediction of adsorption equilibria in such mixtures. The parameters obtained by the sorption analysis characterize adsorption

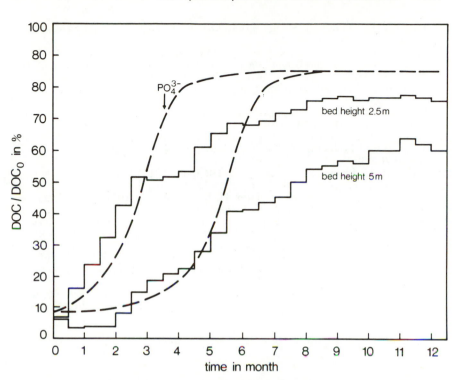

Figure 12. GAC column concentration profiles (dimensionless DOC effluent concentration) of the groundwater referred to in Figure 11 (GAC is F 300). Influent concentration is in the range of 6.3–7.0 g DOC/m³, empty bed contact times were 30 and 60 min. Key: ⌐ , experimental results; and – – calculated with film-diffusion model.

behavior and can be used for a more realistic modeling of GAC performance in fixed-bed adsorbers or for other types of activated-carbon treatment plants.

Literature Cited

1. Jossens, L.; Prausnitz, J. M.; Fritz, W.; Schlünder, E. U.; Myers, A. L. *Chem. Eng. Sci.* **1978,** 33, 1097–106.
2. Singer, P. C.; Chen-Yu, Y. In "Activated Carbon Adsorption of Organics from the Aqueous Phase," Suffet, I. H.; McGuire, M. J., Eds., Ann Arbor Science, Ann Arbor, Mich., Vol 1, pp. 167–89.
3. Fritz, W.; Merk, W.; Schlünder, E. U.; Sontheimer, H. In "Activated Carbon Adsorption of Organics from the Aqueous Phase," Suffet, I. H.; McGuire, M. J., Eds., Ann Arbor Science, Ann Arbor, Mich., Volume 1, pp. 193–211.
4. Crittenden, J. C.; Weber, W. J., Jr. *Environ. Eng. Div. ASCE,* **1978,** 104(EE6) 1175–95.
5. DiGiano, F. A.; Baldauf, G.; Frick, B. R.; Sontheimer, H. *Chem. Eng. Sci.* **1978,** 33, 1667–73.

6. Haberer, K.; Normann, S.; Schredelseker, F. *Z. f. Wasser- und Abwasser-Forsch.* 3/4 **1977**, *314*, 82–7.
7. Frick, B. R.; Bartz, R.; Sontheimer, H.; DiGiano F. A. In "Activated Carbon Adsorption of Organics from the Aqueous Phase," Suffet, I. H.; McGuire, M. J., Eds., Ann Arbor Science, Ann Arbor, Mich., 1980, Volume 1, pp 229–42.
8. Frick, B. R. Ph.D. Dissertation, University of Karlsruhe, Federal Republic of Germany, 1980.
9. Schuliger, W. G. "Carbon Adsorption Handbook," Ann Arbor Science, Ann Arbor, Mich., 1978, p 55.
10. Kölle, W. Use of Macroporous Ion Exchangers for Drinking Water Purification. Special Problems of Water Technology, Vol. 9 (1979) U.S.EPA 1979 600/9-76-030
11. Myers, A. L.; Prausnitz, J. M. *AIChE J.* **1965,** *11*, 121–7.
12. Radke, C. J.; Prausnitz, J. M. *AIChE J.* **1972,** *18*, 761–8.
13. Baldauf, G.; Frick, B. R.; Sontheimer, H. J. *Vom Wasser* **1977,** *49*, 315–30.
14. Kidnay, A. J.; Myers, A. L. *AIChE J.* **1966,** *12*, 981–6.
15. Sontheimer, H. "German Experiences in the Use of Activated-Carbon Treatment," presented to U.S. EPA-NATO CCMS Conference on Adsorption Techniques, Washington, D.C., 1979.
16. Hölzel, G. "Laboratory Activated Carbon Test Methods for Water Utilities," presented to U.S. EPA-NATO CCMS Conference on Adsorption Techniques, Washington, D.C., 1979.
17. "Optimierung der Aktivkohleanwendung bei der Trinkwasseraufbereitung," Report No. 12, Engler-Bunte-Institut der Universität Karlsruhe, Federal Republic of Germany, 1979.

RECEIVED for review August 3, 1981. ACCEPTED for publication March 16, 1982.

Discussion II
Modeling and Competitive Adsorption Aspects

Participants

Katherine Alben, New York State Department of Health
Georges Belfort, Rensselaer Polytechnic Institute
Andrew Benedek, McMaster University
Bernd R. Frick, University of Karlsruhe
F. K. McGinnis III, Shirco, Inc.
Massoud Pirbazari, University of Southern California
Michael R. Rosene, Calgon Corporation
E. Shpirt, New York State Department of Health
Chi Tien, Syracuse University
Walter J. Weber, Jr., University of Michigan

BENEDEK: I'll ask Dr. Rosene to please comment on what he means by percentage of breakthrough concentration.

ROSENE: The percentage of breakthrough concentration that we plot on the vertical axis is really just the effluent concentration divided by the influent concentration. So it is a percent breakthrough. Generally that's the way it's described.

BENEDEK: In other words, C over C_0.

ROSENE: That's correct. If it reads 90, that means we've got 90 percent of the influent value coming up the column.

BENEDEK: I'm surprised that you got very rapidly to hundred-percent exhaustion.

ROSENE: With the minicolumn, as you are aware, we use a reduced particle size, which does enhance the adsorption rate considerably to get equilibrium data. It works very satisfactorily, and we've reproduced it with batch isotherms and gotten the same results.

I couldn't help but notice Dr. Frick's paper[1] about the plateauing

---

[1]Chapter 11 in this book.

0065-2393/83/0202-0269$06.00/0
© 1983 American Chemical Society

effect and the effect of the multicomponent system. We have seen data that would also suggest that the plateau you see in a multicomponent system like that may not be a kinetic effect at all but simply a continuing adsorption of the strongly adsorbed fraction of that multicomponent mixture.

BENEDEK: The reason I asked that question is you've used phenol, and so have we. We used a Calgon carbon, and you obviously didn't.

ROSENE: We did find that if you use a reduced particle size you don't have any difficulty in obtaining equilibrium.

PIRBAZARI: This is a question to the panel involving a comment. Most of the competitive adsorption that we saw today is competition between a specific compound and a background of humic substances.

Actually in a water treatment plant the carbon filter doesn't see the humic substances in the raw water. It sees them after the water's treated. So wouldn't you think, as far as practice is concerned, the background solution should first be treated—whether it's a softening or coagulation or whatever—and then be subjected to any kind of adsorption or modeling studies?

WEBER: I'll start off the response to that, Massoud. Yes, you are certainly correct, at least from the absolute point of view. If one is developing coefficients, mass-transfer coefficients, you expect to be able to transfer to real situations. That is certainly the case because we know that the larger fractions of the humic acids will, number one, be separated in pretreatment steps such as coagulation.

We know further that humic acids are entirely capable of complexing with metal ions that are typically used in coagulation operations which will, indeed, affect their adsorption capacity. So you're right. To be rigorous, one should pretreat the solutions of humic acids so that they contain compounds that are more representative of what a carbon column would see.

ROSENE: I couldn't agree more. It's necessary that you use the water that you're going to attempt to model from the real system. If you're going to model what's going to happen when you go on a carbon filter in a water treatment plant, you'd best get the water that you're doing your pilot tests on, or your isotherm tests on, from the influent to that filter, rather than, say, the raw water source, because other treatment effects can impact on the organic level and on the subsequent competitive effect that you get.

FRICK: We always use humic acids selected or isolated from the raw waters we have to treat. So I think we have the right conditions to simulate competitive adsorption which can occur in activated carbon filters for water treatment.

SHPIRT: My first question is to Mike Rosene on actual dispersion coefficient. Did you try to incorporate it in Peclet number and find relation with Reynolds?

The second question is how far in your range of flow rate are you from a plug-flow reactor?

ROSENE: The correlation with a Peclet's number and Reynolds number is really given in Ebach's dissertation[2]. He goes to great lengths to correlate the dispersion phenomena that he measures with the Peclet's number and with the Reynolds number and to show the ranges for which the correlation is valid.

In terms of a plug-flow reactor, we're getting plug-flow behavior; we're not getting laminar flow developed—profiles developed in the pack bed or anything. I did test that out with a dye, and we're getting essentially plug-flow response as far as that goes; we're not getting wall effects or anything.

WEBER: What Mike has presented here is in reality contradictory to the findings of a number of other investigators with respect to the effects of dispersion. We at Michigan have examined this phenomenon in some depth, both experimentally and in terms of the characterization of Peclet numbers that are typical of flow ranges up to 20 gallons per minute per square foot, which is roughly two and a half times the highest rate that Mike used. Our conclusion is that the actual dispersion is negligible in activated carbon adsorbers at any realistic flow rates.

I think others have found this, too. Andy Benedek, I believe you've worked with that, and I believe, Fran DiGiano, you've done some work with dispersion effects and probably a number of others. So I'm curious about Mike's findings.

I would suggest that in the last few slides, where Mike showed how he utilized the reduction in flow rate to represent an increase in the effects of axial dispersion, the data may be a bit misleading in the sense that I believe, Mike, you were using only a pore-resistance model and neglecting the external film transport. But of course as you reduce flow rate, in addition to potentially impacting dispersion effects, you will even more sharply increase the effects of extra particle transport, which would have precisely the same tendency to spread out the breakthrough curve.

So I really question whether axial dispersion is significant in activated carbon applications. We have included it in MADAM; MADAM has all of the dispersion terms. It's just that when you put in the numbers, the model is totally insensitive to axial dispersion. Most systems we work with are totally insensitive to ranges of dispersion.

[2]Ebach, E. A. "The Mixing of Liquids Flowing Through Beds of Packed Solids," Ph.D. Dissertation, University of Michigan, 1957.

ROSENE: I did run different bed depths, as I had indicated earlier, and we saw the phenomenon in terms of a sharpening of the mass-transfer zone, as I indicated, which is not explained by film transfer. And I might point out that as you lower the flow rate, the dispersion phenomenon is really not overly sensitive to flow rate, and what you're really reducing is the impact of the intraparticle, which is really the predominant slow step in most cases. You are giving more and more time to achieve intraparticle adsorption, and the axial dispersion is there all along but is usually masked.

I want to point out that we looked at single-component systems, so it's most likely that you will see this phenomenon with a single-component system. As you get to multicomponent interactions you begin to slow down the rate of intraparticle adsorption and then the flow rate at which you begin to see the phenomenon, and you have to go to a low flow rate in order to see it because intraparticle diffusion will otherwise become lower and lower.

So the more components you have in the system, the more your dynamics of intraparticle mass transfer slow down. So you really wouldn't expect to see any kind of dispersion phenomenon at very high flow rates because it would be easily masked by intraparticle mass transfer.

It would only be at the very low flow rates that we would see the dispersion phenomenon coming into play. And to determine whether it's a film effect or a dispersion effect you run your varying bed depths, and you see what happens to the breakthrough curve.

TIEN: I'd like to add another comment. I think the discovery that the dispersion effect is important is contrary to the large body of experimental evidence available, not only in adsorption columns but also in the general study of the dispersion effect in a packed bed, which was studied extensively in the fifties and sixties by a large number of chemical engineers.

I think, for example, when you showed two breakthrough curves— one corresponding to 4.8 gallons per minute, the other one corresponding to only 3—you mentioned the fact that you have the same contact time. So this means that for large flow rates, essentially, you have a bed which is 16 times as high as a smaller one, and the fact that you have a very sharp breakthrough curve is simply a confirmation of the earlier work by people in the forties.

Essentially, that is, the local equilibrium condition is approached when you have the long bed. If you have the equilibrium condition you know the solution which is given by the H transform has a block change. That is what has happened.

And I think you mentioned in fact a mass-transfer zone, and I think if you look at the historical paper by Rosen, which appeared in the *Journal for Applied Physics* in 1952, it will show such a thing doesn't exist. That is, what is defined as mass-transfer zone, in fact, increases its length as a square proportional to the square root of time. So such a thing really doesn't exist. I just wonder, how can you use it in an approximate manner to do your design calculation? I wonder whether you can elaborate on that.

ROSENE: As I pointed out, the mass-transfer zone representation that we used is an approximation, and as long as your mass-transfer zone as defined in my equation is not a significant portion of the total bed, like 90 percent of it or something, then the approximation is fairly good, and you can use that measurement as an indication.

I might point out that the MTZ, as we defined it, is generally interpreted as being the result of the adsorption limitation. How fast can you adsorb? The assumption that is explicit in the dispersion model is that you are not limited in your adsorption; in other words, you have attained instantaneous equilibrium within the column. Now, for this phenomenon to be important, your intraparticle mass transfer has got to be fairly rapid or the time that you have to obtain equilibrium has got to be long enough.

So if you're at very low flow rates you do achieve this condition, and the dispersion phenomenon is a real phenomenon, and the chemical engineers who did the studies back in the late fifties—Ebach was among them—established a very credible correlation to show that there is, indeed, a strong dispersion effect. You get a spreading in the column, and it's a function of the interstitial velocity and the effective particle diameter.

McGINNIS: A question to the panel in general: Over the last two days there have been a number of papers presented that have commented on the effects of carbon particle size on adsorption kinetics. In the U.S. we generally use a granular carbon which has a distribution of particle sizes. In Europe, as Herr Frick showed, the Norit extruded carbon is used, and it has a much more uniform particle size, if you will. It's very narrowly distributed—about 0.8 millimeters in this case.

I'd be interested in comments on the effects of that distribution, if any, versus uniform carbon particle size on performance, particularly in terms of the long-term "slow adsorption" effects.

FRICK: We mostly used shaped carbons, not irregular carbons like Norit or the B-turn from the Lurgi Company. We had good agreement when we predicted breakthrough curves, and we did not have that good agreement between measurement and prediction when we used a commercial carbon or other carbon products or carbon qualities.

WEBER: There's no question about the effects of particle size, as I mentioned earlier this morning. It's well established that the rate is either dependent upon the inverse of the diameter or dependent upon some higher power approaching second-order dependency if the control is intraparticle.

That simply falls out of the appropriate equations describing the rates. Ideally, one should use a smaller particle as conditions of efficient operation permit. The ultimate extension of this would be powdered carbon, but conditions of efficient operation do not necessarily obtain with powdered carbon. So it would be ideal to use as small a granular carbon as possible—possible being defined by the conditions of headloss one suffered—whether or not the carbon has to play the dual role of filtration, in which case a perfectly uniform carbon would not be ideal.

So there are many practical factors that enter into it. As it turns out, the ranges of particle sizes that are commercially available in the typical U.S. carbons are quite good, and using an average particle diameter one can obtain reasonably good modeling prediction.

TIEN: I think your question, whether a slow adsorption has any connection with the particle size, really cannot be answered unless one knows something about the structural composition of the pores inside the carbon particles. I believe that Andy's recent publication[3] is barely beginning to explore that problem because you can think in terms of large pores and small pores or pores following a given spectrum of sizes, but that is not enough in describing the intraparticle diffusion.

You have to understand the sequence. That is, does the diffusion take place from large pores to small pores, in a sequential manner, or are they processes which are going on in a parallel manner?

If you consider that there is a structural preference of the various-sized pores, this can presumably have something to do with the preparation of the particle in which the interaction relates to size. So, therefore, it is conceivable that there may be an optimum size at which the slow adsorption you find could be reduced to a minimum.

BENEDEK: Let me just make one short comment to Dr. McGinnis' question. What you really were getting at is the fact that there's a variation in particle size within a given batch of carbon that would cause a pseudo-slow adsorption effect. Is that what you're asking?

McGINNIS: Well, I guess my speculation was that when you have a distribution in particle size, perhaps the smaller particles become spent, if you will, very quickly, and it's the large particles that are really contributing to the so-called slow adsorption.

BENEDEK: Yes. I think you're right. There is an effect of that kind, but when you look at our data, you find that is probably a small portion of the slow adsorption effect because there's ten- and twenty-fold increases in the time required to exhaust the slow adsorption capacity versus what I call fast adsorption capacity. And that could not occur for the relatively small variation in particle sizes that you normally have in a given batch.

BELFORT: I have two questions. One of them is to Dr. Frick. I like the idea of using probes. I think it's a very interesting way of doing it. However, I notice that the three probes you've used are all aromatic in

[3]Benedek, A. and Y. Richard, "Advances in Biological Treatment," presented at the International Water Supply Association Conference, Paris, France, September 1–4, 1980.

nature, and I'm wondering if you people are looking at any other probes that are aliphatic and whether that's of interest at all. I just would like to hear your comment on that.

FRICK: The type of compounds we used as tracer compounds has the problem of the analytical models we have, and so we are using aromatics because we can detect them by UV measurements. We tried it with chloroform and trichloroethylene, and we measured it with GC, but that's a very expensive job.

BELFORT: That's right. I just think it would be interesting because, after all, some of the effects we would like to know about have to do with aliphatic compounds.

My other question was really a comment. It's something I noticed in certain papers, and that is, out of these dynamic models we get D_S. I've noticed that it can vary quite substantially between different models and essentially for the same compound.

For example, Dr. Ishizaki's D_S[4] was very much higher; it was 10^{-7}, whereas if I remember correctly, the D_S from your paper,[5] Dr. Weber, was of the order of 10^{-9} for the same compound. I'm wondering whether the panel has any comments on whether that means anything.

TIEN: I think, Georges, what you say is quite true. I think if you only consider those parameters as a way of fitting your data, that's one thing. You just pick up whatever values fit the data the best.

Now, if you want to attach any kind of single significance to it, I think you have to be careful. For example, if you take your batch kinetic data and you take an initial rate, presumably that would give you the case of F_Y. But nevertheless, there are also correlations available which give you the mass-transfer coefficient from a particle added to liquid. The most up-to-date work in this area is by Ken Smith at MIT.

I would suggest to you that you can extract value from your experimental data, but perhaps you should also compare the value that you have extracted from the experiment with the correlation.

Now, referring to the intraparticle diffusion coefficient, essentially you have either a D_{pore} or a D_S. Our preference is to obtain what you call the K_{SI}, but that is a particle phase-transfer coefficient from either D_S or D_{pore}.

On the other hand, you can get D_{pore} by knowing the diffusivity in the bulk phase and knowing the size of the pore. That simply is a hydrodynamic retardation effect. So you can go through the cycle to do the confirmation—to see whether at least you obtain some kind of agreement on the order-of-magnitude basis.

---

[4]Chapter 6 in this book.

[5]Chapter 7 in this book.

ROSENE: By looking at different adsorbents you can see quite significant changes in the calculated effect of diffusivities, and Dr. Weber's work and my work were fairly close together—about 10^{-9}, I believe. And I believe Dr. Ishizaki's work was on Filtrasorb 200, and it was treated material, undergoing different treatment. The defect in the pore structure of the adsorbent itself could result in quite significant differences in the calculated effective diffusivity.

ALBEN: If you're going to use minicolumns to study competition between chloroform and humics—the way I see it, an advantage of a minicolumn is that you get early breakthrough. It would be very nice if you could continue collecting data. Some of your profiles level off and don't reach complete breakthrough. But if you're looking for competition, it would be nice to have the full breakthrough curve way past saturation of the column and see if there's any displacement in that sense either, especially for chloroform.

BIOLOGICAL/ADSORPTIVE INTERACTIONS

Comparison of Adsorptive and Biological Total Organic Carbon Removal by Granular Activated Carbon in Potable Water Treatment

S. W. MALONEY, K. BANCROFT, and I. H. SUFFET

Drexel University, Environmental Studies Institute, Philadelphia, PA 19104

P. R. CAIRO

Philadelphia Water Department, Research and Development, Philadelphia, PA 19107

The mechanism of long-term quasi-steady-state removal of total organic carbon by granular activated carbon is explained in terms of two current theories: biological total organic carbon removal and slow adsorption kinetics. Eight parallel columns indicated that granular activated-carbon contactors are reproducible for total organic carbon, dissolved oxygen, pH, and alkalinity changes across the column when conditions are identical. Carbon regeneration may be assisted by bacteria but is not caused by bacteria. Temperature and total organic carbon removal were not related over a short time, but temperature and dissolved oxygen removal were strongly related.

R ECENT ADVANCES IN ANALYTICAL CHEMISTRY led to the discovery of nanogram per liter to microgram per liter concentrations of undesirable and potentially harmful low molecular weight nonpolar organic chemicals in raw water supplies and finished drinking water. One viable technique for the removal of a broad spectrum of organic chemicals is activated-carbon adsorption. Granular activated carbon (GAC) provides a vast surface area for adsorption of nonpolar organic chemicals. The major drawback of activated carbon is high operating costs associated with regeneration of spent carbon.

0065-2393/83/0202-0279$07.00/0

The purpose of the GAC is to remove trace organics of health and organoleptic concern. These trace organic compounds are only a small portion of the total organic carbon (TOC), so monitoring TOC removal does not give a measure of health-related trace organic removal. However, comparison of single solute adsorption isotherms in the presence and absence of natural background organics (1) showed that competition for adsorption sites may occur between natural organics in water and organics of health concern. Therefore, reducing the load of natural TOC onto the GAC may extend the useful life of the GAC for removal of health-related organics.

Many researchers investigating GAC are studying mechanisms that may increase the carbons' useful life. One parameter often used to evaluate carbon contactor performance is TOC removal. Typically, GAC contactors exhibit a large initial capacity for TOC followed by partial breakthrough to a level of quasi-steady-state removal (2,3). The objective of the research reported here was to develop an understanding of the mechanism of this long-term quasi-steady-state removal. Two current theories used to explain the long-term removal are biological TOC removal (4–6) and slow adsorption kinetics (7, 8). The use of micro-organisms on the GAC to help remove TOC is of particular interest as is increasing the biodegradability of naturally occurring TOC which appears to occur by pre-ozonation (9).

Biological Regeneration

The occurrence of large numbers of bacteria on GAC filters is well documented (2). In chlorinated treatment plants, bacterial densities are much higher in the effluent than in the influent of a carbon contactor. This increase has been attributed to removal of chlorine and other toxic materials by the top of the GAC column, concentration of organics on the carbon surface, adsorption of bacteria by GAC, and a "sheltered" environment in the macropores of the GAC which shields bacteria from shear forces. These same factors apply in a nonchlorinated system.

Enhanced biodegradation of organics on GAC is proposed to occur due to acclimation of the microbes to the adsorbed substrate (5). In this proposed complex process, GAC adsorbs recalcitrant organics as well as bacteria. Thus, the carbon keeps the bacteria in contact with higher concentrations of refractory organics than in the bulk liquid. Maintaining contact between the bacteria and adsorbed recalcitrant organics allows the bacteria to acclimate to the recalcitrant organics. Biological re-generation emphasizes that the solute must be adsorbed first (4, 6). The bacteria then use a portion of the retained organics, opening sites on the GAC for further adsorption of other organics.

The action of GAC in adsorbing organics decreases the aqueous phase concentration of organics, and, logically, this change would inhibit acclimation of bacteria to adsorbed substrate. Thus, acclimation has been referred to as a common fallacy in activated-carbon adsorption theory (10). Typically, in a real situation, the influent concentration of organic compounds is variable. Some of the time, carbon is adsorbing the organic in question, but when the influent concentration drops to a low or zero value, the GAC may release the organic. In this case, carbon acts as a buffer for bulk phase concentration of an organic that has a rapidly varying influent concentration, and acclimation of bacteria is possible on GAC; on an inert substance such as sand, acclimation is less likely due to the sporadic appearance and disappearance of the organic in the aqueous phase.

Improved biodegradability of naturally occurring TOC has been observed following ozonation (9). Therefore, the interaction of ozonation and GAC unit processes may be engineered to maximize the long-term TOC removal capability. It is proposed that carbon bedlife is prolonged by converting a portion of the recalcitrant organics to biodegradable organics in ozonation (4). The attached microbes then convert the biodegradable portion to biomass, carbon dioxide, and waste products. The TOC that is converted to biomass or carbon dioxide does not take space on the GAC. Other biological end products may or may not adsorb, but the net result is a lesser load to the GAC.

Slow Adsorption Kinetics

The long-term removal of TOC by GAC also can be explained by a dual rate kinetic model (7, 8). Two distinct stages of adsorption have been observed during batch equilibrium studies. In the first stage, rapid adsorption to 50%–80% of the total carbon capacity is observed within a few hours. The remaining adsorptive capacity is exerted very slowly. The dual rate kinetic model takes both stages into account and predicts the quasi-steady-state removal so commonly seen on GAC for lumped parameters such as TOC.

This model of activated carbon considers, as a first approximation, two types of pores. Macropores are defined as those pores in which diffusion rates are unhindered by the pore walls, and micropores are those pores with radii of comparable size to the diffusing species (7). Because of strong multidirectional adsorptive forces in micropores and the restricted diffusion rates caused by the proximity of the pore walls, transport rates are considerably lower than in the large pores. The rapid initial uptake rate in batch isotherms and large initial removals encountered in column studies are assumed to occur in the macropores. The slow approach to equilibrium after initial uptake observed in batch studies and the quasi-steady-state removal seen in column studies are attributed to the micropores.

Experimental

A method for differentiating between adsorptive and biological removal was required in which little modification of the actual operation of a carbon contactor occurred. The method chosen was temperature control of the contactor influent. By using this method, the contactors were operated at actual field conditions, because the temperatures were controlled within the temperature range normally encountered at the water treatment plant. Furthermore, the contactors were subjected to varying influent temperatures such as encountered in a full-scale facility. A temperature differential was maintained between the contactors by chilling the influent to the cold contactor during the summer and early fall and by heating the influent to the warm contactor during the late fall and winter. Figure 1 shows expected results. The difference between the warm and cold effluent gives an estimate of the bacterial contribution.

The temperature dependency of most reactions is represented by Arrhenius' law (11):

$$k = k_0 e^{-E_a/RT} \tag{1}$$

where k is reaction rate at temperature T, E_a is the activation energy, R is the universal gas law constant, and T is the absolute temperature.

The rate of the removal reaction determines the magnitude of the removal as the reaction is limited by the contact time. The overall removal rate is actually the sum of many individual rates for various compounds that make up the TOC.

The key to the reaction rate is E_a. Physical mechanisms such as diffusion and adsorption have small activation energies, up to 5 kcal/mol K. These reactions show little temperature dependency. Previous work with activated carbon indicates that normal temperature variations have only minor effects on adsorption (12).

For a 10°C temperature increase, a general rule of thumb for biological systems states that the reaction rate should double. This rate corresponds to an activation energy of 10–20 kcal/mol K. Stephenson (13) measured the effect of temperature on *Escherichia coli* using various single substrates and found an average ratio of substrate utilization rates of 2 for the 23°C–33°C range in a growth limited system. In an (activated sludge) substrate limited system, the ratio of rates has been measured as low as 1.35 (14). In a study similar to the one reported here, parallel GAC contactors treating wastewater effluent were operated at different temperatures (5°C and 25°C). The results of that study (15) indicate that temperature can be used to control bioactivity on GAC as the warm contactor exhibited greater removal capacity for TOC than the cold contactor. Average influent TOC concentration was 20 mg/L for the carbon contactors.

In this parallel column study, all columns were receiving an identical influent. The objective of starting all columns on ambient influent was to develop a bacterial population acclimated to warm temperatures and then "shock" the bacteria on one set of columns with a sudden temperature change. This procedure avoids the possibility of selecting for cold-adapted microorganisms in the cold columns.

Biological activity was monitored by substrate utilization and the modified plate count (*16*) on the GAC and in the influent and effluent. Organic removal was monitored by TOC, macroreticular resin (MRR) extraction, and carbon core analysis. This paper deals only with temperature effects on TOC and dissolved oxygen (DO) removal in the carbon adsorption process.

DO removal was monitored with an Orion Probe (#97-08). TOC was monitored on a Dohrman 54-D low level analyzer. Alkalinity and pH were monitored for 13 weeks. However, no change was observed in the pH or alkalinity across the contactor (7 weeks at ambient temperature, 6 weeks in ambient-chilled mode) even though dramatic effects were observed in the DO removal when the temperature was lowered in two columns (after 7 weeks).

DO was sampled in a standard BOD bottle and was allowed to overflow for a sufficient time to replace the bottle volume twice. This method of overflowing was employed because the chilled sample appeared to be picking up oxygen from the air and consistently had a higher DO than the ambient sample. After the overflowing technique was adopted, the influents to the cold and warm columns registered the same DO concentration, indicating that absorption of oxygen from the atmosphere was no longer a problem.

Pilot Plant Facility.   The source of influent water was the pilot plant at the Philadelphia Torresdale water treatment plant located on the Delaware River estuary in northeast Philadelphia. The complete pilot plant has been described elsewhere (*17*). Figure 2 shows a schematic of the pilot plant. In general, raw

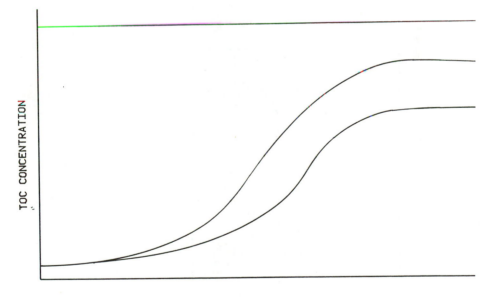

Figure 1. Conceptual diagram of total organic carbon (TOC) removal as a function of temperature. Reading the curves from top to bottom: influent, cold effluent, and warm effluent.

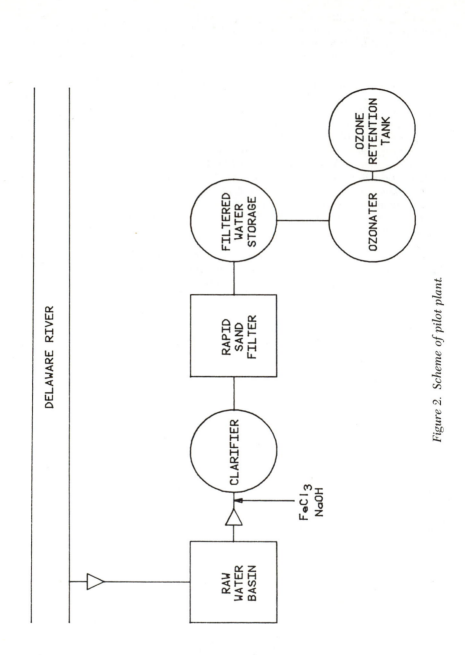

Figure 2. Scheme of pilot plant.

water is pumped from the Delaware River into a raw water basin, then to an upflow clarifier with chemical addition ($FeCl_3$ and NaOH) en route from the raw water basin to the clarifier. Water from the clarifier flows through a rapid sand filter into backwash tanks. The pilot plant water is not disinfected to this point. Water from the backwash tanks is pumped through an ozone contactor into a retention tank which allows for dissipation of ozone residual.

Minicolumn Operational Characteristics. Water from the ozone retention tank is pumped through a small sand filter to eight parallel GAC mini-columns. Figure 3 shows a schematic of the minicolumn apparatus and Table I gives its physical characteristics. Table II summarizes the modes of minicolumn operation. For the first 7 weeks of the experiment, all columns received ambient influent (Mode 1). During this period, Columns 1–4 were taken off line at 2-week intervals, and carbon samples were analyzed for bacterial densities and chemical analyses (not reported on in this paper). At the beginning of Week 8, the influent temperature to two of the remaining four columns was reduced by passing water through a cooling bath in 6.1 m (20 ft) of 0.6-cm (¼-in.) stainless steel tube (Mode 2). Two remaining columns (one cold and one warm) were taken off line 2 weeks later and analyzed as described.

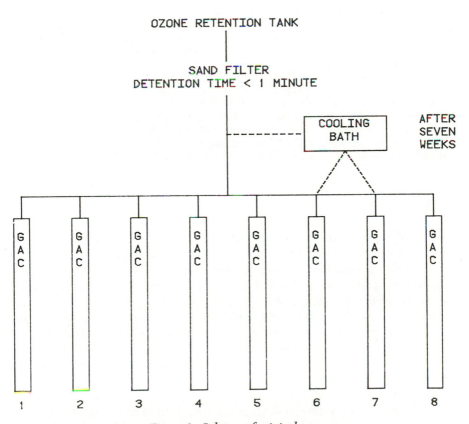

Figure 3. Scheme of minicolumn.

Table I. Minicolumn Physical Characteristics

Diameter	4 cm (1.6 inches)
Length	102 cm (40± inches)
Material	Teflon
Flow rate	80 mL/min
EBCT[a]	15 min
Carbon type	Filtrasorb 400
Carbon depth	91 cm (36± inches)

[a]Empty Bed Contact Time.

The original intent of the study was to observe differences in TOC removal after bacterial populations had been established. No difference in TOC removal as a result of temperature control was observed during Mode 2 (Table II). This finding led to two possible conclusions:

1. The bacterial contribution to TOC removal in an activated-carbon contactor was too small (<100-μg/L difference) to be measured by TOC removal.

2. The adsorptive capacity of the carbon was removing those materials that were not degraded (i.e., masking the effect of lowered bacterial removal). In this case, lower bacterial activity in the cold contactor would be masked by adsorptive capacity in the short run. In the long run, however, the cold contactor would be loaded faster and eventually a difference in TOC removal would be seen as TOC would break through the cold contactor more rapidly.

Based on the second possible conclusion, it was decided to continue the warm–cold comparison (Mode 2). The TOC values continued to show no difference well into the fall (up to 167 days) when maintaining a temperature differential between the warm (ambient) and cold (chilled) contactors became impossible due to falling influent temperatures. Therefore, the mode of operation of the minicolumns was changed from ambient-chilled to ambient-warmed (Mode 3).

Results

GAC Contactor Reproducibility. At the onset of this study, eight columns were operated in parallel under identical conditions (Mode 1). During this period, GAC contactors were reproducible for TOC removal, DO removal, pH, and alkalinity. Figure 4 shows the pH of the influent and effluents of two contactors. After the temperature change (Mode 2 at 52 days), and there is no effect on the pH. During the first 90 days, there was no consistent trend in pH value change across the contactor. The ozone–GAC process does not appear to affect pH. Figure 5 shows alkalinity values for the influent and effluents from two contactors. Alkalinity values were much more variable over time than pH, but there is again no trend in change across the GAC contactor.

Figure 6 shows dissolved oxygen levels in the influent and effluents from the warm and cold contactors. Figure 7 shows TOC levels for the

Table II. Minicolumn Operational Characteristics

Mode	Date of Operation	Days	Temperature Range		Comments
			Warm Column	Cold Column	
1: Ambient	6/2/80–7/23/80	1–52	24–30°C	24–30°	Temperature differential for most of this period was greater than 15°C
2: Ambient-chilled	7/23/80–11/16/80	53–167	15–31°C	6–17°C	No difference in TOC Large difference in DO
3: Ambient-warmed	11/16/80– 4/30/81	168–325	27–32°C	6–17°C	DO removal very high on warm column TOC values diverge about 80 days after temperature change

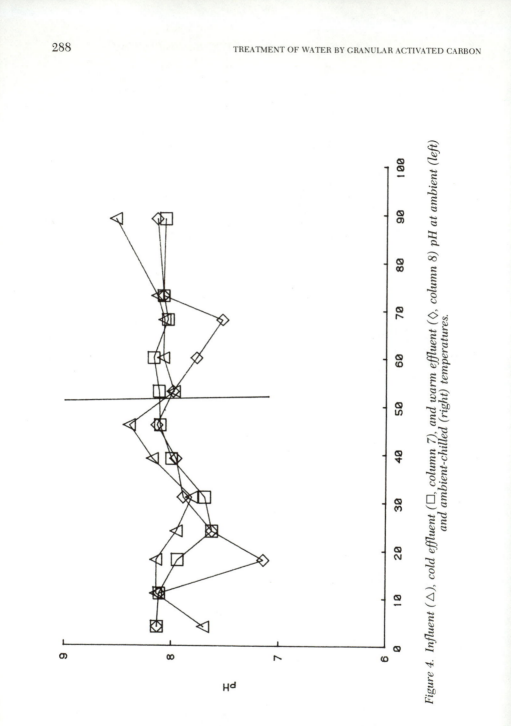

Figure 4. Influent (△), cold effluent (□, column 7), and warm effluent (◇, column 8) pH at ambient (left) and ambient-chilled (right) temperatures.

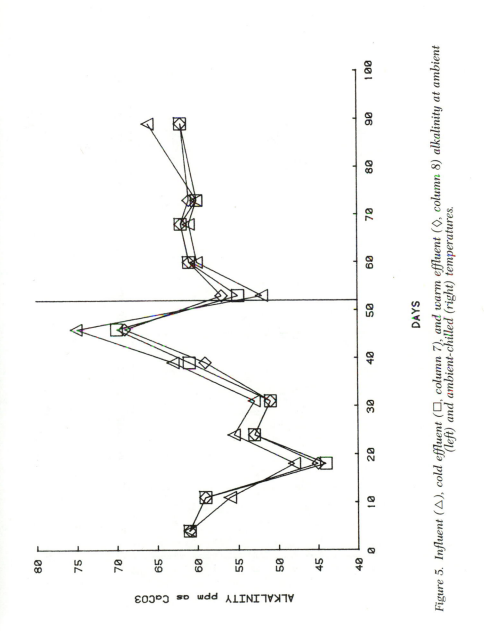

Figure 5. Influent (△), cold effluent (□, column 7), and warm effluent (◇, column 8) alkalinity at ambient (left) and ambient-chilled (right) temperatures.

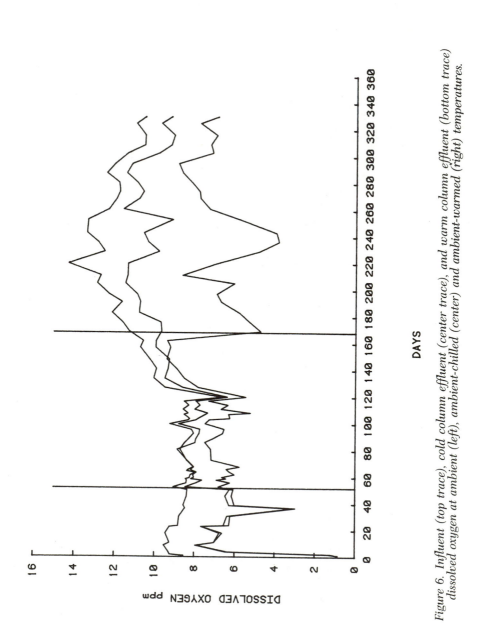

Figure 6. Influent (top trace), cold column effluent (center trace), and warm column effluent (bottom trace) dissolved oxygen at ambient (left), ambient-chilled (center) and ambient-warmed (right) temperatures.

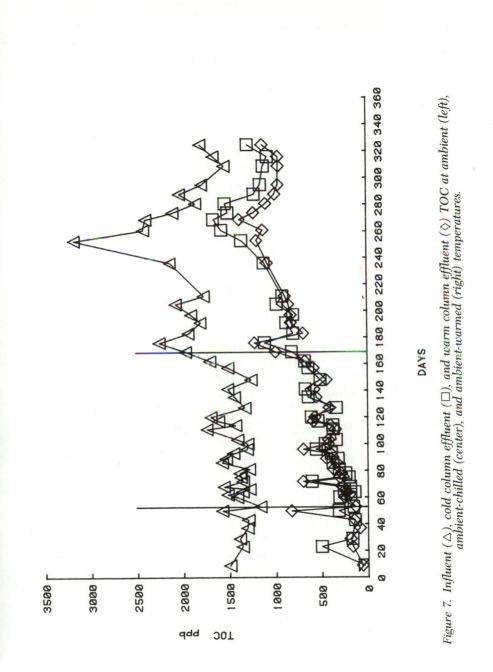

Figure 7. Influent (△), cold column effluent (□), and warm column effluent (◇) TOC at ambient (left), ambient-chilled (center), and ambient-warmed (right) temperatures.

same columns. During the ambient operation of the system (Mode 1), DO and TOC values are almost identical. There are two points on the TOC plot that are greatly different, but they occur early in the study when the columns were producing water very low in TOC. Errors on the TOC analyzer and in sample collection appear more pronounced when dealing with very clean water.

Figure 8 shows the TOC mass loading for the warm and cold GAC contactor. Figure 8 includes the two points on Figure 7 where the TOC was not identical in the early portion of the study. However, the high values did not occur on the same column and thus canceled each other; therefore, the TOC mass loading curve has points that are nearly identical for most of the study. The divergence in TOC mass loading values occurs only very late in the third mode of the study, approximately 80 days after changing to an ambient-warmed mode of operation (Figure 7).

Figure 9 shows the bacterial loading to the GAC contactors. The influent to the minicolumns is on the same order of magnitude as the effluent and up to two orders of magnitude greater than the bacterial density in the ozone retention tank. Previous studies (2, 18) showed that bacterial numbers increase dramatically across carbon contactors. This increase has been attributed in part to chlorine removal by GAC and in part to the proposed enhancement of microbial activity on GAC. The results from this study indicate that, in the absence of a disinfectant stable for a long time such as chlorine, bacterial densities increase dramatically in a very short time without contact with GAC. Referring to Figure 3, the only difference in treatment between sampling at the influent to the minicolumns and sampling at the ozone retention tank is passage through a sand filter with a hydraulic retention time less than 1 min. Thus, the bacteria do not require GAC on which to grow when no disinfectant residual is present.

Discussion

The results presented up to this point in combination with material presented elsewhere (16) and short-term experiments conducted in parallel to the long-term comparison of warm versus cold contactors suggest that:

1. Bacteria contribute only a small portion to the long-term TOC removal exhibited by GAC.

2. Bacteria–GAC interaction, proposed to be synergistic through the action of acclimation to adsorbed substrate, is not a significant mechanism.

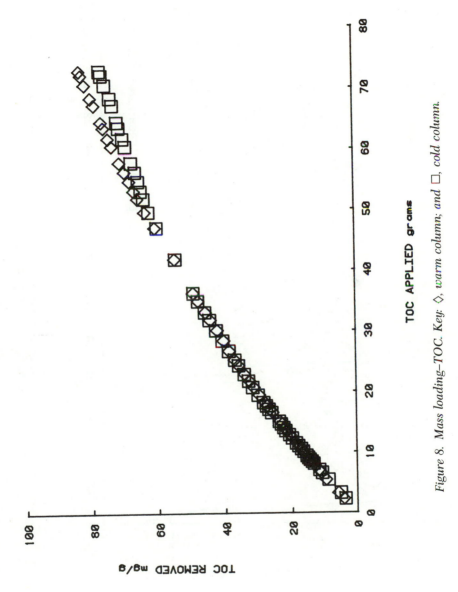

Figure 8. Mass loading–TOC. Key: ◇, warm column; and □, cold column.

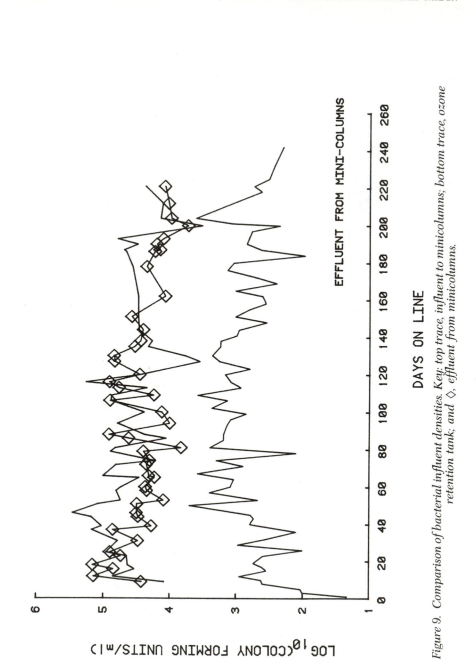

Figure 9. Comparison of bacterial influent densities. Key: top trace, influent to minicolumns; bottom trace, ozone retention tank; and ◇, effluent from minicolumns.

First, a portion of the results of the parallel work done at the City of Philadelphia pilot columns (*16*) will be discussed to help interpret the results of this study. Among the many comparisons being evaluated in that study, two sand contactors, with empty bed contact times (EBCT) of 15 min each, were operated in series using water from the ozone retention tank as influent. These contactors are operating in parallel to the mini-columns reported here. The results pertinent to this study are as follows. The first sand column exerts a DO demand of a fairly constant magnitude but always less than carbon and removes approximately 15% of the TOC applied to the sand column over the study period. The second sand column in series exerted very little DO demand and removed approximately 2% of the influent TOC to the first column over the study period. This value is below the limit of reproducibility (<100 µg/L) of the TOC analyzer. From the data, it is concluded that biodegradable TOC generated by ozonation of refractory TOC is easily attacked by the bacteria and an EBCT of 15 min on a sand medium is sufficient to remove the biodegradable fraction present.

It is also concluded that the portion being degraded in the sand column must have been produced by ozonation. This conclusion is based on the observation that degradation takes an EBCT time of 15 min (or less). Prior to ozonation, the pilot plant water was processed by an upflow clarifier with a hydraulic retention of 300 min and sand filtration with an EBCT of 36 min. (These high retention times are due to the pilot plant operating below design capacity.) Thus, if solutes in the river are biodegradable within a 15-min EBCT, they (the biodegradable solutes) would degrade in the pilot plant sand filter. The observed biodegradation in the pilot sand contactors indicates that biodegradable materials are being formed in the ozonation process.

DO uptake has long been an indirect measure of biological activity. GAC demonstrates a large initial demand for DO (*19*) which diminishes drastically after 80 bed volumes. In a study dealing only with O_2–GAC interactions (*19*), the DO demand did not reach zero, but it exhibited the same quasi-steady-state level common to TOC–GAC adsorption. No mention was made of attempts to maintain sterility in the O_2–GAC study. However, an increase in surface acidity of the activated carbon was observed suggesting that a chemical reaction was occurring between the DO and GAC. The GAC exhibited oxygen demand for 1700 h at which time the column studies were terminated (*19*).

DO uptake has been used as a measure of bioactivity in GAC studies (*20*), with the assumption that the chemical demand of GAC for DO had been exhausted. These studies were carried out in BOD respirometers and indicate higher DO demand at the surface of an ozonated column than in a nonozonated column.

Data from this study indicate that DO uptake is a poor estimator of bioactivity. Referring to Figure 6, the DO in the effluent of the two contactors is almost identical during the ambient operation (Mode 1). Immediately after the change to an ambient-chilled system (Mode 2), DO removal becomes nearly zero in the cold column. As the influent temperature decreases, the influent DO increases. The DO uptake by the cold column also increases. If biological activity were primarily responsible for DO removal, the DO uptake should not increase during the ambient-chilled mode of operation because temperatures are falling from 31°C to 17°C and bioactivity should be less. The fact that DO uptake increases during Mode 2 from Day 110 to Day 160 suggests that the principal mechanism in DO uptake is a concentration-dependent chemical reaction. Biological reaction rates have been shown to be independent of oxygen concentration when concentrations are above 2 mg/L (21).

Although DO removal does not appear to be an absolute measure of bioactivity, it can be used when some method is devised to estimate the chemical demand of GAC. One such method involves using a parallel GAC column that is "new" relative to the column being studied. This technique does not remove biological oxygen demand, but it estimates increased oxygen demand due to any enhancement of bioactivity on GAC, assuming that the time required to establish a stable microbial population on the "new" column is not exceeded. Figure 10 shows the results of such an experiment conducted over a short period spanning the change from ambient-chilled (Mode 2) to ambient-warmed (Mode 3) operation.

In Figure 10, the "new" column was operated at ambient temperatures. Because the experiment spans the change in mode of operation, the first part of the graph shows the warm column effluent and new effluent. During this comparison, the new column is removing about the same or a little more oxygen than the older loaded (with TOC) column. This finding is not surprising because of the large initial oxygen demand of GAC. However, once the temperature was increased on the warm and cold columns, the comparison of new to loaded GAC at ambient temperatures involves the cold TOC-loaded column and the new column shown in diamonds and triangles, respectively. The cold column, now operating at ambient temperatures, consistently removed more oxygen than the new column. The last data points were collected after the new column had been on line for 90 days. This figure also suggests that acclimation is not a significant mechanism, because 90 days is sufficient time for the bacteria to reach a stable population. (In one report (18), a stable population was reached after as little as 14 days.) After a stable population is developed, the effect of acclimation should be exhibited by both columns. The consistent difference in DO removal between the loaded and new GAC columns (operated at ambient conditions) indicates that some mechanism other than acclimation probably is occurring. However, these data are too few to draw firm conclusions and more work is needed. It is

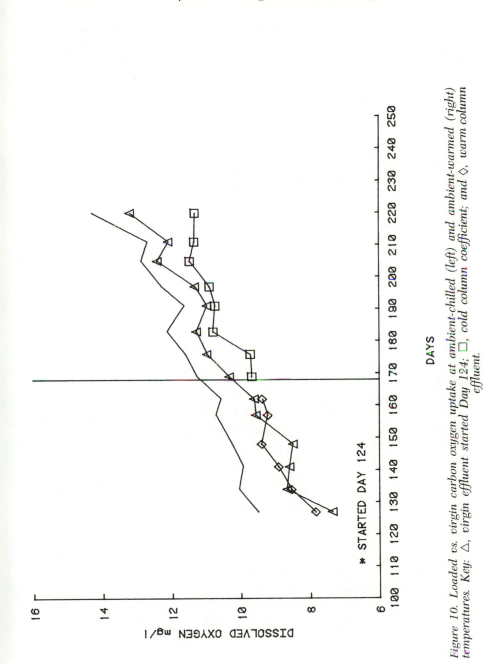

Figure 10. Loaded vs. virgin carbon oxygen uptake at ambient-chilled (left) and ambient-warmed (right) temperatures. Key: △, virgin effluent started Day 124; □, cold column coefficient; and ◇, warm column effluent.

hypothesized that adsorbed material cannot be acted on by bacteria. In fact, GAC may adsorb biodegradable organics before bacteria can act on it, thus gaining no advantage of the ozonation. The biodegradable materials may not be acted on by microorganisms until a change in equilibrium conditions releases them to the aqueous phase.

Another way to look at DO uptake as a function of temperature is to evaluate the data from one column. Data for the warm column are presented in Figure 11 with DO removal plotted as a function of temperature. Data for ambient operation were evaluated using linear regression and 95% confidence interval was plotted around the regression line. Data for warmed ambient operation are shown in diamonds on the plot. All data for the warmed ambient operation lie above the confidence interval for the regression line. This result is another indication that the mechanisms for DO removal changed when the mode of column operation was changed.

The increased DO consumption on the loaded column versus the new column can be interpreted as storage of biodegradable materials, followed by release caused by a temperature induced shift in equilibrium. Thus, biological activity can assist in partial regeneration of the GAC. However, when the storage of biodegradable material is considered in light of the pilot plant sand column results (16) discussed previously, it is somewhat contradictory that biodegradable materials would be available for adsorption. Recall that almost all TOC removal that occurred in the sand columns took place in a 15-min EBCT. The minicolumn used in this study also had a 15-min EBCT, so biodegradable materials entering the column should have been converted to inorganic carbon, unless adsorbed material is unavailable to bacteria. This result again leads to the conclusion that acclimation is not an important mechanism, because bacteria do not appear able to utilize adsorbed substrate.

TOC data did not echo the marked effect of temperature differences seen in the DO data. Referring to Figure 7, little difference in effluent TOC values is seen until after Day 240 (Mode 3), even though the increased oxygen demand started at Day 170. Thus, the effect of temperature on TOC removal is delayed. This finding could be the result of GAC's adsorptive capacity masking lowered bacterial removal, but this effect is unlikely because the quasi-steady-state removal had been reached by Day 120. If the bacterial contribution to TOC removal were significant, the masking effect should have been exhausted, at least partially, before Day 240. On the other hand, if the GAC's residual capacity, not measured in batch isotherms, is great enough to mask effects of bacterial activity for the period observed in this study, the bacterial contribution to the long-term TOC removal is very small compared to the effect of slow adsorption kinetics.

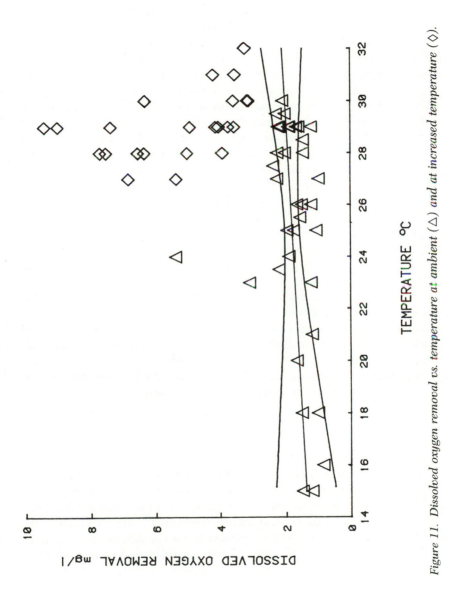

Figure 11. Dissolved oxygen removal vs. temperature at ambient (△) and at increased temperature (◇).

A second theory to explain the observed TOC data is a temperature induced shift in equilibrium. Raising the temperature reduces the equilibrium capacity of GAC. When this occurs, materials previously adsorbed may desorb. If these materials are biodegradable, then the effect of desorption would not be seen in the effluent, because the bacteria convert them to inorganic carbon. The effect of desorption would be seen in DO uptake even though the effluent TOC does not show it. This process opens up space on the GAC to remove additional organics. The warm column was raised to a higher temperature, and, therefore, the change in equilibrium may have been more pronounced. Thus, more capacity could have been opened on the warm column than on the cold column, and the warm column would be expected to have a greater TOC removal than the cold column for a short period. This capacity would be exhausted at some time, and the warm and cold columns would have the same TOC removal. This sequence would explain the large initial difference in TOC removal followed by a slowly decreasing difference in TOC removal. The time required for the warm and cold contactors to return to near identical operation also would be affected by the slow adsorption kinetics phenomenon. Thus, the large initial difference in TOC removal would be due to activity at the surface of the carbon, and additional biodegradable materials may diffuse out of the restricted pores at a much slower rate.

The fact that the magnitude of the difference in TOC removal between the warm and cold column is diminishing with time makes the "masking effect" of bacterial action by GAC an unlikely explanation. If masking were a true mechanism, then once adsorption breakthrough occurred, the difference in TOC removal should be constant. This fact also indicates that acclimation is an unlikely mechanism, because apparently the organics must be in the aqueous phase for bacteria to use them.

Summary

The data from this study are typical of GAC treatment of natural surface waters. TOC removal is quite high initially and then levels off at an apparent steady-state removal. Eight parallel columns operated in this study indicate that GAC contactors are reproducible for TOC, DO, pH, and alkalinity changes across the column when conditions are identical.

Temperature and TOC removal are not related over a short time. Changes in temperature cause a slight apparent regeneration. This regeneration appears to be assisted by bacteria but not caused by bacteria. It is theorized that GAC stores biodegradable materials that can be used by bacteria once the materials are released to the aqueous phase. Acclimation of bacteria to adsorbed substrate does not appear to be a significant mechanism in this study.

Temperature and DO removal are strongly related. DO removal increases with increasing temperature. DO removal also increases with increasing influent concentration. Temperature does not affect pH and alkalinity. The pH and alkalinity of the water do not change across the contactors.

Acknowledgment

This research was completed under a cooperative program between the City of Philadelphia Water Department, William Marrazo, Commissioner and Drexel University, Philadelphia, Pa., supported by the U.S. EPA (Grant CR 806256-02), Cincinnati, Oh., Project Officer Keith Carswell.

Literature Cited

1. Frick, B., Bartz, R.; Sontheimer, H.; and DiGiano, F. A. In "Activated Carbon Adsorption of Organics from the Aqueous Phase"; Suffet, I. H.; McGuire, M. J., Eds.; Ann Arbor Science: Ann Arbor, Mich., 1980; Vol. 1, p. 229.
2. Suffet, I. H., *J Am. Water Works Assoc.* 1980, 72, 41.
3. Wood, P. R.; Demarco, J. In "Activated Carbon Adsorption of Organics from the Aqueous Phase"; McGuire, M. J.; Suffet, I. H., Eds.; Ann Arbor Science: Ann Arbor, Mich. 1980; Vol. 2, p. 115.
4. Rice, R. G.; Miller, G. W.; Robson, C. M.; Kuhn, W.; In "Carbon Adsorption Handbook", Cheremisinoff, P. N.; Ellerbusch, F., Eds.; Ann Arbor Science: Ann Arbor, Mich. 1978; p. 485.
5. Ying, W., and Weber, W. J., Jr., In "Proceedings of the 33rd Industrial Waste Conference, Purdue University"; Ann Arbor Science: Ann Arbor, Mich., 1979; p. 128.
6. Eberhardt, M., In "Translation of Reports on Special Problems of Water Technology"; Sontheimer, H., Ed., EPA - 600/9-76-030, 1976; Vol. 9—Adsorption, p. 331.
7. Peel, R. G.; Benedek, A. *J. Environ. Eng. Div., ASCE,* 1980, 106, 797.
8. Mallevialle, J., "The Interaction of Slow Adsorption Kinetics and Bioactivity in Full Scale Activated Carbon Filters: The Development of a New Predictive Model", Chemviron Award Committee Submission, Feb. 1980.
9. Stephenson, P.; Benedek, A.; Malaiyandi W.; Lancaster, E. A. In "Ozone: Science and Engineering"; Pergamon: Elmsford, N.Y., 1979; Vol. I, p. 263.
10. Benedek, A., In "Activated Carbon Adsorption of Organics from the Aqueous Phase"; McGuire, M. J.; Suffet, I. H.; Eds.; Ann Arbor Science: Ann Arbor, Mich., 1980; Vol. 2, p. 273.
11. Levenspiel, O., "Chemical Reaction Engineering", 2nd ed.; Wiley: New York, 1972; p. 21.
12. Weber, W. J., Jr. "Physiochemical Process for Water Quality Control", Wiley-Interscience: New York, 1972; p. 231.
13. Stephenson, M., "Bacterial Metabolism", MIT Press: Cambridge, 1949.
14. Kerberger, G. J., Norman, J. D.; Schroeder E. D.; Busch, A. W. "B.O.D. Progression in Soluble Subtrates-VII-Temperature Effects", in *Proceedings, 19th Industrial Waste Conference,* Purdue University, 1964.
15. Maqsood, R.; Benedek, A. *JWPCF,* 1977, 49, 2107.
16. Philadelphia, "Research Investigation on the Removal of Trace Organic Compounds by Combined Ozonation and Adsorption in a Biologically Activated Carbon Process," City of Philadelphia's Final Report to EPA, in press, 1982.

17. Radziul, J. V., *Water Sew. Works,* **1977**, *124*, 76.
18. McElhaney, J. B.; McKeon, W. R.; "Enumeration and Identification of Bacterial Populations on GAC," in *Proceedings of the 1978 Am. Water Works Assoc. Water Technology Conference,* Louisville, Ky., 1978.
19. Prober, R.; Pyeha, J. J.; Kingdom, W. E. *AIChE J.* **1975**, *21*, 1200.
20. Benedek, A., *Ozonews,* **1979**, *6*, 1.
21. Kalinske, A. A., *JWPCF,* **1976**, *48*, 2472.

RECEIVED for review August 3, 1981. ACCEPTED for publication March 9, 1982.

Pilot Scale Evaluation of Ozone–Granular Activated Carbon Combinations for Trihalomethane Precursor Removal

WILLIAM H. GLAZE—University of Texas at Dallas, Graduate Program in Environmental Sciences, Richardson, TX 75080

JAMES L. WALLACE, DOUGLAS WILCOX, K. R. JOHANSSON and K. L. DICKSON—North Texas State University, Denton, TX 76203

BOBBY SCALF and ROGER NOACK—Henningson, Durham, and Richardson, Dallas, TX 75230

ARTHUR W. BUSCH[1]—Oligodynamics Corporation, Dallas, TX 75225

The biological activated-carbon process for removing tri-halomethane precursors in drinking water was investigated at a pilot water treatment research facility. A series of carbon columns was used that allowed the regular monitoring of water chemistry and biological parameters. It was found unlikely that the events taking place on the GAC could be explained by purely physical processes, and a largely micro-biological mode is indicated for the last several weeks of the study.

THIS REPORT DESCRIBES a pilot water treatment research facility that was operated to evaluate the so-called biological activated-carbon process for removal of trihalomethane (THM) precursors in a drinking water source. The pilot plant included a package water treatment plant with the capability of supplying 10 gallons/min (GPM) of settled/filtered water of low turbidity. This water provided the test medium for the evaluation of granular activated carbon (GAC), with and without preozonation, for THM precursor removal. Various water chemistry and biological parameters were monitored on a regular basis to determine to what extent

[1]Current Address: University of Texas at Dallas, Graduate Program in Environmental Sciences, Richardson, TX 75080.

0065-2393/83/0202-0303$06.00/0

precursor levels were removed by various unit processes in the plant and how treatment could be improved by control of process variables such as ozone contact time. Most importantly, the project was designed to determine to what extent GAC columns maintained sustained removal capabilities past the time of expected exhaustion and hopefully to determine the role microorganisms played in the purification process.

Pilot Plant Facility

The entire pilot plant with the exception of the raw water intake line [5-cm (2-in.) black steel] and pumps was housed in a 12.2-m (40-ft) drop-frame trailer (similar to a moving van). The trailer was located within the gates of the Thomas L. Amiss Water Treatment Plant in Shreveport, La, and utilized raw water taken directly from Cross Lake, the principal source of water for the city.

The pilot plant (Figure 1) had the capability for preoxidation of raw water with ozone, chlorine dioxide, and perhaps other oxidants, but in this study no preoxidant was used. Raw water was first filtered through a Sinflex strainer [0.08-cm (0.03-in.) perforations] and then flowed directly to a 10 GPM-rated Neptune-Microfloc "Waterboy" package treatment plant with four chemical feed options, tube settler, and multimedia filter. This unit was operated near rated capacity with the product water ultimately being split between three treatment trains, each with two carbon columns in series. The train with Columns 5 and 6 was a "control" train in that no ozone contacted the water prior to entering the carbon columns. Trains with Columns 1–4 received water which had been oxidized in the ozone contact basin. This contact basin utilized a high-speed turbine injector designed by Kerag (Switzerland) and manufactured by Howe-Baker (Tyler, TX). At 6.0 GPM, retention time in the contact basin was approximately 3.3 min. Ozone was generated from dry cylinder oxygen using a Howe-Baker "ESCOZONE E4" generator capable of producing up to 20 g/h of ozone. Water from the ozone contact basin (6.0 GPM) was split into a 2.0-GPM stream to Columns 1 and 2 and a 4.0-GPM stream was sent to an ozone detention basin. Retention time in the ozone detention basin was an additional 30 min, after which 2 GPM of water was pumped to Columns 3 and 4.

A total of eight carbon columns was included in the pilot plant, each 30 cm (12 in.) diameter 304 SS, schedule 12 pipe. With 1.2 m (4 ft) of GAC, each carbon column had an empty bed contact time (EBCT) of 12 min at 2.0 GPM. In the first phase study, six carbon columns were operated in pairs (1 and 2; 3 and 4; 5 and 6). Each pair had its own backwash storage basin receiving water from the respective column pairs, a backwash pump, a feed pump, and fittings for air sparging as a part of the backwash procedure. Before and after each column a pressure gauge was

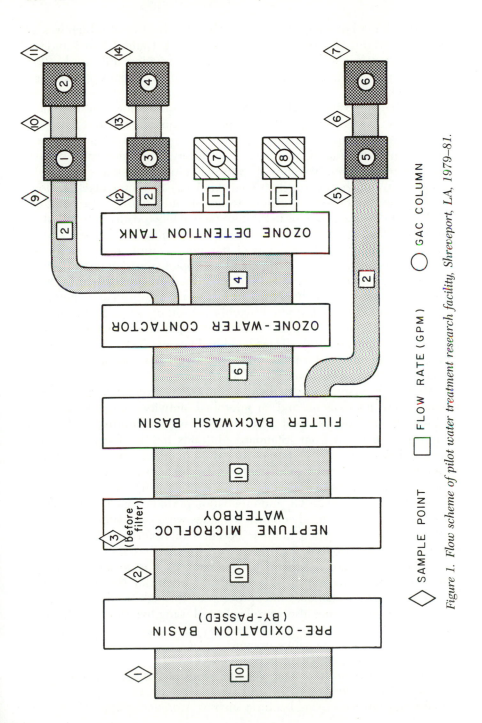

Figure 1. Flow scheme of pilot water treatment research facility, Shreveport, LA, 1979–81.

mounted in-line to provide a basis for initiation of backwash, which was a manual procedure. Also between Columns 3 and 4, 1 and 2, and 5 and 6, in-line injectors for oxygen enrichment were provided. Filtrasorb 400 carbon was provided gratuitously by Calgon Corp. as a means of assisting this research.

Weekly composite samples were taken from the pilot plant by timer controlled valves located at the sampling points shown in Figure 1. Samples were taken at 4-h intervals; they were not of equal volume, however, since the sample points were at different pressures and therefore delivered different volumes of water when opened for a set interval. Water was transferred through 0.3 cm (1/8 in.) diameter black Teflon tubing to glass containers in refrigerators positioned close to the point of sampling. On the date composite samples were collected, discrete samples also were taken at each point by manually opening the valves. These samples were used for the analysis of inorganic carbon, ozone residual, dissolved oxygen, ammonia, nitrate, and microbiological parameters. Trihalomethane formation potential (THMFP), total organic carbon (TOC), color, UV absorbance, conductivity, and pH were measured on composite samples. THMFP was determined by chlorination of samples at 20-mg/L dose at ambient temperature and at a buffered pH of 6.5. Incubation was for 3 days, at which point the THM yield had leveled. TOC and inorganic carbon were determined using the Dohrmann DC-54 analyzer following the manufacturer's directions.

Microbiology studies were divided into two components: a microbial identification component and a microbial activity component. The methods of analysis for total plate counts, total coliforms, and fecal coliforms were taken from "Standard Methods for the Examination of Water and Wastewater" (1). Some procedural modifications were implemented in the total plate count method, i.e., changes in culture media (from tryptone glucose yeast agar to soil extract agar), incubation temperature (from 35°C to 28°C), and incubation period (from 48 h to 7 days (2). Total coliform and fecal coliform densities were calculated according to the membrane filtration method outlined in Reference 1 and EPA Document No. 600/8-70-017 (3).

Isolation of major groups of microorganisms was accomplished by plating on soil extract agar. The dominant organism (colonies) were subcultured, gram stained, and processed through a determinative scheme. Isolated colonies also were evaluated using five routine selective media:

1. McConkey: gram negative organisms
2. Actinomycete isolation agar: *Actinomycetes*
3. *Pseudomonas* isolation agar: specific types of Pseudomonands such as *P. fluorescens*
4. Burkes: nitrogen fixing bacteria such as *Azotobacter sp.*

   5. Cooke rose bengal agar: most types of fungi

Cell counts of organisms associated with the carbon particles were determined by sonication of the carbon sample for 1 min at 60 watts in a phosphate buffer to dislodge the microorganisms. An aliquot of the phosphate buffer containing dislodged microorganisms was then plated on soil extract agar and incubated for 5–7 days.

Pilot Plant Operational Parameters

The Shreveport pilot research facility was operated from Dec. 29, 1979 to July 23, 1981. This report describes the first 60 weeks of operation extending to March 6, 1981. During this period, the following operational conditions were used. The alum dose was 50 mg/L, nominal, and the sodium hydroxide dose was selected to increase water pH and alkalinity so as to optimize raw water turbidity removal. The ozone dose was 6.3 mg/L from Dec. 29, 1979 to Feb. 10, 1980; it was 2.0 mg/L to Apr. 10, 1980, and then was 2.5 mg/L to March 6, 1981; transfer efficiency was 75–85%. The GAC column parameters were 2.0 GPM or 2.5 gal/ft^2-min, 12 min of EBCT for each column, and downflow mode.

As an example of raw water characteristics, the average values for summer and winter periods are given in Table I.

Pilot Plant Performance

The data obtained from the first 60 weeks of operation of the pilot plant are too extensive to analyze in detail. Therefore, a summary of principal conclusions is given. More detailed discussions and analysis of the data will be published elsewhere (4, 5).

Pretreatment. The Neptune-Microfloc "Waterboy" pretreatment unit performed well, giving low turbidity water (<1 NTU) consistently. Figure 2 shows THMFP of the raw and settled/filtered effluent from the Waterboy. It is of particular interest that the efficiency of THMFP removal was approximately 20–30% in the winter months, but as high as 50% in the summer. Whether this reflected the improved performance of the Waterboy per se or a change in the natural organic influent is not clear.

Figure 2 also shows the variability of the raw water source, Cross Lake, La. This lake is surrounded by wooded areas and its backwaters are marshy and heavily loaded with natural aquatic humus. No significant anthropogenic pollution of the lake was evident.

TOC values in the lake and settled/filtered water showed the same features as THMFP, but there were some periods in the spring and summer where disproportionately high THMFP values were observed. It is possible that these periods of high THMFP/TOC ratios reflected production of organics in the lake.

Table I. Raw Water Characteristics
(Average Values for Summer and Winter)

Characteristic	Summer	Winter
TOC, mg/L	11	7
THMFP, μg/L	350–450	150–200
Alkalinity, mg/L	35	18
Color, Pt units	70	30
pH	7.5	6.5
Temperature	30°	10°

Ozonation. Figure 3 shows that oxidation with low doses of ozone was not an effective method for the reduction of THM precursors. Only 5–15% reduction of THMFP was observed with settled/filtered water as the test medium, even during the initial 6-week period when the (transferred) dose was 6.3 mg/L (5).

GAC Filtration. Figure 4 is a typical performance plot of the GAC column pairs operated in tandem in the pilot plant. Plots of this type are not as instructive as effluent/influent plots (Figure 5) and plots of cumulative removal versus time (Figure 6). These figures illustrate the sustained removal of THMFP during the test period. Plots of TOC removed have essentially the same features as Figures 5 and 6. Even more instructive are plots such as Figure 7 (6, 7). The slopes of these plots represent the fraction of substrate removal at any time, and, as such, they are particularly useful in discussing mechanisms that operate in GAC columns. Before entering into such a discussion, however, two other types of data will be presented.

Figure 8 shows cumulative inorganic carbon (INC) production across Columns 3 and 4 and 5 and 6, receiving ozonated and nonozonated water, respectively. INC production became significant and nearly constant from Week 20 to Week 44, at which point production dropped precipitously. It was just at this time that the very mild Louisiana winter became quite cold and lake water temperatures dropped.

Table II shows total plate counts of water samples taken before and after GAC Columns 3 and 5. It is clear that the ozonation process which preceded GAC Column 3 resulted in substantial disinfection of the water entering the column. By comparison, Column 5 received significantly higher levels of bacteria, particularly during the first 5 months of operation.

Adsorption and Microbiological Mechanisms of Organic Carbon Removal. The data accumulated during the 60 weeks of operation of the Shreveport pilot research facility, however instructive, are not sufficient to resolve unequivocally the question of the relative roles of physical and microbiological processes on GAC columns (6–15). Nonetheless, the data

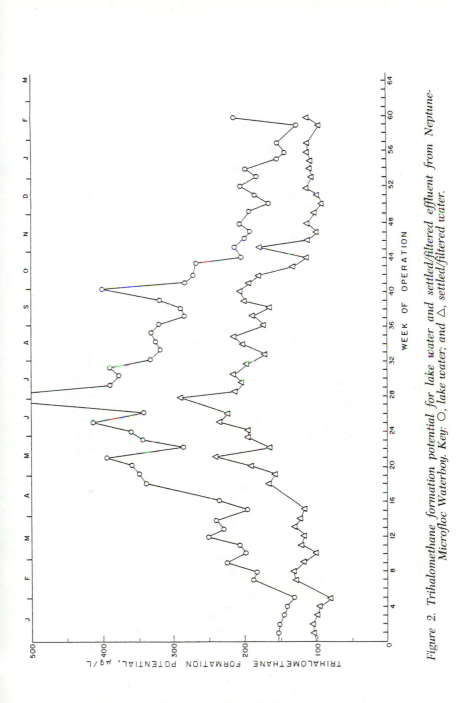

Figure 2. Trihalomethane formation potential for lake water and settled/filtered effluent from Neptune-Microfloc Waterboy. Key: ○, lake water; and △, settled/filtered water.

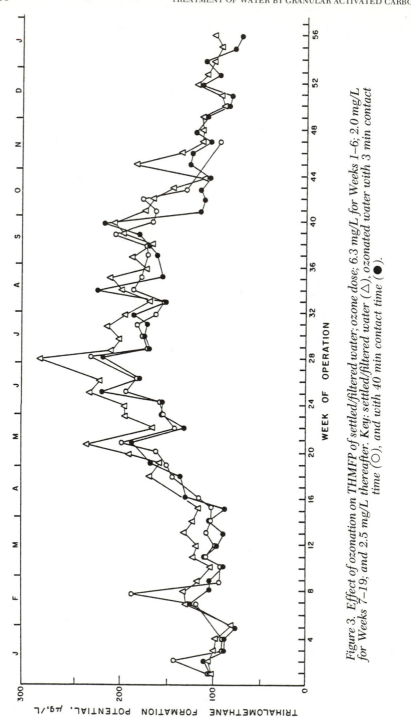

Figure 3. *Effect of ozonation on THMFP of settled/filtered water; ozone dose; 6.3 mg/L for Weeks 1–6; 2.0 mg/L for Weeks 7–19; and 2.5 mg/L thereafter. Key: settled/filtered water (△), ozonated water with 3 min contact time (○), and with 40 min contact time (●).*

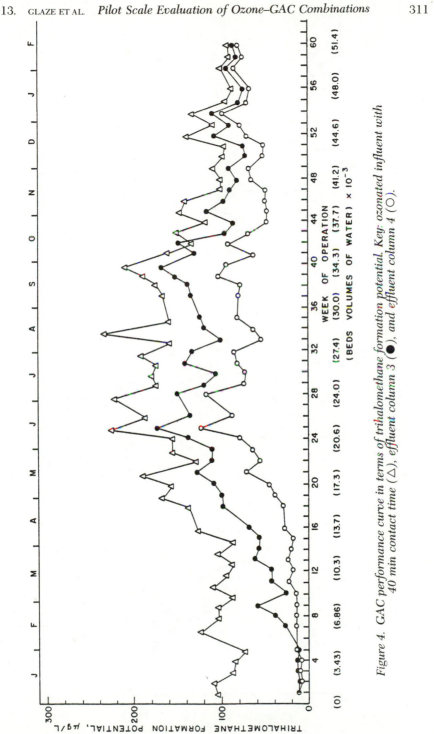

Figure 4. GAC performance curve in terms of trihalomethane formation potential. Key: ozonated influent with 40 min contact time (△), effluent column 3 (●), and effluent column 4 (○).

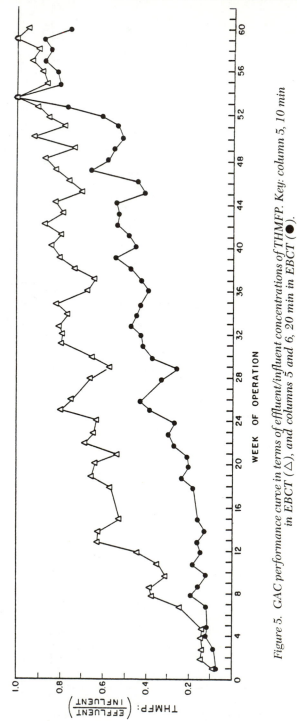

Figure 5. GAC performance curve in terms of effluent/influent concentrations of THMFP. Key: column 5, 10 min in EBCT (△), and columns 5 and 6, 20 min in EBCT (●).

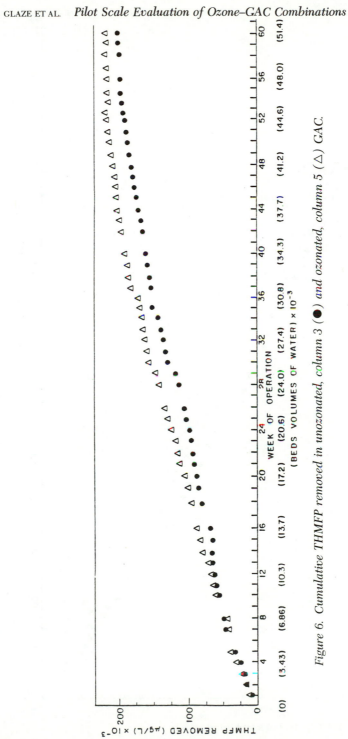

Figure 6. Cumulative THMFP removed in unozonated, column 3 (●) and ozonated, column 5 (△) GAC.

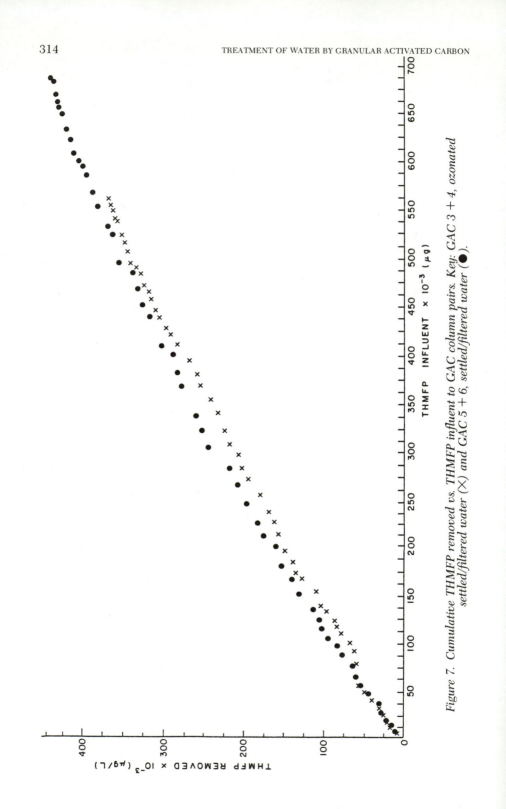

Figure 7. Cumulative THMFP removed vs. THMFP influent to GAC column pairs. Key: GAC 3 + 4, ozonated settled/filtered water (×) and GAC 5 + 6, settled/filtered water (●).

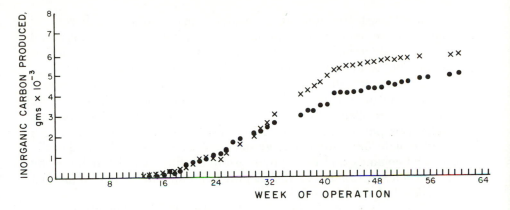

Figure 8. Cumulative inorganic carbon production across GAC columns receiving ozonated and unozonated influent water (values expressed as INC in kg carbon). Key: column 3, ozonated (✕) and column 5 settled/filtered water (●).

Table II. Total Plate Counts (Number/mL) of Water Samples Taken Before and After GAC Carbon Columns

	GAC Column 5 (Unozonated Water)		(Ozonated Water—40-min Contact)	
Date	Before	After	Before	After
Jan. 4, 1980	1527	1066	0	494
Jan. 23, 1980	2790	3720	0	1465
Feb. 12, 1980	145	150	5	90
Feb. 26, 1980	2600	2300	8	1.5×10^4
March 26, 1980	1×10^4	5000	0	3500
May 6, 1980	6050	3050	1500	2350
June 3, 1980	1200	5150	336	4610
July 8, 1980	3315	4100	290	1.3×10^4
Aug. 12, 1980	2500	550	30	1100
Sept. 15, 1980	400	1600	78	3400
Oct. 6, 1980	130	750	9	550
Nov. 12, 1980	1700	300	240	1500
Dec. 8, 1980	1920	7550	2	1275

Note: Total plate counts with soil extract nutrient agar, 28°C, and 7-day incubation.

provide circumstantial and indirect evidence that tends to corroborate the position of Sontheimer and coworkers who attribute a significant removal capacity to microbial action (13, 14). On the other hand, Figure 7 is precisely of the form predicted by the Benedek dual diffusion model (7–9). The figure and the corresponding plot of cumulative TOC removed versus cumulative TOC influent show a period of rapid, macropore diffusion controlled adsorption, which occurred during the first few weeks. This period was followed by a period of slower removal which lasted until the 52nd week or so for the unozonated GAC columns. For the ozonated columns, and to a certain extent for the second column in the unozonated set (Column 6), the removal continued to the end of the first operational phase (Week 62).

This performance is similar to that attributed by Mallevialle (7) to slow micropore diffusion, but a closer inspection of the data reveals that microbiological activity must account in part for the pseudo-steady-state removal observed after Week 20. The important features of the data in this regard are as follows.

While organic carbon removal was proceeding in the initial adsorption period, little inorganic carbon production was observed. For the first 20 weeks, TOC removal averaged 0.54 mol carbon/m^3 GAC-h, and INC production was only 0.04 mol carbon/m^3 GAC-h (unozonated and ozonated GAC columns performed similarly during this period). However, at the outset of the pseudo-steady-state period, the INC production rate increased and eventually reached a steady-state value itself (Figure 8). Cumulative INC production was initially higher in the unozonated columns but the ozonated columns eventually excelled in this regard. Table III sumarizes TOC and THMFP removal rates and INC production rates for the four columns from Week 20 to Week 44.

It is quite unlikely that these INC production data can be rationalized on the basis of endogenous respiration of bacteria which are filtered out of the water. The higher rates of INC production from the ozonated GAC columns, combined with the plate count data in Table II, speak strongly against this argument. Rather, it appears that microbiological removal is proceeding at rates of about the same magnitude as those observed by others (12). Finally, the data in Table II clearly indicate that plate counts are increasing as ozonated water passes through the columns. Biomass accumulation in GAC columns is not then a result of simple filtration of bacteria. There is apparently a proliferation of several species on the surface of the GAC which eventually contribute to the removal of natural carbonaceous substrates.

At the beginning of the period of cooler water temperatures, INC production dropped significantly (Figure 8). Approximately 4 weeks later, there is observed a precipitous breakthrough of TOC and THMFP (Figure 5). Apparently, the available GAC surface area was exhausted and when

Table III. Pseudo-Steady-State Removal Rates and Inorganic Carbon Production Rates in GAC Columns

Parameter	Column 5	Column 6	Column 3	Column 4
TOC, mol carbon/m^3 GAC-h	0.60	0.59	0.44	0.59
INC, mol CO_2/m^3 GAC-h	1.21	0.83	0.83	0.45
THMFP, mmol THMFP as $CHCl_3$/m^3 GAC-h	2.2	2.1	2.1	2.7

Note: Week 2 through Week 44.

microbiological removal ceased, no removal mechanism remained. Again, it is highly unlikely that the events could be explained by purely physical processes, and a largely microbiological mode is indicated for the last several weeks of the study period. Interestingly, the loss of active removal from Week 52 to Week 60 was not as great for the ozonated columns as for the unozonated columns.

Postscript. Following the presentation of this paper, the pilot research facility was operated for an additional 20 weeks. After surface water temperatures rose in 1981, removal of TOC and THMFP resumed at a rate of approximately 0.26 mol carbon/m^3 GAC-h and 1.00 mmol THMFP/m^3 GAC-h. Inorganic carbon production was essentially equal to TOC removal and ozonated GAC columns continued to outperform unozonated columns while in a purely microbiological mode. During this period, additional data were obtained that corroborated the conclusions tentatively presented at the Atlanta ACS Meeting. Details of the full study are presented in Reference 4 and subsequent papers.

Acknowledgment

This work was sponsored by the U.S. Environmental Protection Agency through Cooperative Agreement CR-806157. We are particularly indebted to J. Keith Carswell, Project Officer, Drinking Water Research Division, USEPA/MERL, Cincinnati, Ohio. Analysis of water samples was carried out in part by Ken Carney and Richard Williams.

Literature Cited

1. "Standard Methods for the Examination of Water and Wastewater", 14th ed.; American Public Health Association: Washington, D.C., 1975.
2. McElhaney, J.; McKeon, W. R. "Enumeration and Identification of Bacteria in

Granular Activated Carbon Columns", Proceedings of Am. Water Works Assoc. 6th Water Quality Technology Conference, Louisville, Ky., 1978.
3. "Microbial Methods for Monitoring the Environment: Water and Wastes", EPA-600/8-78-017, Environmental Monitoring and Support Laboratory, U.S. EPA, Cincinnati, Oh., 1978.
4. "Evaluation of Biological Activated Carbon for Removal of Trihalomethane Precursors", Final Report, Cooperative Agreement CR806157, Drinking Water Research Division, U.S. EPA, Cincinnati, Oh., 1982.
5. Glaze, W. H.; Wallace, J. *J. Amer. Water Works Assoc.* submitted.
6. Benedek, A.; Bancsi, J. J.: Malaiyandi, M; Lancaster, E. A. *Ozone: Science and Engineering*, **1980**, *1*, 347–56.
7. Mallevialle, J., "The Interaction of Slow Adsorption Kinetics and Bioactivity in Full Scale Activated Carbon Filters: The Development of a New Predictive Model", Unpublished manuscript submitted to the Chemviron Award Committee, February 1980.
8. Peel, R.; Benedek, A.; Crowe, C. M. *AIChE J.* **1981**, *27*, 26–32.
9. Peel, R; Benedek, A. *J. Environ. Eng. Div., ASCE,* **1980**, *106*, 797–813.
10. Peel, R.; Benedek, A. *Environ. Sci. Technol.* **1980**, *14*, 66–71.
11. Tsezos, M.; Benedek, A. *J. WPCF,* **1980**, *52*, 578–86.
12. DiGiano, F., "The Influence of Microbiological Activity on the Performance of GAC", Presented at Am. Water Works Assoc. Preconference Seminar on Organics, San Francisco, Calif., June 24, 1976.
13. Rice, R. G.; Robson, C. M.; Miller, G. W.; Clark, J. C.; Kühn, W., "Biological Processes in the Treatment of Municipal Water Supplies", EPA-600/2-80-1980, U.S. EPA/MERL, Cincinnati, Ohio.
14. Eberhardt, M.; Madsen, S.; Sontheimer, H., Report No. 7, Engler-Bunte-Institut, Karlsruhe University, Federal Republic of Germany, 1974.
15. Sontheimer, H.; Heilker, E.; Jekel, M. R.; Notte, H.; Vollmer, F. H. *J. Am. Water Works Assoc.* **1978**, *70*, 393.

RECEIVED for review August 3, 1981. ACCEPTED for publication April 12, 1982.

14

Interaction of Adsorption and Bioactivity in Full-Scale Activated-Carbon Filters: The Mont Valerien Experiment

F. FIESSINGER and J. MALLEVIALLE

Société Lyonnaise des Eaux et de l'Eclairage, Laboratoire Central, 38 rue du Président Wilson, 78230 Le Pecq, France

A. BENEDEK

McMaster University, Department of Chemical Engineering, Hamilton, Ontario, Canada L8S 4L7

The water purifying efficiency of a two-step system was investigated. The first process was a physicochemical process with coagulation, flocculation, and sedimentation, followed by rapid sand filtration. The second process was slow sand filtration. Different types of carbon were evaluated, and different filtration velocities in the second filtration stage were tested. The use of very slow filtration velocities and comparison with higher rates made it possible to study the mechanism of removal of organic matter. The relative importance of adsorption and biodegradation for the removal of organic matter using granular activated carbon also was assessed.

T HE PLANT AT MONT VALERIEN PRODUCES potable water out of the Seine river water withdrawn at Suresnes immediately downstream from the City of Paris (Figure 1). There the water is not appreciably more polluted than upstream from the city (Table I) because wastewaters from the area are collected and sent to a large secondary treatment plant at Acheres a few kilometers below Suresnes. Urban runoff, however, may contribute to the degradation of water quality when it flows through the French capital. Consumers' demand for a water of perfect quality and the scope of new

0065-2393/83/0202-0319$06.00/0

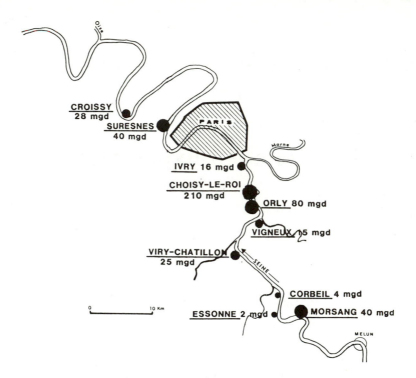

Figure 1. Potable water treatment plants in the Paris area.

Table I. Main Characteristics of the Seine River Water at the Suresnes Bridge over the Period of Experiment (Feb. 79 - Feb. 80)

Parameter	Average	Maximum	Minimum
Temperature (°C)	14.0	22.5	4.0
Dissolved oxygen (mg/L)	8.6	13.0	3.3
Turbidity (Ftu)	24.3	92	2.5
pH	7.75	8.20	7.21
NH_4 (mg/L)	0.52	1.5	0.2
N Total K_j (mg/L)	1.3	3.6	0.6
PO_4 (mg/L)	0.87	1.85	0.3
Cd (μg/L)	< 0.5	2	< 0.5
Pb (μg/L)	18.0	110	3
Hg (μg/L)	< 0.3	< 0.3	< 0.3
COD–$KMnO_4$ (mg/L O_2)	4.46	10.0	2.2
UV extinction ($A_{25\ 4nm}^{1m}$)	10.42	25.5	1.94
Fluorescence (mV)	2.82	5.54	1.66
TOC (mg/L)	3.45	5.5	2.5
Total organic chloride (μg/L)	44.3	154	11

EEC water quality standards (*1*) prompted the need for the transformation of the old Mont Valerien plant into an advanced one, capable of removing trace pollutants effectively.

The plant at Mont Valerien consists of two different lines. The first (Process A) consists of a slow sand filtration process with a capacity of 25 MGD (100,000 m^3/day). This line, built in 1904, includes coarse gravel strainers and rougheners preceding slow (biological) sand filters where the filtration velocity is about 5 m/day (0.2 m/h). The second line (Process B) is a more recent physicochemical process with coagulation, flocculation, and sedimentation in an upflow "pulsator" clarifier, followed by rapid sand filtration at a velocity of 5 m/h. This line, built in 1960, has a capacity of 15 MGD (50,000 m^3/day).

These two processes have been operated in parallel but could, through simple modifications, be operated in series as well. The filtration velocity could be raised appreciably so that the total flow rate would be the same.

The objective of the study was to evaluate the efficiency of Process B followed by Process A in a full-scale experiment. Slow sand filters could be topped by a granular activated-carbon (GAC) layer and ozonation could be introduced between the two processes, i.e., between rapid sand filtration and slow GAC/sand filtration.

Different types of carbon were evaluated and different filtration velocities in the second filtration stage were tested. The GAC empty bed contact time (EBCT) was, however, kept the same in all filters.

The use of very slow filtration velocities and comparison with higher ones made it possible to study the mechanism of removal of organic matter. The second objective was to study the relative importance of adsorption and biodegradation for the removal of organic matter in GAC. This constitutes the core of the present paper.

Experimental

Flow Chart. Figure 2 shows the flow schematic for the study. The performances of the original two lines are measured at sampling points 0 and 6.

No prechlorination was applied during the experiment. The only chemical application consisted of coagulant and coagulant aide [average dosage of 40 ppm $Al_2(SO_4)_3 \cdot 18H_2O$ and 0.5 ppm of sodium alginate]. The "pulsator" clarifier was operated at a low upflow velocity of 2 m/h. The first rapid sand filter 6, which fed all others in the experiment, was operated at about 5 m/h with air scouring and backwashing approximately every 30 h.

For this study, the following unit processes were tested:

1. Slow sand filtration at 0.625 m/h (15 m/day) in 50-m^2 filters.
2. Slow GAC filtration at 0.625 m/h. Norit PKST 1/4 - 1 GAC, 15 cm, was put on top of 65 cm of sand in 50-m^2 filters. Norit PKST is a nonreactivatable peat-based broken carbon, of low cost with a total surface area of 800 m^2/g (manufacturer's data).

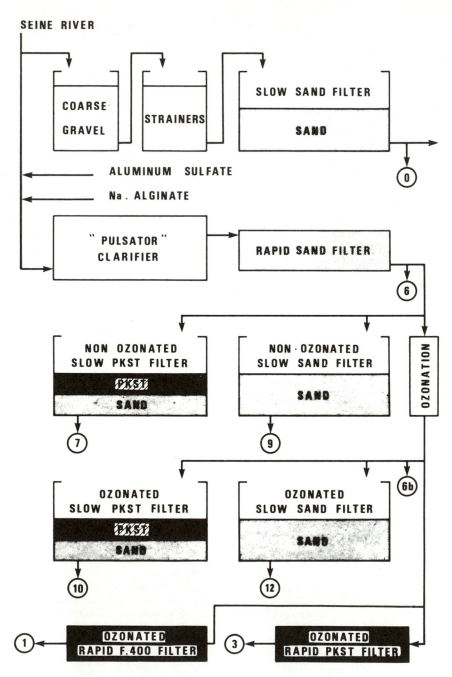

Figure 2. Flow schematic for the Mont Valerien experiment.

3. Ozonation and slow sand filtration with same operating parameters as 1. Ozonation was carried out in a two-chamber contactor with a total contact time of 10 min. The average ozone dosage applied was 1.4 ppm with the residual after the second chamber being kept constant at 0.25 ppm.
4. Ozonation and slow PKST filtration. Operating parameters for the filtration and ozonation steps were the same as in 2 and 3.
5. Ozonation and rapid GAC filtration. Two different carbons were tested in 25-cm diameter pilot columns, Chemviron F-400 and Norit PKST. Chemviron is the European branch of Calgon Corp. and their F-400 carbon is widely used in water treatment. It is a high quality coal-based broken carbon which can be reactivated. Its total surface area is higher than 1200 m²/g. The filtration velocity was 8.3 m/h, and the bed depth was 2 m, yielding a EBCT of 14.4 min in both rapid and slow filters for each carbon.

Filter Operating Conditions and Media Characteristics. Table II summarizes the operating conditions for the filters. Control of the flow rates ensured constant operating conditions for all filters. The EBCT was equal to 14.4 min in all GAC layers. Total EBCT (GAC + sand layers) was considerably longer in the slow filters than in the rapid ones. Previous studies incidated that the pilot sized rapid filters would correctly reproduce the performances of larger ones. Therefore, it is assumed a comparison between these small rapid filters and the large industrial (50 m²) slow ones is valid.

The small rapid filters were backwashed approximately every month for about 1 h (bed expansion of about 30%); the large slow filters did not have to be backwashed during the year of experiment. Manipulation of the thin (15 cm) GAC layer could thus be avoided.

Table III represents the filter media characteristics. Rather similar particle sizes were sought for the different media, and the choice of the two carbons was primarily dictated by this consideration.

The particle size distribution curves were established and from them the average sizes were drawn.

The average external surfaces per volume of bed could thus be calculated. This characteristic was important for the interpretation of bioactivity measurements in the filter beds.

Water Quality Parameters. Most water quality parameters from the EEC potable water standards (1) were measured weekly at different stages of the

Table II. Filter Operating Conditions

Filter	Sampling Point Number	Filtration Velocity (m/h)	Filter Surface (m²)	Bed Height and Medium	EBCT (min)
Rapid sand	6	5	50	0.8 m sand	9.6
Slow sand	9 (no ozone)	0.625	50	0.8 m sand	76.8
	12 (ozone)				
Slow PKST	7 (no ozone)	0.625	50	0.15 m PKST	14.4
	10 (ozone)			+0.65 m sand	62.4
Rapid F-400	1	8.3	0.05	2 m F-400	14.4
Rapid PKST	3	8.3	0.05	2 m PKST	14.4

Table III. Filter Media Characteristics

Filter	Effective Size (10%) (cm)	Uniformity Coefficient (60%/10%)	Average Diameter (50%)(cm)	Bulk Density (g/cm³)	Particle Density (g/cm³)	Average External Surface per Volume of Bed (cm²cm³)
Sand (slow and rapid)	0.048	2.60	0.109	1.14	2.54	58.84
Norit PKST 1/4-1	0.042	1.40	0.055	0.25	0.54	111.11
Chemviron F-400	0.066	1.52	0.093	0.41	0.75	63.83

processes. Table I, which represents the main characteristics of Seine river water, includes only a part of those parameters. Lumped organic parameters and more particularly TOC are discussed in detail.

TOC measurements were carried out on a TOCSIN 2 apparatus. Organic carbon is converted to CO_2 and eventually to methane which is measured in a flame-ionization detector. The detection limit of the method is as low as 0.1 mg/L. No prefiltration of the samples was made.

Chemical oxygen demand (COD) was determined with $KMnO_4$ in hot acidic medium. The result is expressed in mg/L oxygen.

UV absorbance was measured by light extinction at 254 nm using a 10-cm cell and expressed for a 1-m cell. To be able to calculate and report cumulative removal profiles, the UV absorbance values were converted into μg/L of fulvic acid (Figure 3).

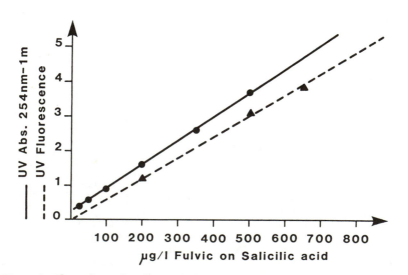

Figure 3. Plots of UV absorbance and fluorescence in terms of μg/L of fulvic and salicylic acid concentrations. Key: ●, fulvic acid; and ▲, salicylic acid.

UV fluoresence also was measured, with an excitation wavelength of 320 nm and an emission wavelength of 405 nm. The fluorescence was converted into $\mu g/L$ of salicylic acid (Figure 3). Expressed as such, these two parameters could be included in the material balance.

Biological examinations were conducted on samples from various depths of the filter beds including total viable cell (plate) counts, nitrifying bacteria counts, ATP measurements, and scanning electron microscopic observations. ATP measurements were made with the bioluminescence method; 100 mg of GAC was treated with dimethyl sulfoxide, then MOPS was added to the solution, and the sample was quickly frozen. Readings were made with a Dupont 760 biometer.

The frequency of measurements for most parameters was once a week and the duration of the experiment was approximately 1 year from February 1979 to January 1980.

Results and Discussion

Inorganic Parameters.   Most inorganic pollutants and in particular heavy metals present at low concentrations were effectively removed during the first stage clarification process. The only noteworthy parameter is ammonia. It was effectively removed most of the time through biological oxidation in the sludge blanket of the clarifier and in the rapid sand filter, except for a short period in January 1980. Table IV indicates the results (eight measurements) for the different unit processes. When the level of ammonia in the raw water increases and the temperature decreases, GAC filters, particularly slow ones without ozonation, seem to be more efficient.

Organic Lumped Parameters.   BREAKTHROUGH CURVES.   Figures 4 and 5 exhibit the typical patterns of breakthrough curves in terms of lumped parameters for such filters. Initially, the removal is rather high but, after a few months of operation, effluent concentrations increased rapidly and, thereafter, a relatively steady removal, between 10 and 20% of the feed (filter 6), continued to occur throughout the experiment. The comparison of these two figures indicates the rapid filters are slightly more efficient than the slow ones. It is difficult, however, to make clear conclusions from such figures; therefore, the results were averaged over the period of operation.

AVERAGE REMOVALS.   Table V summarizes the average percentages of removal for the 1-year experiment. These are averages of 45–50 data points and represent the reduction across each unit process, with the exception of the first two, which represent the reduction across the overall process up to the considered sampling point. The first two lines represent the reduction observed on the original processes of the plant. In the table, rapid sand filter 6 is the feed for nonozonated filters and rapid sand filter + ozone 6b is the feed for ozonated slow and rapid filters.

TOC.   The initial large reduction is probably due to the removal of particulate organics. Table V indicates that the efficiency, for TOC

Table IV. Average Ammonia Concentration in January 1980
(Eight Measurements)

Effluent from	NH_4 (mg/L)
Rapid sand filter	0.31
Slow sand filter	0.28
Slow PKST filter	0.10
Ozonated slow PKST filter	0.11
Ozonated rapid PKST filter	0.30

Note: Mean water temperature was 5°C.

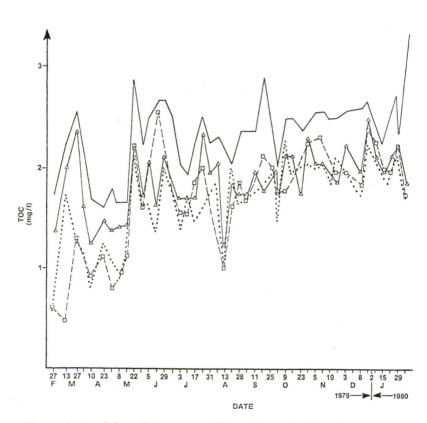

Figure 4. Breakthrough curves at Mont Valerien for slow filters. Key: ——, rapid sand filter 6; □ nonozonated slow PKST filter 7; · · · ·, ozonated slow PKST filter 10; and △, ozonated slow sand filter 12.

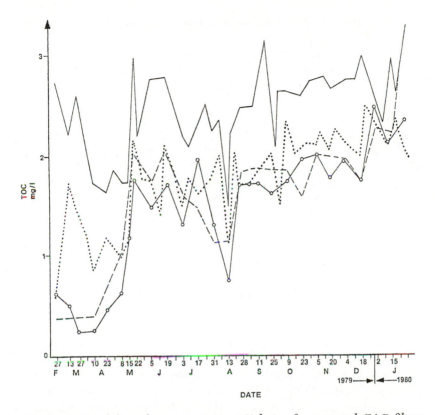

Figure 5. Breakthrough curves at Mont Valerien for ozonated GAC filters. Key:——, rapid sand filter 6; ····, ozonated slow PKST filter 10; ○, ozonated rapid F-400 filter 1; and - - -, ozonated rapid PKST filter 3.

Table V. Average Percent Organic Reduction across Unit Processes

Filter	Sampling Points	TOC	COD by KMnO₄	1m A254nm	Fluorescence
First Stage					
Conventional slow sand	0	33	—	—	—
Rapid sand	6	32	70	63	35
Slow sand	9	10	17	10	9
Slow PKST	7	25	45	46	49
Ozone	6b	5	25	41	54
Second Stage					
Ozonated slow sand	12	12	12	0	12
Ozonated slow PKST	10	24	36	38	48
Ozonated rapid PKST	3	34	46	52	60
Ozonated rapid F-400	1	39	58	56	67

reduction, of the conventional slow sand filtration process is very similar to the physicochemical one.

During the second stage of treatment, an additional 10% of TOC removal is brought about by slow sand filtration 9. GAC filtration 7 removes 25% which is 15% more than the comparable sand filter 9.

Ozone itself, at the relatively low doses used, removes only 5% of the TOC. The ozonated sand filter appears to remove a little more than the nonozonated one. The ozonated GAC slow filter seems to remove less organic material than the nonozonated one, but the difference is very small.

Rapid GAC filters remove appreciably more organics than the slow ones. For the same Norit PKST carbon, the additional TOC reduction is about 10% and, with the best F-400 carbon, the average TOC reduction for the 1-year experiment is almost 40% which is considerable given the 14.4-min EBCT. As an example of the improved water quality, the average TOC concentration in the effluent of rapid sand filter 6 was 2.33 mg/L (standard deviation 0.59) whereas the average TOC concentration in the water produced by the best filter 1 was 1.34 mg/L (standard deviation 0.42).

COD, ABSORBANCE, AND FLUORESCENCE. The other lumped parameters showed somewhat different percentages, particularly with ozonated waters. It has been shown (2) that ozonation decreases the size of the organic particles/molecules and therefore influences spectrophotometric measurements like UV absorbance or fluoresence. All data, however, are consistent with TOC values. The comparison made on the basis of TOC is confirmed using these additional parameters.

During the second stage, the slow sand filter 9 removed a limited amount of organic matter (through biodegradation ?). Ozonation does not appear to enhance this removal.

GAC is very efficient, particularly, if used in contactors with a rapid filtration velocity. Ozonation, on the other hand, seems to have a detrimental influence on overall GAC performance. However, the combination of the two processes, ozonation + GAC filtration, has a net positive effect over GAC filtration alone. Here again, these results are in agreement with previous observations (3).

The higher efficiency of the rapid GAC filters also is in agreement with accepted adsorption theories (4–6): higher liquid film velocities outside the carbon leads to higher liquid film transfer coefficients. In the long term, these slow and rapid GAC filters with the same EBCT will remove the same amount by adsorption but the kinetics are apparently faster in the case of the rapid filters.

These differences in favor of rapid filters also suggest that adsorption plays a major role in the removal of organics by GAC even for an extended time period.

Cumulative Removals.   The cumulative organic removals in a filter were plotted as a function of the cumulative organics applied. These curves are easier to interpret than the breakthrough curves and shed light on the respective influence of adsorption and biodegradation.

TOC.   The first plot in Figure 6 represents the cumulative removal for the two slow sand filters (nonozonated and ozonated). The slope of these curves is almost constant throughout the operating period indicating that a small fraction of the TOC was removed by these slow sand filters right from the start of operation. Toward the end of the period of operation, a slightly greater removal of TOC occurred in the pre-ozonated slow sand filter 12, possibly indicating that the nature of the water changed and ozone had a more significant effect on organic removal.

The effluent from these slow sand filters was biologically stable, i.e., biodegradation was not observed on storage. The operation of these filters was not limited kinetically and therefore the same amount of TOC should be removed by bacteria in both sand and carbon beds. It was thus assumed that the fraction removed by the slow sand filters represents the total amount of biodegradable organics in each type of feed. All other cumulative plots were therefore prepared by subtracting from each data point the corresponding sand filter effluent concentration, e.g., the removal in the pre-ozonated slow sand filter 12 was subtracted from the TOC applied and removed in the case of the pre-ozonated slow GAC filter 10. The resulting curves were assumed to represent removal by adsorption only.

For the rapid GAC filter, the TOC removed was corrected for the difference in contact time (multiplied by 14.4/76.8) before subtraction from the TOC removed in the pre-ozonated rapid GAC filters 1 and 3. The very rough assumption was thus made that biodegradation was proportional to the overall EBCT.

Results are presented in Figures 7 and 8. On the whole, the curves appear to be clearly adsorption based. Their slope is constantly diminishing, particularly in the case of the rapid filters where more rapid exhaustion due to greater adsorption is expected. As shown in Figure 8, the filters were fully exhausted—slope near zero—before the end of the experiment. This finding also indicates that the kinetics of biodegradation are not significantly faster on carbon than on sand. GAC does not enhance biodegradation.

The fact that the slopes for the slow filters 7 and 10 (Figure 7) are not zero at the end may be explained by the fact that they had not reached their total adsorptive capacity (because of slower kinetics) at the end of

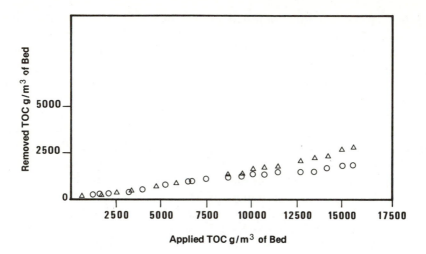

Figure 6. Cumulative plots for slow sand filters. Key: ○, slow and filter 9 and △, ozonated slow sand filter 12.

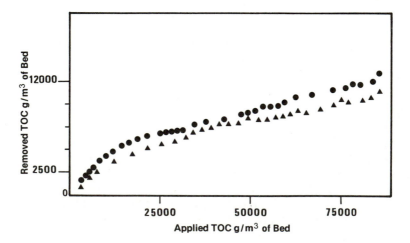

Figure 7. Cumulative plots for slow PKST filters. Key: ●, slow PKST filter 7 and ▲, ozonated slow PKST filter 10.

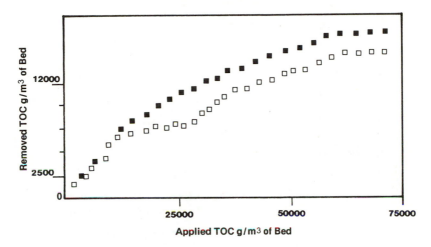

Figure 8. Cumulative plots for rapid GAC filters. Key: ■, ozonated rapid F-400 filter 1 and □, ozonated rapid PKST filter 3.

the experiment. Eventually, they too would be expected to show a slope of zero.

COD, FLUORESCENCE, AND UV ABSORBANCE. Similar calculations were made for the other lumped parameters. These results are presented in Figures 9–11. The same general conclusions can be drawn and are best illustrated by Figure 9 for COD using $KMnO_4$. The case of UV absorbance is more difficult to interpret since the cumulative curves have an appreciable slope at the end of the experiment (Figure 11). One hypothesis is that GAC, because of its rough surface, retains more particulate organic matter than sand (as measured by UV).

Biological Examinations. TOTAL CELLS COUNTS. Cell counts are summarized in Table VI; they constitute geometric means for eight measurements and indicate similar values in all filters. Counts seem to be slightly higher in ozonated GAC filters and more particularly in the rapid ones.

Bacterial counts were compared with scanning electron microscope observations; numerous discrete bacteria were seen at the surface of the GAC particles, particularly in the vicinity of holes and crevices. Breakage of particles confirmed that bacteria were restricted to the outer surface of the particles.

ATP MEASUREMENTS. ATP measurements of the samples collected at the surface of the filters were taken in femtograms (10^{15}g) per gram of filtration medium.

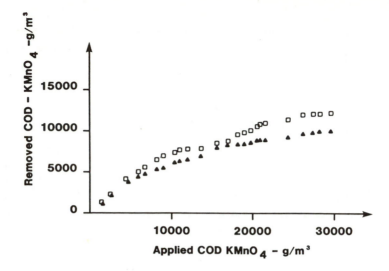

Figure 9. Cumulative plots for COD removal. Key: □, ozonated rapid PKST filter 3 and ▲, ozonated slow PKST filter 10.

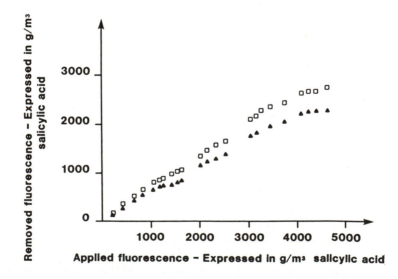

Figure 10. Cumulative plots for fluorescence reduction. Key: □, ozonated rapid PKST filter 3 and ▲, ozonated slow PKST filter 10.

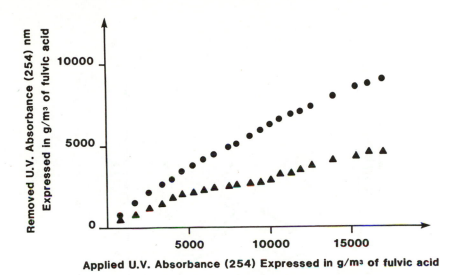

Figure 11. Cumulative plots for UV reduction. Key: ●, *slow PKST filter 7 and* ▲, *ozonated slow PKST filter 10.*

Table VI. Mean Geometric Total Viable Cell Counts (20°C) within the Filter Beds (cells per cm³ of media)

Filter	Sampling Point	*Filter Bed Depths*					
		Surface	15 cm	15 cm	20 cm	50 cm	75 cm
Slow sand	9	sand 2×10^6	sand —	sand —	sand 2.8×10^5	sand —	sand —
Slow PKST	7	PKST 3.3×10^6	PKST 1.4×10^6	sand $8. \times 10^5$	sand 2.8×10^5	sand 4.2×10^4	sand —
Ozonated slow sand	12	sand 2.5×10^6	sand —	sand —	sand 7.2×10^5	sand —	sand —
Ozonated slow PKST	10	PKST 9.3×10^6	PKST 5.2×10^6	sand 3.7×10^5	sand 3.6×10^5	sand 1.4×10^5	sand —
Ozonated rapid PKST	3	PKST 9.1×10^6	PKST —	PKST —	PKST —	PKST —	PKST 3.1×10^6

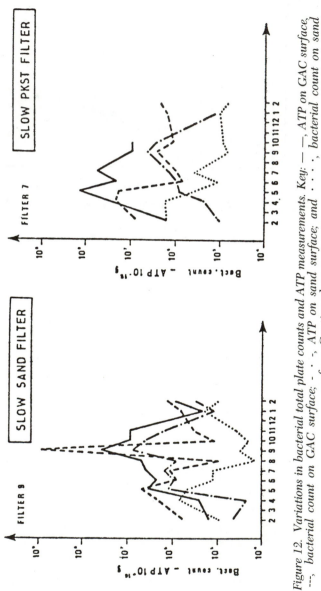

Figure 12. Variations in bacterial total plate counts and ATP measurements. Key: — —, ATP on GAC surface; - -, bacterial count on GAC surface; - · -, ATP on sand surface; and · · · ·, bacterial count on sand surface. Continued on next page.

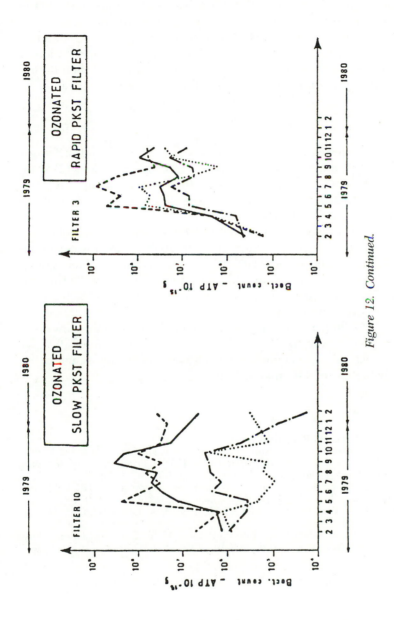

Figure 12. Continued.

ATP concentrations are higher in GAC than in sand (10-fold). They seem slightly higher after pre-ozonation and after rapid filtration.

Figure 12 also represents the total bacterial counts and clearly indicates that there was correlation with ATP concentrations.

Respirometric measurements also were performed but resulted in somewhat erratic values. Bacteria are present in the GAC beds in larger numbers than in the sand but no evidence could be drawn of their higher efficiency in the carbon than in the sand, for the removal of organics.

Conclusions

The Mont Valerien experiment also was used as a first industrial test for Benedek's (6) new adsorption predictive model. Predictions (2) were in good agreement with measurements, thus substantiating the slow adsorption theory. This finding brings further evidence that GAC does not significantly stimulate biodegradation.

After a year of experiment at Mont Valerian, the general following conclusions can be drawn. Slow sand filters treating clarified water in the second stage remove an average 10% TOC primarily through biodegradation. Slow GAC filters remove an additional 15% TOC primarily through adsorption. The total reduction is thus 25%. Ozone removes 5% but does not enhance GAC performance.

Rapid GAC filters remove at least 10% more TOC than the slow ones. The total reduction with the best carbon reached an average of 40 % for the whole year of experiment. Biodegradation does not seem to be significantly higher in GAC than in sand.

Literature Cited

1. Fiessinger, F. *Aqua* **1980,** 9, 199.
2. Mallevialle, J. "Ozonation des Substances de Type Humique dans les Eaux." Second I.O.A. Symposium, Montréal (May 1975). *Proceedings of the 2nd International Symposium on Ozone Technology,* IOA, 47 1976.
3. McGuire, M.; Suffet, I. H.; Radziul, J. V. *J. Am Water Works Assoc.* **1978,** 70, 565.
4. Wilson, E. J.; Geankoplis, C. J. *In. Eng. Chem. Fund. 1966,* 5(1), 9.
5. Crittenden, J. C.; Weber W. J. *ASCE J1 Environ. Eng. Div.* **1978a,** 104(EE2), 185.
6. Benedek, A. Slow Adsorption Phenomenon. ACS. Environ. Chem. Div. Symposium on Activated Carbon, Atlanta, (March 1981).

RECEIVED for review August 19, 1981. ACCEPTED for publication July 1, 1982.

15

Microbial Population Dynamics on Granular Activated Carbon Used for Treating Surface Impounded Groundwater

R. M. DONLAN and T. L. YOHE

Philadelphia Suburban Water Company, Bryn Mawr, PA 19010

A series of granular activated-carbon (GAC) columns was used to evaluate the microbiological aspects of GAC. Pathogen retention and growth on GAC, types and numbers of microorganisms developing on GAC, and microbial sloughing from GAC into effluents were three areas investigated. Predominant bacteria isolated were similar to previous studies. It was also found that, once established, the numbers of microorganisms in influents and effluents are similar until environmental changes cause fluctuations in the steady-state.

D URING THE LAST FEW YEARS, an increasing number of groundwaters have been found to be contaminated with organic chemicals. One affected water is Philadelphia Suburban Water Company's Upper Merion Reservoir, located in King of Prussia, Pa.

Upper Merion is a groundwater-fed reservoir, and it is an important source of water for Philadelphia Suburban Water, one of the country's largest investor owned water utilities, supplying water to approximately 850,000 people. Although the source is groundwater, the impoundment takes on some characteristics of a surface supply. Treatment is therefore more extensive than would be the case with "normal" groundwater supply.

In the spring of 1979, low levels of trichloroethylene (TCE) were detected in this reservoir by both the EPA and our own laboratory. It was mutually agreed that daily testing should be conducted while laboratory and pilot column studies were performed to find the most cost effective way to remove TCE and other organics from the reservoir water. Both granular activated carbon (GAC) and aeration were included in the study.

The GAC pilot columns were constructed of glass, stainless steel, and Teflon. They were located in the basement of the treatment building. The feed to these columns was untreated water from the reservoir. The design parameters used for this study are shown in Table I. Three carbon types were evaluated with one column serving as a replicate. The glass columns all had 10-cm (4-in.) inside diameters and were charged to a bed depth of 91 cm (36 in.). The empty bed contact time for this 6-month study was 5 min. During the study, approximately 50,000 bed volumes passed through the columns. Column 4 was taken off-line after 142 days to develop postrun analytical methods. Column 3 was left on-line an additional 4 months to verify data anomalies. Table II provides the actual run parameters of this study.

Included in the GAC pilot column study were provisions to evaluate the microbiological aspects of GAC. Specifically, three areas were pursued in these microbiological investigations:

1. Pathogen retention and growth on GAC.
2. Types and numbers of microorganisms developing on GAC.
3. Microbial sloughing from GAC columns into effluents.

Experimental

Sampling Procedures.   All influent and effluent samples for microbiological analysis were obtained in sterile glass bottles. Of 260 total plate count samples, 42 replicates were collected. These replicates were used to determine confidence intervals for the observed results. Similar replication also was conducted for total coliform and *Pseudomonas aeruginosa* determinations. Samples were transported immediately to the laboratory where they were refrigerated and analyzed within 24 h. At the completion of the pilot study, GAC columns were disassembled and core samples of each column were taken. Each 91-cm core was divided into six sections. The top 15-cm section was subdivided into a 10-cm lower portion and 5-cm upper portion. Each core section was sampled using a sterile spatula, and samples were placed into sterile "whirl-pac" bags. Samples were then transported to the laboratory, refrigerated, and tested within 6 h.

Table I. Pilot Column Design

Parameter	Specification
Carbon types	WV-G(2)
	F-400
	ICI HD-1030
Diameter	10 cm (4 in.)
Bed depth	94 cm (37 in.)
EBCT	5 min
Run time	6 months
Bed volumes	50,000

Table II. Pilot Study Run Parameters

Parameter	Col. 1 WV-G	Col. 2 WV-G	Col. 3[a] F-400	Col. 4[b] HD-1030
Run length (days)	176	176	300	142
Bed volumes treated	50,100	49,600	85,000	39,900
Liters	371,000	368,000	630,000	296,000
Empty bed contact time (min)	5.0	5.1	5.1	5.1
Surface loading (gpm/sq. ft.)	4.5	4.4	4.4	4.4
Head loss (feet of water)	50	55	56	16
pH	influent average 7.48, range 7.20–7.65			
average	7.52	7.53	7.54	7.47
range	7.2–8.8	7.2–8.8	7.2–8.8	7.2–7.7
Alkalinity (as $CaCO_3$)	influent average 156, range 152–166			
average	161	161	161	160
range	155–177	155–172	156–175	155–166
Dissolved oxygen (mg/L)	influent average 4.0, range 0.5–8.2			
average	3.4	3.3	3.3	4.2
range	0.1–7.5	0.2–7.5	0.1–7.6	0.2–7.7
Turbidity FTU units	influent average 0.7, range 0.2–1.18			
average	0.46	0.47	0.46	0.50
range	0.27–0.60	0.17–0.60	0.20–0.62	0.20–0.74
Total organic carbon (mg/L)	influent average 0.67, range 0.5–1.24			
average	0.45	0.43	0.42	0.48
range	0.2–1.18	0.18–1.33	0.22–0.98	0.24–0.84
Temperature °C	influent average 13.0, range 9.5–21.0			

[a]Chemical values reported only for 176-day period.
[b]All values represent only 142 days of sampling.

Influent and Effluent Analyses. For total plate count analyses, samples were diluted in sterile buffered water. Two plates of standard plate count agar (Difco Laboratories, Detroit, MI) were planted for each dilution. The pour plate procedure was used. Plates were incubated at room temperature for either 6 or 15 days.

Colonies of predominant organisms on total plate counts were isolated on standard plate count agar (spc). Isolates were identified according to the methods of Shayegani et al. (1). Other sources also were consulted to aid in identification (2–15).

Total coliforms were enumerated with the membrane filter procedure (6), and verification was performed on both typical and atypical colonies (7). Verification results indicated that only typical (green sheen forming) colonies were coliforms.

P. aeruginosa also was enumerated with the membrane filter procedure (6). All colonies were verified on milk agar (6).

Two media were tested for their efficacy in enumerating total enteric bacteria: MacConkey's agar (Difco) and Tergitol-7 agar (Difco) supplemented with 1% triphenyltetrazolium chloride. For each, 10 mL of sample was filtered through a 0.45-μm filter which was then placed on the agar medium. MacConkey's agar plates were incubated at 35°C for 24 h; Tergitol-7 agar plates were incubated at room temperature for 48 h. After incubation, all colony types were counted. Verification of colonies on both media was performed by isolating colonies on standard plate count (spc) agar and identifying through the API-20E system (Analytab Products, Plainview, NY).

Enteric pathogens were enumerated according to the procedures of the "Microbiological Methods for Monitoring the Environment" (7). This procedure is shown in Figure 1. Two liters of each sample was filtered through a 0.45-μm membrane filter. The filter was then divided in half and each half placed into a separate flask of selenite broth (Difco). One flask was incubated at 35°C for 4 days,

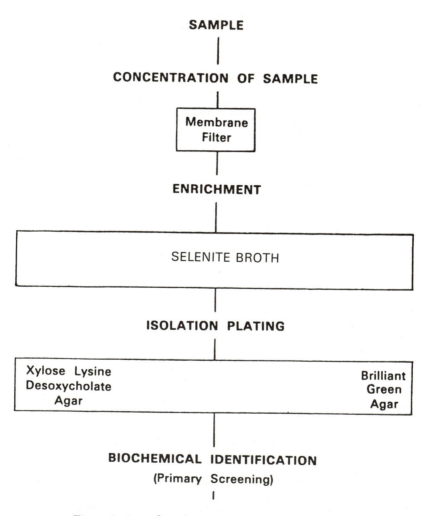

Figure 1. Procedure for enumerating enteric pathogens.

and one was incubated at 41.5°C for 4 days. After 48 h and again after 4 days, a subculture from each selenite broth was streaked onto two xylose lysine desoxycholate (XLD) agar plates (Difco) and two brilliant green lactose bile (BGLB) agar plates (Difco). One XLD plate and one BGLB plate were incubated at 35°C, and one XLD plate and one BGLB plate were incubated for at 41.5°C. BGLB plates were incubated for 48 h, and XLD plates were incubated for 24 h. Suspect colonies were then picked, isolated on spc agar, and identified through the API-20E System.

Core Analyses. Total plate counts of cores were performed by placing a 0.5-g core sample into a sterile Pyrex tissue grinder, adding 1.0 mL of sterile buffered water, and grinding to a fine consistency. To this was added 4.0 mL of sterile buffered water. The diluted sample was then drawn out of the tissue grinder with a pipet and placed into a sterile tube. The subsample, now a 1:10 dilution of the original sample, was mixed on a Thermolyne maxi-mix for 5 min. The subsample was then diluted out in sterile buffered water and planted on agar in duplicate. Plates were incubated for 6 days and counted. Predominant isolates on total plate counts were identified as described previously.

Biological activity (CO_2 production) was determined by a modification of the technique of Stotzky (8) as shown in Figure 2. Carbon dioxide produced by the sample is drawn through the 0.05 N NaOH forming sodium carbonate. The sodium carbonate formed is treated with barium chloride to form barium carbonate. Excess sodium hydroxide present is titrated with 0.05 N HCl with phenolphthalein as the indicator. From the amount of sodium carbonate formed, the amount of CO_2 produced was calculated.

Sample flasks and bubblers were acid washed, rinsed in deionized water, and sterilized. A 50-g core sample plus 150 mL of filter sterilized influent was then added to the flask and mixed. As a negative control, two flasks prepared as above were subsequently sterilized for 30 min in an autoclave. Carbon dioxide collectors and bubblers were acid washed and rinsed as above. Then 15 mL of 0.05 N NaOH was added to the collector.

The flask and collector were then attached to the aeration system. The pressure regulator was set at 50 lb/in.², and needle valves were adjusted to obtain a continuous flow. The flow rate was ultimately adjusted by the pinch clamp so that all flasks had similar rates. Flasks were then purged for 30 min, collectors were changed, and flow rates were readjusted. The flasks were then covered with foil to block out light. The experiment was carried out until a trend of CO_2 production was established, with collectors being changed every 24 or 48 h. At the completion of the experiment, the solution in the collector was poured into a beaker. To this was added 5 mL of 2 N barium chloride and 2 drops of phenolphthalein. The solution was titrated using 0.5 N HCl. At least one blank of 0.05 N NaOH was run with each set of titrations. Both the hydrochloric acid and sodium hydroxide solutions were standardized against a phthalic acid solution.

Results

Pathogen Retention and Growth on GAC. Table III shows the predominant bacteria isolated from influent source water, GAC column effluents, and GAC cores. A total of 64 isolates was identified. Several have previously been isolated from clinical specimens (2, 3). None of the isolates was in the family Enterobacteriaceae.

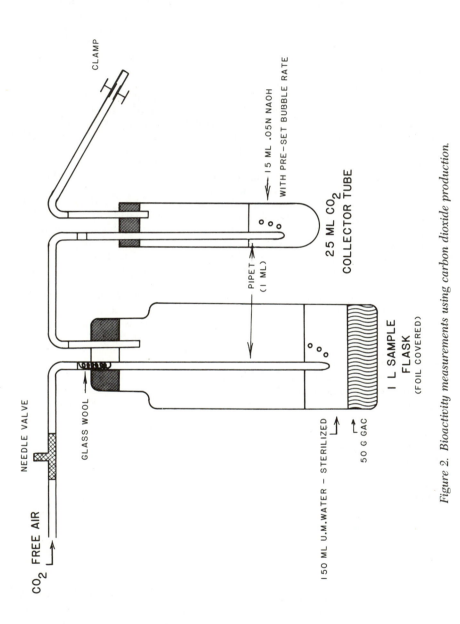

Figure 2. Bioactivity measurements using carbon dioxide production.

Table III. Predominant Isolates by Location

Influent	Effluent	GAC Bed
Pseudomonas acidovorans	Pseudomonas cepacia	Group Ve Biotype 1
Group IIk Biotype 1	P. putida	Group Ve Biotype 2
Group Ve Biotype 1	Group IIk Biotype 1	Group IIk Biotype 1
Acinetobacter	Group Ve Biotype 2	Alcaligenes denitrificans
calcoaceticus var.	Group IIf	Achromobacter xylosidans
LWOFFI	Acinetobacter	Biotype IIIa
Alcaligenes denitrificans	calcoaceticus var.	Bacillus cereus
Bacillus spp.	LWOFFI	Bacillus spp.
Chromobacterium spp.	Micrococcus spp.	Micrococcus spp.
Actinomycete	Actinomycete	fungus
	fungus	

The results from the total coliform enumerations are shown in Table IV. The mean numbers of total coliforms in the column effluents were not higher than the number in the influent water. The 95% confidence interval per determination was within −1 to +2 of the measured value.

No pathogenic enteric bacteria, i.e., *Salmonella spp.* or *Shigella spp.*, were isolated from 2-L influent or column effluent samples from 10 determinations. Two selective media were tested for their efficacy in selectively enumerating total enteric bacteria. These media were MacConkey's agar and Tergitol-7 agar. Both MacConkey's agar (9) and Tergitol-7 agar (10) have been used by other investigators for specifically enumerating enteric bacteria in water. Verification of colonies on both influent and effluent plates indicated that both MacConkey's and Tergitol-7 agars were not selective enough and cultured nonenteric bacteria. Hence, the number of total enteric bacteria present in influent and effluents was not determined.

Table V shows that *P. aeruginosa* was rarely present in either influent or column effluent samples. When present, its numbers were low. In addition, *P. aeruginosa* was not among the predominant isolates on the column core samples (Table III). This finding shows that *P. aeruginosa* would not be a problem if GAC were used to treat this water.

Table IV. Total Coliform Counts

Sample	Average	Range	n
Influent	8.1	0–33	48
Column 1	5.6	0–26	48
Column 2	4.5	0–23	48
Column 3	4.8	0–20	48
Column 4	5.2	0–25	41

Note: 95% C.I. per determination ($X - 1$ to $X + 2$).

Table V. *Pseudomonas aeruginosa*

Number of Samples	Number Positive	Average Number/100 mL Positive
175	9	2

Types and Numbers of Microorganisms on GAC Columns. The types of microorganisms predominant on GAC columns and column effluents were similar to the types predominating in the influent water (Table III). Most bacterial isolates may be placed into one of two broad groups: gram negative rods which are unable to ferment glucose and gram positive rods which form spores.

The first grouping includes members of the genera *Pseudomonas*, *Acinetobacter*, *Alcaligenes*, *Achromobacter*, Group *IIk*, Group *Ve*, and Group *IIf*. Most of these organisms normally are found in soil and water, although many also have been isolated from clinical specimens (2, 3). The second grouping is the genus *Bacillus*. Because of its ability to form resistant endospores, this genus may be isolated from numerous sources (2) and is common in soil and water. The remainder of the bacterial isolates, *Chromobacterium*, *Micrococcus*, and the actinomycetes, also are commonly found in water.

Table VI compares the mean total plate counts for influent, effluent, and columns. The mean total plate counts of GAC column sections averaged approximately 6.8×10^5 microorganisms/g carbon. The mean total plate count of influent water was 1.3×10^3/mL of water. This represents greater than a 2 log increase in number indicating that sizeable microbial populations developed on GAC. Column 3, which showed a higher mean total plate count within the GAC bed, was left on-line an additional 4 months after termination of the sampling program. Figure 3 shows the total plate count versus column depth for Columns 1, 2, and 4. Column 3, which was taken off-line at a later date, showed similar trends. Counts for all four columns were always highest in the top 15 cm of the column.

The total plate count procedure does not enumerate all microorganisms present on GAC. The direct microscopic procedure was attempted to determine the efficiency of the total plate count procedure. Unfortunately, results of this test were inconclusive, but it is a promising area for future research.

The biological metabolic activity of core sections from Columns 1, 2, and 3 is shown in Figures 4, 5, and 6, respectively. All core sections tested were significantly higher in CO_2 production than negative controls. This result implies that biological populations are actively metabolizing on the GAC columns.

Table VI. Plate Counts by Location

Sample	CFU[a]/mL Average	Range	n
Influent	1274	63–2550	45
Effluents			
Column 1	1110	65–4310	45
Column 2	1103	59–3950	45
Column 3	1193	48–5940	45
Column 4	1162	27–2665	38
Cores 10^5 CFU/g			
Column 1	6.27	1.61–18.8	7
Column 2	5.95	1.18–8.67	7
Column 3	8.73	6.13–12.4	7
Column 4	6.28	1.26–14.1	7

Note: 95% C.I. for influent/effluent determinations ($X - 216$ to $X + 56$).
[a]Colony forming units.

To test the ability of influent organisms to grow in a low nutrient environment, such as is found on GAC, influent water was filter sterilized and inoculated with a seed of influent water. As a negative control, filter sterilized Milli-Q water (Milli-Q System, Millipore Corp., Bedford, MA) was inoculated with an influent seed. The positive control contained filter sterilized influent amended with 10 ppm glucose plus added influent seed. As can be seen in Figure 7, all flasks showed a logarithmic increase in total plate count. The positive control increased 1 log higher than the other samples initially and then decreased. Counts were higher in the negative control than in the unamended samples. These results show that these organisms can grow in environments with very low nutrient concentrations.

It has been shown previously (*11*) that total plate counts of GAC effluents may be increased by extending the incubation time. For this reason, a 6-day incubation period was used for all total plate count analyses. To investigate the possibility that counts would continue to increase with an even longer incubation, plates were counted after both 6 and 15 days of incubation. The percent increase in total plate count when a 15-day incubation was used is shown in Table VII. The mean increase in total plate count of influent and column effluents was between 50% and 71%. This result indicates that an even longer incubation time may be necessary to enumerate the microbial populations associated with GAC more accurately.

Microbial Sloughing from GAC Columns into Effluents. Figure 8 shows influent and column effluent total plate counts over a 100-day

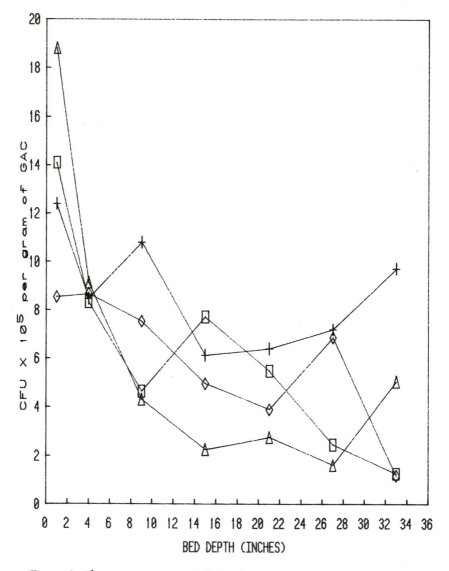

Figure 3. Plase counts versus bed depth. Key: △, Column 1 (WV-G, 176 days); ◇, Column 2(WV-G, 176 days); +, Column 3 (F-400, 300 days); and □, Column 4 (HD 10×30, 142 days).

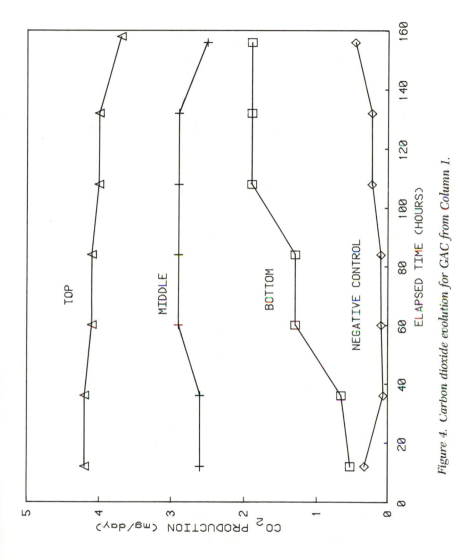

Figure 4. Carbon dioxide evolution for GAC from Column 1.

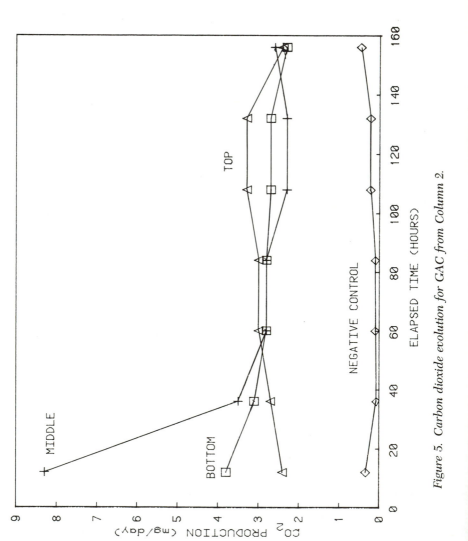

Figure 5. Carbon dioxide evolution for GAC from Column 2.

COLUMN 3   ACTIVITY

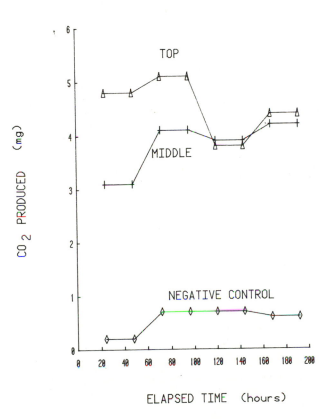

Figure 6. Carbon dioxide evolution from Column 3.

period of the study from replicate GAC Columns 1 and 2. After an initial 20-day period when effluent counts were significantly higher than influent counts, the effluent levels were generally equal to or below the influent. Occasionally, certain column effluent counts would increase significantly above the influent, but the mean influent count for the entire study was higher than each of the four mean column effluent counts (Table VI). This study shows that large numbers of microorganisms (as compared to the influent) are not being sloughed off the GAC columns.

Testing was conducted to determine whether hydraulic fluctuations would affect effluent bacterial water quality. With flow variations of approximately ± 15%, no significant increase in effluent bacterial counts was observed.

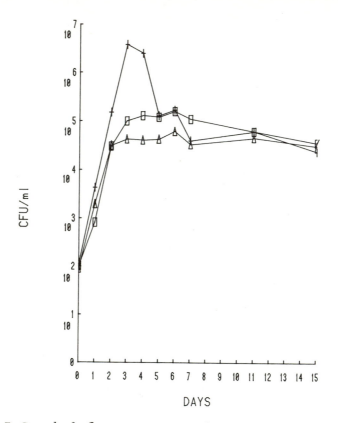

Figure 7. Growth of influent organisms on influent water. Key: □, DI water; △, average influent; and +, influent and glucose.

Table VII. Percent Increase in Total Plate Counts When 15-Day Incubation Was Used

Day	Influent (%)	Column 1 Effluent (%)	Column 2 Effluent (%)	Column 3 Effluent (%)	Column 4 Effluent (%)
107	11.1	3.74	5.03	1.83	8.40
112[a]	23.3	30.0	13.9	29.9	ND
114	204	6.29	19.1	24.9	7.35
119[b]	82.9	161	161	205	154
128[c]	16.4	77.7	126	126	100
133	81.0	93.9	93.5	124	87.8
140	6.60	ND	1.22	ND	1.32
142	9.67	7.69	2.59	5.82	3.28
154[b]	15.3	76.9	37.5	48.4	ND
$\bar{X}$	50.0	57.2	51.1	70.7	51.7

[a]14-day total incubation.
[b]16-day total incubation.
[c]7-day initial incubation.

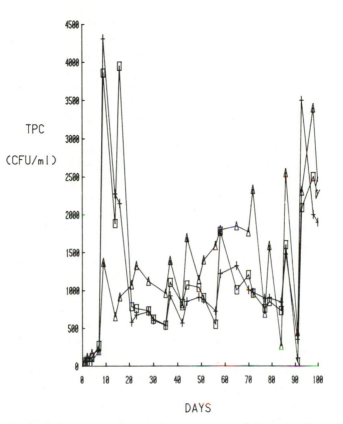

TPC

(CFU/ml)

DAYS

Figure 8. Total plate counts for 100 days. Key: △, *influent;* +, *Column 1; and* □, *Column 2.*

Discussion

The predominant bacteria isolated from influent source water, GAC column effluents, and GAC column cores (Table III) are similar to the types found by other investigators on GAC column cores and column effluents. Miller et al. (*12*) cited a study which showed that the predominant bacterial species on carbon columns were nonpathogenic organisms, found naturally occurring in soils and waters. McElhaney and McKeon (*11*) had similar results, but they noted that some isolates could become opportunistic pathogens. In this light, it is important to note that "almost any bacterial species can produce disease in humans" (*3*). An increasing number of species of glucose nonfermenting gram negative rods are being isolated from clinical specimens. These organisms are often opportunistic pathogens, infecting the already debilitated host (*3*). But, in spite of this, the numbers of bacteria in the effluents are similar to the numbers in the influent which indicates that the bacteriological quality with regard to pathogenic organisms, including the opportunistic pathogens, is equivalent in influent and column effluents. We found the numbers of total coliforms in the effluents to be slightly lower than the numbers in the influent (Table IV). Fiore and Babineau (13) reported similar results with GAC filtered well water. They found that total coliforms did not colonize the filter system and grow to levels above that found in the influent water. Brewer and Carmichael (9) found that levels of both total coliforms and total enterics were similar or lower in effluents as compared to influents. Very little data have been collected on the survival of pathogens or coliform indicator bacteria on GAC columns. These data demonstrate that coliform organisms are not thriving on GAC columns; passage of this source water through GAC does not increase their numbers.

Enteric pathogens, e.g., *Salmonella spp.* and *Shigella spp.*, were not detected in influent or column effluents. This result is to be expected as the source water was surface impounded groundwater with little surface run off.

We were unsuccessful in enumerating the total enteric bacterial population. Both MacConkey's agar (9) and Tergitol-7 agar (10) were found to cultivate nonenteric organisms from influent and effluents. Our results demonstrate that these two media are not suitable for specifically enumerating enteric populations.

P. aeruginosa, unlike the enteric organisms, is an aquatic bacterium, readily isolated from surface waters (14). *P. aeruginosa* infections are a major threat to compromised hosts, especially burn patients (15). Since water may serve as a vector for this organism, its multiplication on GAC and sloughing into the GAC effluents are important. We are not aware of other studies which have attempted to enumerate specifically *P. aeruginosa* on GAC. Our results show that, at least for surface impounded groundwater, it does not multiply in large numbers on GAC.

The types of bacteria isolated from GAC columns, influent, and column effluents (Table III) are representative of the types normally found in surface and groundwater (16, 17). These types have also been isolated from GAC cores and column effluents (11, 12).

Significant populations of microorganisms did develop on the GAC columns, as evidenced in Table VI. This finding confirms the data of other investigators (11, 12, 18) that bacteria will colonize GAC columns and develop significant populations. Our data show that populations were highest in the top 15 cm of the columns.

Biological activity per se can only be measured in metabolic terms, e.g., oxygen uptake, carbon dioxide evolution, enzyme activity, or heat evolution (8). For this reason, biological activity was estimated by measuring total CO_2 evolution from GAC cores. The data demonstrate that populations were metabolically active.

A growth experiment was performed to determine growth rates of influent organisms on sterilized influent source water. Results imply that these microorganisms can, in fact, grow in the very low nutrient source water. Van Der Kooij et al. (*19*) showed that *Aeromonas hydrophila* could grow in tap water when the glucose concentration was as low as 17 $\mu g/L$. Favero et al. (*20*) found that *P. aeruginosa* could grow to relatively high numbers rapidly in distilled water.

A 6-day incubation time was used for all total plate count analyses. Because of data from other authors (*11, 12*), an experiment was attempted to determine maximal incubation time. The data (Table VII) demonstrate that a 15-day incubation will yield higher counts than a shorter incubation. An even longer incubation time might yield even higher counts. Different incubation conditions using different media must be investigated and compared to direct microscopic counts to determine the most efficient means of estimating biomass on GAC.

This investigation was different from other studies (*9, 11*) that used disinfected water as the feed to GAC columns. In such studies, influent total plate count populations were generally quite low, and a major concern was whether bacterial regrowth would occur on GAC columns and cause higher numbers of bacteria to slough into the effluent water. The untreated source water for this study contained much higher bacterial populations than in other studies that used chlorinated source water. Our concern was whether bacteria might grow on GAC and slough off in numbers significantly higher than in the influent, placing an increased loading on further stages of treatment. The data show that, over the 176-day period of the study, the mean total plate counts were actually higher in the influent than in column effluents (Table VI). We did notice a sharp rise in effluent counts initially (Figure 8) and occasionally thereafter noticed a trend similar to the results of McElhaney and McKeon (*11*). It appears that initially the effluent counts increase because the "pseudo-steady-state" (*18*) has not yet been established. Once established, numbers in the influent and effluents are similar until environmental changes cause fluctuations in the steady-state. In addition, changes in flow rate through Column 3 have no noticeable effect on effluent bacterial levels. We conclude from these data that bacterial sloughing is negligible based on the fact that effluent numbers of bacteria are normally lower than the numbers in the influent water.

Acknowledgments

This study was conducted by the Philadelphia Suburban Water Company. The authors wish to acknowledge Kenneth E. Shull, Vice

President of Treatment, Quality Control and Research, for his leadership and support of this work. We are also indebted to Edward Crist, Robert Plowman, and Gene Guthan for their assistance in various aspects of the study. Thanks also to Dr. I. H. Suffet and the Drexel University staff for assisting in the preparation of the figures for the manuscript.

Literature Cited

1. Shayegani, M. et al. *Health Lab. Sci.* **1977,** *14,* 83.
2. "Bergey's Manual of Determinative Bacteriology," 8th ed., Williams and Wilkins, Baltimore, Md., 1974.
3. "Manual of Clinical Microbiology," 2nd ed., American Society for Microbiology, Washington, D.C., 1974.
4. "Anaerobic Laboratory Manual," 3rd ed., Virginia Polytechnic Institute and State University Anaerobe Laboratory, Blacksburg, Va., 1975.
5. "Difco Manual," 9th ed., Difco Laboratories. Inc., Detroit, Mich.. 1953.
6. "Standard Methods for the Examination of Water and Wastewater," 14th ed., American Public Health Association, Washington, D.C., 1976.
7. "Microbiological Methods for Monitoring the Environment—Water and Wastes," Environmental Monitoring and Support Laboratory, Office of Research and Development, U.S. EPA, Cincinnati, OH. 1978.
8. "Methods for Studying the Ecology of Soil Microorganisms, IBP Handbook No. 19," International Biological Programme, Blackwell Scientific Publications, London, England, 1971.
9. Brewer, W. S.; Carmichael, W.W. *J. Am. Water Works Assoc.* **1979,** *71,* 738.
10. Evans, T. M. et al. *Appl. Environ. Microbiol.* **1978,** *35,* 376.
11. McElhaney, J. B.; McKeon, W. R. "Enumeration and Identification of Bacterial Populations on GAC," In Proceedings of the 1978 American Water Works Association Water Quality Technology Conference, Louisville, Ky., 1978.
12. Miller, G. W. et al. In "Activated Carbon Adsorption in the Aqueous Phase," McGuire, M. J., and Suffet, I. H., Eds., Ann Arbor Science, Ann Arbor, Mich., 1980, Volume 2, pp. 323–47.
13. Fiore, J. V.; Babineau, R. A. *Appl. Environ. Microbiol.* **1977,** *34,* 541.
14. Hoadley, A. W. In "*Pseudomonas aeruginosa:* Ecological Aspects and Patient Colonization," Young, V. M. Ed. Raven Press, New York, N.Y., 1977, pp. 31–57.
15. Holder, I. A. In "*Pseudomonas aeruginoas:* Ecological Aspects and Patient Colonization," Young, V. M. Ed., Raven Press, New York, N.Y., 1977, pp. 77–95.
16. Lamka, K. G. et al. *Appl. Environ. Microbiol.* **1980,** *39,* 734.
17. Geldreich, E. E. et al. *J. Am. Water Works Assoc.* **1972,** *64,* 596.
18. Benedek, A. In "Activated Carbon Adsorption in the Aqueous Phase," McGuire, M. J.; Suffet, I. H. Eds., Ann Arbor Science, Ann Arbor, Mich., 1980, Volume 2, pp. 273–302.
19. Van Der Kooij, D. et al. *Appl. Environ. Microbiol.* **1980,** *39,* 1198.
20. Favero, M. S. et al. *Science* **1971,** *173,* 836.
21. Safe Drinking Water Committee of the National Research Council, "Drinking Water and Health," National Academy of Science, Washington, D.C., 1980, Volume 2.

RECEIVED for review August 3, 1981. ACCEPTED for publication March 29, 1982.

Mathematical Description of the Microbial Activity in a Batch System of Activated Carbon, Phenol, and *Arthrobacter* Strain 381

JOHN G. DEN BLANKEN

Delft University of Technology, Department of Civil Engineering, Laboratory of Sanitary Engineering, Stevinweg 4, 2628 CN Delft, The Netherlands

The batch system consisted of Norit powdered activated carbon (PAC), phenol, and Arthrobacter *strain 381 (A. 381) in a recording constant-pressure respirometer at 20°C and pH 7.5–8.0. The consumed phenol was oxidized completely by A. 381 to CO_2, water, and biomass. The yield factor (Y) was 2.24×10^8 bacteria/mg phenol and the maximum growth constant (μ_{max}) was 0.024 h^{-1} for attached bacteria, while $K_s = 2.4 \times 10^{-4}$ mg phenol/L at a phenol concentration of 4.7 mg/g PAC. When the phenol concentration was higher than 10 mg/g PAC, the phenol carbon consumption regeneration velocity was 0.010 g/h by 10^{11} attached bacteria. Chemical regeneration of PAC was not important when phenol was the only adsorbed substrate under the described experimental conditions.*

D URING THE LAST 10 YEARS, attention has been given to the microbial activity in granular activated-carbon (GAC) filters to determine its contribution to the removal of organics and water quality (*1–3*). A good practical measure for the microbial activity in steady-state GAC filters in water treatment is oxygen consumption (ΔO_2). With known conversion factors, one can estimate how much organic carbon has been degraded to CO_2 (*1*).

This study collected data about the adsorption of a specific type of bacteria and organic compound(s) on activated carbon and measurement of the bacteriological activity in a model batch system (ground Norit

0065-2393/83/0202-0355$06.00/0

RBW 0.8 Extra-*Arthrobacter* strain 381-phenol) and describes the different processes with simple, mathematical models.

Experimental

The studied processes are presented in Figure 1. The oxygen consumption of the different processes also was studied. Because so many simultaneous processes take place in a filter during water treatment, it is practically impossible to study each process separately; a mathematical model was chosen. Processes 1–4 (Figure

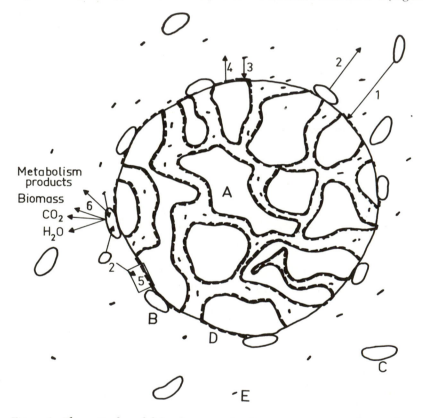

Figure 1. Theoretical model for the system in the respirometer vessels: 1 is the adsorption of bacteria, 2 is the desorption of bacteria, 3 is the adsorption of a phenol molecule, 4 is the desorption of a phenol molecule, 5 is chemical oxidation of phenol, 6 is endogenous respiration, 7 is bacterial degradation of phenol, and 8 is growth of bacteria. Key: A, particle of activated carbon; B, adsorbed bacterium; C, free bacterium; D, adsorbed phenol molecules; and E, free phenol molecule.

1) were studied by determination of adsorption isotherms, and Processes 5–7 were studied by measurement and calculation of the oxygen consumption that takes place as a result of these processes. Oxygen consumptions were measured with a recording constant-pressure respirometer (Figure 2) with large vessels. The vessels are immersed in a constant-temperature water bath and are provided with magnetic stirrers. When oxygen is consumed, a lower pressure develops and is measured by a low-range pressure transducer. Pure oxygen is then supplied to the vessels by burettes. Carbon dioxide produced by the bacteria is adsorbed in a small vessel containing concentrated KOH solution (Figure 3). This apparatus was developed in the laboratory.

The oxygen consumption by Process 7 is equal to the difference between the total oxygen consumption minus the oxygen consumption by the nonbacteriological processes and the endogenous respiration in the respirometer vessels.

Ground Norit RBW 0.8 Extra (particle size 74–104 μm) was used as adsorbent, and phenol was used as the carbon source. Phenol was very soluble in water, simple to analyze, adsorbed well on activated carbon, and was biodegradable (3). In GAC filters, the number of phenol oxidizers was larger than $5.8 \times 10^3/$ cm^3 GAC (1). KNO_3 was added as nitrogen source in a ratio phenol/ $KNO_3 = 1:1.28$.

Arthrobacter globiformis strain 381 (A.381) was the phenol degrading bacterium, which is a typical representative of the *Coryneform* group (4). This

Figure 2. Front of the respirometer

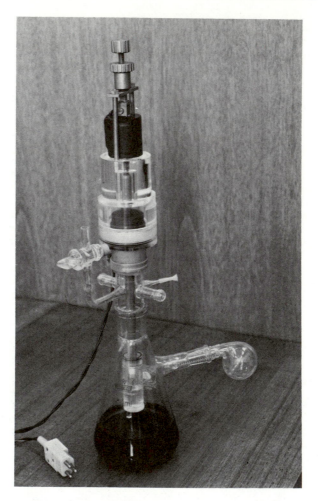

Figure 3. Respirometer vessel with circulation pump.

group of bacteria was present in large numbers in GAC filters in water treatment
(*1*) and utilized phenol at high concentrations in the medium.

Because adsorption and bacteriological processes are pH and temperature
dependent, these parameters were kept at 7.5–8.0, with 0.4 *M* Tris(hydroxy-
methylaminomethane) buffer, and 20°C, respectively. This pH corresponds with
that of drinking water. In a respirometer vessel with 300 cm³ solution and about 2
g ground carbon, the pH is kept reasonably constant by addition of 10 cm³ 0.4 *M*
Tris solution and 16.2 cm³ 0.2 *M* HCl. The presence of this buffer had no effect on
the growth and phenol degradation of A.381.

Activated carbon was always wet sterilized (15 min at 121°C), and the contents of the vessels were stirred continuously at 475 rpm for a good gas exchange.

The bacterial suspensions of A.381 were cultured in yeast extract (Oxoid L21, 1.0 g/L) and phenol solution (pH 7.2), shaken during 24 h at 20°C, filtered, washed three times, and centrifuged with ¼-strength Ringer solution. Bacterial growth was determined by counting the colony forming bacteria after 2–5 days at 25°C on yeast extract (7.0 g/L)–bactodextrose (10.0 g/L) agar.

All glassware was cleaned with hot (80°C) soap solution (Extran, liquid 2%), and then rinsed with tap water and at least three times with demineralized water.

It was necessary to determine what part of the dosed phenol had remained at the end of the experiments. The quantity of phenol remaining in the liquid was determined directly with a gas chromatograph, while the activated carbon was extracted in a Soxhlet with chloroform for 22 h at 75°C.

Mathematical Model for Oxygen Consumption in Respirometer Vessels

During the experiments in the respirometer vessels, several processes consume oxygen: interaction of oxygen with the carbon surface, chemical oxidation of free and/or adsorbed phenol, endogenous respiration by free and/or attached bacteria, and consumption of phenol by attached bacteria. The total theoretical oxygen consumption is equal to the sum of the consumptions of these processes, so that:

$$-\left(\frac{dO_2}{dt}\right)T = \left(\frac{dO_2}{dt}\right)ad + 0.17[u(\mathrm{Ph}) + u(a) - v_2(\mathrm{H}) - v_2(h)]$$

$$+ eB + \left(\frac{dO_2}{d\mathrm{Ph}}\right)\left(\frac{k_3 a^{k2} B}{K_s + k_4 a^{k2}}\right) \tag{1}$$

The different constants will be explained in the next paragraphs. The terms in this equation were determined, because no useful data are available in literature.

For the development of the mathematical model for the phenol consumption by bacteria, the well-known simple model of Monod (5) will be used. Other models, such as those of Williamson (6) and Robertaccio (7), are not useful for the description of the situation in filters in water treatment; no biofilm is produced and the water contains a low substrate concentration (2). The model of Robertaccio is only valuable for high substrate concentrations.

Higher substrate concentrations than commonly occur in water treatment were used during the experiments, because otherwise the measured oxygen consumptions would remain much too low. Higher concentrations also allow the bacteria to enter the logarithmic growth phase. The model of Monod applies only to the logarithmic phase:

$$\mu = \mu_{max} \frac{C_s}{K_s + C_s} \tag{2}$$

where μ is the growth rate constant (h^{-1}), μ_{max} is the maximum growth rate constant (h^{-1}), C_s is the concentration of the limiting substrate (mg/L), and K_s is the substrate concentration (mg/L), at which $\mu = \frac{1}{2} \mu_{max}$.
Then:

$$\frac{1}{\mu} = \frac{K_s + C_s}{\mu_{max} \, C_s} = \left(\frac{K_s}{\mu_{max}}\right)\left(\frac{1}{C_s}\right) + \frac{1}{\mu_{max}} \tag{3}$$

Then μ_{max} and K_s are calculated from a Lineweaver–Burk plot.

Results

Adsorption of A.381 on GAC RBW 0.8 Extra. Equilibrium is reached after about 30 min of shaking, and the adsorption data at 20°C follow a Langmuir-type adsorption isotherm (Figure 4):

$$\frac{a}{m} = \frac{525\ C}{1 + (1.84 \times 10^{-5})C} \tag{4}$$

where a/m is the number of bacteria per gram of activated carbon and C is the number of bacteria per cm^3 in the surrounding liquid.

The external surface is 81 cm^2/g, on which a maximum of 2.84×10^7 bacteria can attach at 20°C. From these data, one can calculate that only a small part (2.16%) of the external surface is covered with bacteria.

Studies with a scanning electron microscope confirmed this computation (2). This finding means that no bacterial biofilm is present on the activated carbon surface and that the bacteria do not hinder the adsorption of phenol (no results published) or other organic compounds. During desorption experiments, it appeared that only 10–30% of the adsorbed bacteria could desorb.

When ground activated carbon (particle size 74–104 μm) was used, almost all dosed bacteria attached to the carbon particles (to about 5×10^9 bacteria/g carbon), while the number of free bacteria was very low (< 100/cm^3). Because of this, an adsorption isotherm could not be determined.

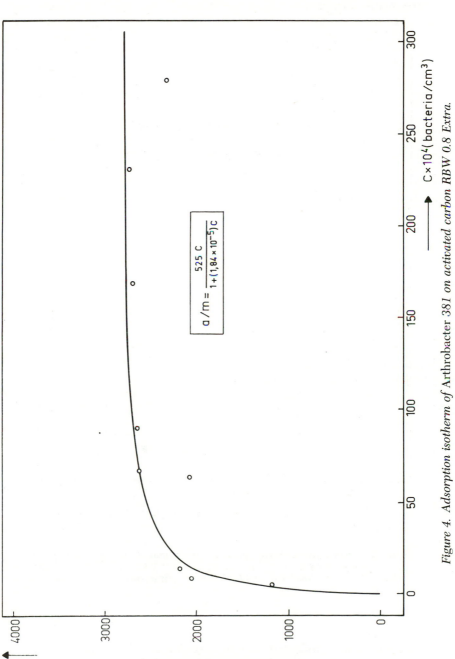

Figure 4. Adsorption isotherm of Arthrobacter 381 on activated carbon RBW 0.8 Extra.

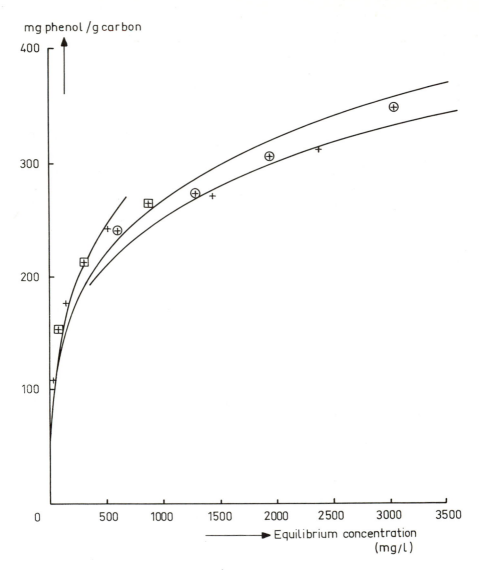

Figure 5. Adsorption isotherm of phenol on ground RBW 0.8 Extra (74–104 μm). Key: a/m = 44.67 C$^{0.27}$ (left curve), a/m = 44.67 C$^{0.26}$ (center curve), and a/m = 44.67 C$^{0.25}$ (right curve).

Adsorption of Phenol. The adsorption of phenol on the ground activated carbon followed the Freundlich-type adsorption isotherm (Figure 5). Because phenol concentrations much lower than 500 mg/L were used in the respirometer vessels, the equation:

$$\frac{a}{m} = 44.67 \ C^{0.27} \tag{5}$$

will be used for the calculation of the adsorption equilibrium.

Oxygen Consumption in Vessels by Nonbacteriological Processes. To determine the quantity of oxygen used by A.381 in the presence of activated carbon and/or phenol, it is necessary to know how much oxygen the other nonbacteriological processes require. Therefore, the oxygen consumption caused by adsorption of oxygen on activated carbon and the chemical oxidation of free and adsorbed phenol were determined.

The contents of the vessels are given in Table I, and the results of the oxygen consumption studies are presented in Figure 6. From the measurements, it appeared that no oxygen was adsorbed on activated carbon, by which the term $(dO_2/dt)_{ad} = 0$ (*see* Equation 1). The particles were not ground further by the magnetic stirrers. Oxidation of free phenol did not occur, so that the terms $u(Ph)$ and $v_2(H)$ are also equal to zero.

Only adsorbed phenol (a) is partly oxidized probably to hydroquinone (h):

$$C_6H_5OH + \tfrac{1}{2}O_2 \underset{v_2}{\overset{v_1}{\rightleftarrows}} C_6H_4(OH)_2$$

Table I. Contents of the Respirometer Vessels to Measure the Oxygen Consumption by Nonbacteriological Processes

Vessel No.	Activated Carbon (g)	Phenol Solution[a] (cm³)	Ringer (cm³)	Buffer[b] (cm³)	Total Volume (cm³)
1	1.5	—	280	20	300
2	—	10	270	20	300
3	1.5	10	270	20	300
4	3.0	—	280	20	300
5	—	20	260	20	300
6	3.0	20	260	20	300

[a]Phenol solution: 75 g/L.
[b]Tris buffer.

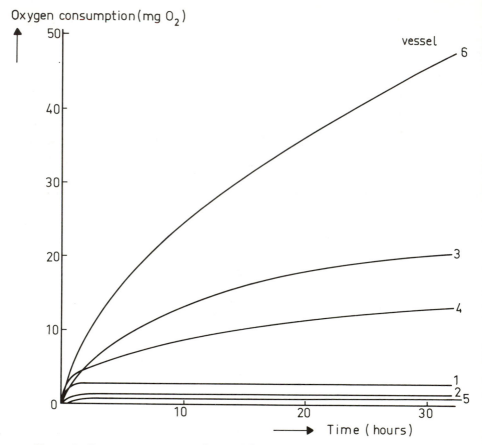

Figure 6. Oxygen consumption (mg O_2) by nonbacteriological processes in the respirometer: Vessel 1 contains 1.5 g ground activated carbon, Vessel 2 contains 750 mg phenol, Vessel 3 contains 1.5 g carbon + 750 mg phenol, Vessel 4 contains 3.0 g carbon, Vessel 5 contains 1500 mg phenol, and Vessel 6 contains 3.0 g carbon + 1500 mg phenol in 300 cm^3 Ringer solution.

Then $-(d\mathrm{Ph}/dt) = u(a) - v_2(h)$, in which $u = v_1(O_2)^{1/2}$. The oxygen consumption as a consequence of this process is equal to:

$$-\frac{dO_2}{dt} = \left(\frac{1}{2}\right)\frac{32}{94}\,[u(a) - v_2(h)] = 0.17\,[u(a) - v_2(h)] \qquad (6)$$

where 32 and 94 are the molar weights of oxygen and phenol, respectively. From the oxygen consumption, one can calculate that about half of the adsorbed phenol has been oxidized to hydroquinone after 116 h, because then exactly 90 mg of oxygen has been used per gram of phenol. This value corresponds with $0.09 \times (94/32) = 0.26$ molecule O_2 per molecule of phenol. The oxygen consumption as a consequence of the chemical oxidation of adsorbed phenol is then equal to $-dO_2/dt = 0.17(0.0110a - 0.0085h)$. The corrections necessary for the chemical oxidation of adsorbed phenol are summarized in Table II.

Endogenous Respiration of A.381. The experimental conditions for measuring endogenous respiration by free and/or attached bacteria and the oxygen consumption rate are presented in Table III. Two experimental runs each with six vessels were carried out. Experiment 1 (Vessels 1–6) lasted 36 h; the endogenous respiration also is shown in Figure 7. Experiment 2 (Vessels 7–12) lasted 92 h. The number of viable bacteria decreased during the experiments, and the number dying was lower in the vessels with activated carbon than in the vessels without activated carbon (1, 2, 3, 7, 8, and 9). This finding means that the viability of this strain increases when attached to activated carbon.

It is clear that the oxygen consumption rate in all vessels was low, and calculations indicate that it is not correlated with the average number of bacteria during the experiments ($R = -0.5$).

This finding means that the measured, very low oxygen consumption cannot be caused by endogenous respiration of the bacteria. The endogenous respiration of free and attached bacteria was so small that it was not necessary to correct for it.

Trace organic compounds cannot have served as substrate, because the oxygen uptake by the culture was very low and the viability generally decreased.

Oxygen Consumption as a Result of Bacteriological Phenol Degradation. The contents of the respirometer vessels given in Table IV were used to measure the oxygen consumption by bacteriological processes. In Figure 8, the oxygen consumption at different quantities of dosed phenol is presented, while in Table V the quantity of dosed and consumed

Table II. Calculation of the Correction for the Chemical Oxidation of Adsorbed Phenol in Different Periods (h)

Vessel No.	Dosed Phenol (mg)	Adsorbed Phenol (mg)	Oxygen Consumption Rate (mg O_2/h) in period					
			1–12	*12–25*	*25–50*	*50–100*	*100–150*	*Total*
1	6.25	6.2	0.015	0.007	0.004	0.003	0.002	0.606
2	12.50	12.3	0.030	0.013	0.009	0.006	0.005	1.274
7	18.75	18.0	0.044	0.019	0.013	0.008	0.007	1.806
8	25.00	23.0	0.056	0.025	0.016	0.011	0.009	2.341
13	37.50	31.3	0.076	0.033	0.022	0.014	0.013	3.165
14	50.00	37.7	0.092	0.040	0.026	0.017	0.015	3.782

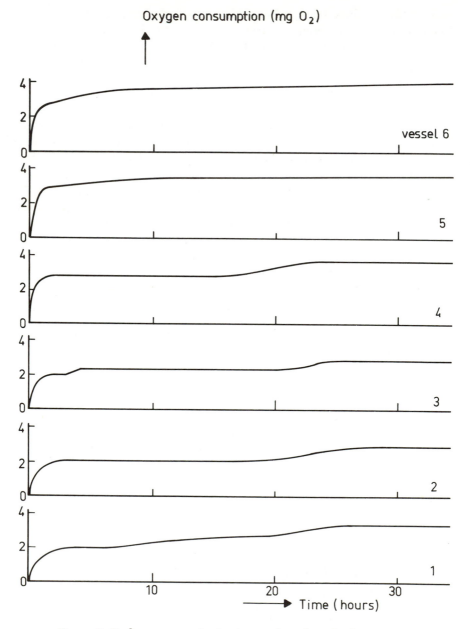

Figure 7. Endogenous respiration in vessels with Arthrobacter *381.*

Table III. Contents and Results of the Respirometer Vessels for Measuring the Endogenous Respiration

Vessel No.	Activated Carbon (g)	Buffer[a] (cm³)	Ringer (cm³)	Number of Bacteria at the start	the end	Oxygen Consumption Rate (mg O₂/h)
1	—	20	279	2.3×10^6	4.8×10^5	0.056
2	—	20	278	4.6×10^6	2.6×10^6	0.035
3	—	20	277	6.9×10^6	3.5×10^7	0.029
4	1.5	20	279	2.3×10^6	6.9×10^5	0.034
5	1.5	20	278	4.6×10^6	1.4×10^6	0.023
6	1.5	20	277	6.9×10^6	2.1×10^6	0.038
7	—	20	279	4.3×10^8	1.1×10^7	0.023
8	—	20	278	8.6×10^8	2.9×10^7	0.024
9	—	20	277	1.3×10^9	6.0×10^8	0.026
10	1.5	20	279	4.3×10^8	3.9×10^8	0.026
11	1.5	20	278	8.6×10^8	8.4×10^7	0.012
12	1.5	20	277	1.3×10^9	1.1×10^9	0.021

[a]Tris buffer.

phenol, the bacterial growth, and the oxygen consumption per vessel (by bacterial and chemical oxidation) are summarized.

The oxygen consumption as a result of the phenol consumption by adsorbed A.381 is equal to the last term of Equation 1, in which:

$$k_3 = \left(\frac{1}{Y}\right) \mu_{max} \left(\frac{1}{k_1 m}\right)^{k_2} \tag{7}$$

$$k_4 = \left(\frac{1}{k_1 m}\right)^{k_2} \tag{8}$$

These parameters can be calculated from the results of the different experiments.

From the results in Table V it appears that the oxygen consumption per mg of consumed phenol is equal to 2.395 mg. This figure agrees well with the theoretically needed quantity of oxygen for a complete oxidation of phenol to CO_2 and water, i.e., 2.38 g oxygen per gram of phenol, $C_6H_5OH + 7O_2 \rightarrow 6CO_2 + 3H_2O$.

The values of k_1 and k_2 follow directly from the adsorption of phenol on the activated carbon used:

$$\frac{a}{m} = k_1 C^{1/k_2} = 44.67 \, C^{0.27} \tag{9}$$

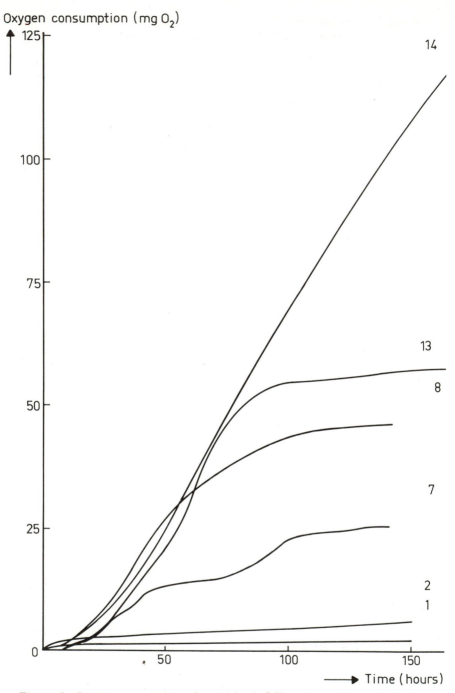

Figure 8. Oxygen consumption (mg O_2) at different quantities of dosed phenol (6.25, 12.50, 18.75, 25.0, 37.5, and 50.0 mg) into vessels with 2.0 g carbon and 10^8 bacteria in 300 cm³ Ringer solution.

Table IV. Contents of the Respirometer Vessels to Measure the Oxygen Consumption by Bacteriological Processes

Vessel No.	Activated Carbon (g)	Number of Bacteria[a]	Buffer[b] (cm³)	Substrate[c] (cm³)	Ringer (cm³)
1	2.0	10^8	20	2.5	267.5
2	2.0	10^8	20	5.0	265.0
7	2.0	10^8	20	7.5	262.5
8	2.0	10^8	20	10.0	260.0
13	2.0	10^8	20	15.0	255.0
14	2.0	10^8	20	20.0	250.0

[a]*Arthrobacter globiformis* strain 381.
[b]Tris buffer; 200 cm³ Tris(0.4 M) + 324 mL HCl (0.2 M).
[c]Phenol (2.5 g/L) and KNO_3 (3.2 g/L) dissolved in demiwater.

Table V. Experimental Results with Activated Carbon, Bacteria, and Phenol in the Respirometer Vessels

Dosed Phenol (mg/vessel)	Length of the Experiment (h)	Consumed Phenol	Bacterial Growth per Vessel	Oxygen Consumption (mg/vessel) Bacteria	Oxygen Consumption (mg/vessel) Chemical
6.25	150	3.35	6.1×10^8	0.97	0.61
12.50	150	10.00	5.6×10^8	4.01	1.27
18.75	140	16.45	5.4×10^9	23.85	1.81
25.00	140	23.00	3.8×10^9	39.61	2.34
37.50	185	34.30	7.8×10^9	54.35	3.16
50.00	185	47.20	1.0×10^{10}	107.72	3.78

giving $k_1 = 44.67$ L/g carbon, $k_2 = 1/0.27 = 3.70$ (no dimension), and $k_4 = (1/k_1 m)^{k_2} = 7.85 \times 10^{-7} (1/m)^{3.70}/L$.

The yield factor Y is calculated from the slope of the graph, which gives the relationship between the data of the bacterial growth and the phenol consumption (Table V). This resulting slope is 2.24×10^8 bacteria per mg of phenol with a correlation coefficient of 0.95. For a number of different phenol concentrations during the period of exponential growth, it is necessary to calculate the associated values of μ, which equal $(dB/dt)/B$.

From a Lineweaver–Burk plot, μ_{max} and K_s are calculated (Equation 3):

$$\frac{1}{\mu} = 0.01 \frac{1}{C_s} + 42.37 \qquad r = 0.93 \qquad (10)$$

Then $(1/\mu_{max}) = 42.37$, so that $\mu_{max} = 0.024$ h^{-1}, and $(K_s/\mu_{max}) = 0.01$, so that $K_s = 2.4 \times 10^{-4}$ mg phenol/L water at a phenol concentration of 4.7 mg/g PAC.

This result means that the bacteria consume phenol already at a maximum rate with a very low concentration on the carbon surface. In essence, the oxygen consumption is linearly dependent on the number of attached bacteria:

$$\left(\frac{dO_2}{dt}\right) B = 0.0313B/m \ \ \text{mg O}_2/\text{g PAC/h}/10^8 \ \text{bacteria} \quad (11)$$

The ultimate model for the oxygen consumption in the vessels with activated carbon, phenol, and bacteria is as follows. As an end result of the measurements and calculations, Equation 12 gives the oxygen consumption rate in units of mg O_2 per g PAC/h in a respirometer vessel, which contains m gram of PAC, on which a mg of phenol, h mg of hydroquinone, and $B \times 10^8$ bacteria per g activated carbon have been adsorbed:

$$-\left(\frac{dO_2}{dt}\right) T = 0.17 \left(0.0110 \ \frac{a}{m} - 0.0085 \ \frac{h}{m}\right) + 0.0313 \ \frac{B}{m} \quad (12)$$

Restrictions are that the quantity of adsorbed phenol is larger than or equal to 10 mg phenol per g activated carbon, while the temperature and the pH have to be 20°C and 8.0, respectively. The quantity of phenol degraded by attached $A.381$ is equal to $0.0313/2.395 = 0.013$ mg phenol/g PAC/h/10^8 bacteria.

From Tables II and V, it can be concluded that chemical regeneration of PAC is not an important factor when phenol is the only adsorbed substrate under these experimental conditions.

Discussion

The theoretical, mathematical model developed has been checked by experimental studies on a model system, which consisted of one type of activated carbon (Norit RBW 0.8 Extra with a particle size 74–104 μm), one type of substrate (phenol), and one single phenol degrading strain of bacteria ($A.381$).

A literature study to determine the necessary constants indicated that the values looked for were unknown or could not be calculated from published results. It was necessary and possible to determine all these constants by experimental studies with an automatic respirometer. The action of chemical oxidation of adsorbed phenol is in agreement with the results of Hoak (8) but contradicts those of Landers (9). The contradiction in the results can be explained by the fact that Landers worked with different types of carbon and at different pH values. After calculating the μ values and the corresponding phenol concentrations, the values of μ_{max} and K_s could be calculated.

In both vessels with activated carbon, as in vessels without it, the bacteriological oxygen consumption per mg of consumed phenol appeared to be equal to the theoretical oxygen consumption of phenol. From this it can be concluded that the phenol consumed has been oxidized completely. It is unnecessary to correct the respiration results for the oxygen consumption as result of the endogenous respiration by free or attached bacteria. In an environment without substrates, the attached bacteria had a greater viability than free bacteria.

When all calculated constants are filled in the theoretical model (Equation 1), a simple model arises that describes the phenol degradation on the carbon surface (*see* Equation 11). The quantity of phenol degraded by bacteria appeared to be linearly dependent on the number of attached bacteria per gram of activated carbon.

The quantity of adsorbed phenol is only relevant at a concentration of less than 10 mg of phenol per gram of activated carbon. The fact that the bacteria can already consume phenol with maximum rate at such a low phenol concentration on the carbon surface is probably affected by the small size of the carbon particles. The conclusion is that, under the experimental circumstances used, the chemical oxidation of phenol is not important for the regeneration of activated carbon, while bacteriological regeneration was very important.

In the batch experiments described, 0.010 g of phenol carbon/h was oxidized by 10^{11} attached bacteria. It is remarkable that this value also was measured in GAC filters with mixed bacterial populations, dosed phenol, and many other organic compounds (1). But the question still arises if extrapolation of the observations is possible to the behavior of other organic compounds on GAC. To answer this question more research of the behavior of other organic compounds is desirable.

Conclusions

Under the experimental conditions used (20°C and pH 7.5–8.0), dissolved phenol was not chemically oxidized, while adsorbed phenol was partly chemically oxidized to hydroquinone. The endogenous respiration of A.381, either free or attached to activated carbon, was immeasurably small.

The yield factor (Y) was 2.24×10^8 bacteria/mg phenol and the maximum growth constant (μ_{max}) was 0.024 h^{-1} for attached bacteria, while K_s was equal to 2.4×10^{-4} mg phenol/L at a phenol concentration of 4.7 mg/g PAC. When the phenol concentration is higher than 10 mg/g PAC, the phenol consumption velocity by attached A.381 was 0.013 mg phenol/g PAC/h/10^8 bacteria.

Chemical regeneration of PAC was unimportant when phenol was the only adsorbed substrate under the described experimental conditions. More research on the behavior and degradation of other organic compounds is desirable.

Literature Cited

1. den Blanken, J. G., H_2O, **1980,** *13,* 502–10, 512 (in Dutch).
2. den Blanken, J. G., Report No. 79–24 (in Dutch), Laboratory of Sanitary Engineering, Delft University of Technology, Delft, The Netherlands, 1979.
3. den Blanken, J. G., "Research on the Bacteria Population, Its Activity in Activated Carbon Filters, and Its Influence on Drinking Water Quality," Annual Report, Laboratory of Sanitary Engineering, Department of Civil Engineering, Delft University, The Netherlands, 1978.
4. Buchanan, R. E.; Gibbons, N. E. "Bergey's Manual of Determinative Bacteriology," 8th ed., Williams and Wilkins, Baltimore, Md., 1974.
5. Monod, J. *Ann. Rev. Microbiol.* **1949,** *3,* 371–94.
6. Williamson, K. J., Ph.D. Dissertation, Stanford Univ., Palo Alto, Calif., 1973.
7. Robertaccio, F. L., Ph.D. Dissertation, Univ. of Delaware, Newark, Del., 1976.
8. Hoak, R. D. *Int. J. Air Wat. Poll.,* **1962,** *6,* 521–38.
9. Landers, H., Ph.D. Dissertation, Univ. of Aken, West Germany, 1974.

RECEIVED for review August 3, 1981. ACCEPTED for publication February 2, 1982.

Discussion III
Biological/Adsorptive Interactions

Participants

Andrew Benedek, McMaster University
Jack DeMarco, U.S. Environmental Protection Agency
John G. den Blanken, Delft University of Technology
Francois Fiessinger, Societe Lyonnaise des Eaux et
    de l'Elairage
Robert A. Hyde, Water Research Centre, England
Michael J. McGuire, The Metropolitan Water District of
    Southern California
Alan A. Stevens, U.S. Environmental Protection Agency
Irwin H. Suffet, Drexel University
Walter J. Weber Jr., University of Michigan

WEBER: We've had some results presented to us today which, to my way of thinking, still leave open the question of whether or not biological activity on carbon columns plays a significant role in water treatment applications and whether or not preozonation enhances the potential role that it might play.

HYDE: We talked about using ozone in conjunction with carbons today. In fact, the only way that the use of ozone in conjunction with carbon can be justified is if it actually extends the service life of a carbon bed. We made some preliminary cost estimations at the Water Research Centre, in England, which indicate just how long this bed life needs to be increased for different doses of ozone.

One of the examples of the carbons that was spoken about this afternoon was one where 15 minutes' empty-bed contact time was used. If we assume that the normal bed life is six months—I'm just taking that as an example—if, in fact, we use one and a half milligrams per liter of ozone before that carbon, in order to make the use of ozone cost-effective the bed life needs to be increased to four years of operation.

If you take the same example but with a normal bed life of only three months, you're asking the carbon to do much more severe duty. Then, if we use one and a half milligrams per liter of ozone, we need to extend the

0065-2393/83/0202-0373$06.00/0
© 1983 American Chemical Society

bed life from three months to something over five months to economically justify using ozone.

DeMARCO: I think one of the important questions in GAC research is, can we, by operating an ozone GAC system, achieve an efficiency far beyond what we could achieve in a straight GAC system? That is, if we're talking about plateaus with a given contact time, say 15 minutes, and, say, an 80-percent plateau or steady-state condition, could we look at a plateau where we were getting somewhere around 10 percent, 20 percent?

I think that's what much of EPA's research was directed toward, and I think the Philadelphia experience, in my opinion, as well as the Miami and the Shreveport experience, were directed initially toward that primary goal. Once those points were established, then we were going to look at the economic aspects of it. Obviously, from the results that you've seen, preliminary though they may be, we have not achieved some of the glowing expectations that we were led to believe by some of the European investigators. I think that's where we are now.

WEBER: What you outline is clearly a different objective than simply extending the effective life of the activated carbon, but I agree with you that the results that have been presented certainly do not clearly indicate that a preozonation enhances the performance of carbon systems.

SUFFET: I'd also like to put some things in perspective.

The work in Philadelphia and the work in Shreveport had oxygen on the columns. The work at Miami did not. It was an anaerobic system, and an anaerobic system works quite differently than an aerobic system, for obvious reasons.

The work in the Netherlands—I don't know if that's water treatment or not. I'm sort of wondering, is that water treatment, or do we have a phenol-acclimated system where we have half TOC and phenol on a one-to-one concentration basis, and we have acclimation to phenol, and are we studying a very mild industrial waste process and not a real drinking water process? That's something that you might look at later. All this work in a bundle eventually will come to the answers that we're looking for.

I think the European work has been somewhat pioneering. We are trying to optimize and understand these systems. One interesting thing is—and this is for the water treatment industry and coming from the chemist's background—I see no problem with having no prechlorination or ozonation until after the coagulation/sand-filtration step.

In Philadelphia we see this to be a very good process for removing a TOC that is biodegradable, and we're using sand and upflow clarification to do it. And then after that point, putting in chlorine, going with carbon or not with carbon—that's a separate question. But I don't see anything wrong with going back to 1905 technology of slow sand filtration or something of this sort for an improvement in removing these biological components.

WEBER: That's a whole other argument, and I'm not at all convinced that slow sand filtration achieves any better biological removal. As a matter of fact, there's much evidence to the contrary, Mel. Why is it that we always hope in the water field that all of our researches in a bundle will somehow come to the results we seek?

den BLANKEN: We have done the different experiments with only tap water without a phenol dose, which I've shown you in the slides. Besides, I showed you the removal of the small phenol dose as a function of bed thickness, and it was the only experiment I showed you with phenol. Jack's experiments were all without phenol.

We used phenol sometimes to get a better correlation between the growth of bacteria and phenol removal. There have been discussions about correlations between bacterial growth and removal of organic carbon. And it was typical in our results that there was never a correlation between bacterial growth and removal of carbon in filters without phenol.

But there was a clearer correlation in films with a phenol dose because it is important that using a gross medium for bacteria—this plate-count agar, for instance—I think you find only the tip of the iceberg. That is, you find only a part of the bacteria in the water, and that is one reason why in some experiments we used a very small phenol dose of about one to three milligrams of phenol per liter. So you cannot call this wastewater treatment; that is a very small phenol dose.

WEBER: There have been so many people who have done definitive studies that would once and for all answer the question of whether ozone has a negative effect on adsorbability, and the long chain of evidence is that in fact, for most compounds that are found in water supplies, direct ozonation adversely affects adsorbability.

I think it's a very valid point that there are so many uncontrollable parameters operative in ozonated, biologically activated carbon systems that what one gets from one pilot study is not necessarily extrapolateable to any other situation unless you totally characterize the matrix of variables that are operative in that system. It's so very important. The oxidant varies. It's a heterolytic oxidant. It goes through a whole stage of degradation processes. Carbon has different reactivity. Temperature has such an effect. The bioorganisms that you establish are dependent upon the substrates, et cetera.

BENEDEK: I've been listening to these talks this afternoon, and it's very exciting. We are beginning to see some comparisons, and we all want to know the same thing: Do we need ozone or do we not need ozone? And to what extent is it good, to what extent is it bad, what level should it be? And so on and so forth.

First of all, I really do think that we should try to use a cumulative plot to compare things because it makes life much easier. Cumulatively, we can plot organic supply as a function divided by the weight of the mass of carbon in the system against what is removed. And we see nice, smooth curves instead of the typical up-and-down, up-and-down kinds of things that many of these curves show. Now, there's nothing wrong with those curves. The difficulty is to compare one system to another.

There's one more thing that we ought to do, and that is, of course, that we have to determine the adsorptivity as well as we can for each water supply, because it's likely to be different. Miami isn't going to be like Philadelphia. Perhaps we can also take the cumulative plots, then, and divide by the adsorptive capacity that one would expect to encounter and thereby determine the additional removal that one obtains.

These figures will then enable us to compare Philadelphia with Miami. The Shreveport pilot study shows that we've got to go long-term. The reason we can't make conclusions very well today is because we have perhaps too little data. We've got to go maybe another year or two to do this, and we've got to do it in the way I'm indicating.

Hopefully, we can do the adsorption isotherm information prior to completing these studies and thereby determine the mechanisms, which are very, very important, because we don't want to go out and spend a great deal of money and wind up building the wrong kinds of plants.

Finally, I also think we should be very careful when we look at oxygen uptake. There are so many pitfalls in that ballgame that it's very tough not to go wrong. The most important one to recall all the time is that oxygen uptake, per se, and CO_2 release may have absolutely nothing to do with organic uptake, that is, substrate from the water. It could be nothing more than endogenous respiration—bacteria being applied in the feed sitting on the carbon and respiring. We should try to deduct endogenous respiration so that conclusions are not drawn purely on that basis.

WEBER: What you've said about oxygen uptake can also be said about carbon dioxide or inorganic carbon generation. The whole notion of respiration needs to be properly accounted for.

BENEDEK: I don't think that anyone is arguing that one can have biodegradation and adsorption. Francois Fiessinger and I are saying that you can also have slow adsorption, which may or may not be an important part of the biodegradation. But the crux of the question—and it's very difficult to answer—is, what is the effect of a carbon as a support for the biodegradation? Are there synergistic effects? Is it like other bacteria that you see here on the surface of the carbon? Are they like contented cows grazing on a luscious Alpine meadow?

To answer that question, I think we have to answer three questions: First of all, where are the bacteria? Are they on the surface, or are they in the interior? If they are in the interior, then obviously they can have a much bigger area to graze on. That intrigued us—Francois and I—and we split some carbon. We didn't find any bacteria inside the carbon that we split, but we did find them outside. So our feeling is that they tend to remain on the outside, and therefore they are acting on the outside.

The next question is, do they have a higher concentration of organics where they are—that is, on the surface? And this is the most confusing of all questions. I fundamentally believe not only that they don't have a higher concentration, I believe that in fact the bacteria see a lower concentration when they are on the carbon. This is obviously not too easy to explain, contrary to what most of you probably have been hearing and learning. If you have an interface which rejects organic compounds such as sand, the molecules hit the wall and bounce right back. On the other hand, in activated carbon they penetrate, and they are adsorbed. Therefore, the interfacial liquid concentration at the surface of the carbon is, in fact, lower than it is in the case of the sand.

So the third question is, is our sand better than activated carbon if you discount the adsorption? That's what Francois and I tried to do. We ran parallel sand and carbon columns, and we found that sand is not better than carbon. Summers had glass columns,[1] and he didn't find that glass was any worse than activated carbon. I think if you look through the literature—and there is a bottomless amount of literature—you find that usually there is no difference. Occasionally there is.

Bacteria attach more easily to activated carbon than to sand. Therefore, in the beginning of an operation your sand is going to be slower to get bacterial activity, and the carbon works better initially. If you wait long enough this effect disappears.

Second, if you prechlorinate and you've got chlorine coming in, sand doesn't do anything to chlorine. Carbon does; it eliminates the chlorine. Therefore, bacteria can grow on carbon if you have chlorine in the feed.

Finally, if you have high concentration fluctuation, there's a tendency for carbon to dampen that. But this really doesn't have very much to do with the bacterial activity.

So my conclusion is that so far we have not found reason to believe that there's a synergism. I desperately would like to find it, and so would many other people. But we haven't managed it.

SUFFET: Dr. den Blanken, do you have a comment? Yesterday you showed some biological materials which appeared to be inside the pores.

den BLANKEN: I have some comment about bacterial activity. To measure the bacterial activity, you have to use bacterial parameters like oxygen consumption or CO_2 production, and I think ATP is not a good indicator for this.

My second point is that we have done some studies in slow sand filters, and we have seen that oxygen consumption remains nearly constant as a function of running time, while in granular activated carbon

---

[1]Chapter 22 in this book.

filters it increases with running time. I discussed in my paper[2] that the number of bacteria increased and there was a mechanism of attachment of specialized bacteria inside activated carbon pores. So I'm convinced that it is a rough method to subtract the activity of sand filtration from granular activated carbon. Benedek showed that you get a constant-level, steady-state situation when you subtract one from the other. But I think that the oxygen consumption always increases with the steady state in organic carbon removal. So I also think that if you give filters a sufficiently long time you will get more activity and you will get desorption from the carbon, which will increase bacteriological activity.

It's a very complex process, and to find out which parameters are the reason for the effects which you see, I think you have to do a lot of fundamental work with model substrates, model bacteria, temperature, and pH, perhaps in batch experiments, to come a to a good conclusion. What we have seen today is an overall effect of different things together. I think, perhaps, that it would be worthwhile to hold a symposium on only the biological aspects of granulated activated carbon.

STEVENS: I think we do have to agree—at least I think most will agree with me—that this is a complicated situation whereby we have more than one mechanism going, and I think all we are quibbling about is how much of each we have at any given time.

I would like to raise a point about something I think Francois concluded[3] (I'm not quite sure)—that the ATP measurement was inappropriate, or not as good as some other measurement of the biological activity. I'm wondering whether that's just because it didn't correlate with the plate counts or some other method of counting the microorganisms. I'd like to raise the possibility here that other methods of counting the organisms are perhaps really inadequate, and it may not be the ATP measurement's fault at all.

FIESSINGER: Well, in the experiment I presented we measured almost everything which could be measured. I measured the ATP because some people yesterday raised the question of the interest of this measurement. But I think the obvious answer on ATP is that you cannot measure fentograms of something in the presence of carbon because you have adsorption; you don't know what is adsorbed or desorbed. Anyway, I don't think the principle of the method is reliable. We did it, but we couldn't interpret it.

Could I say a word on the previous comment? Of course, den Blanken did his experiments with phenol, which is very degradable, and I made my experiment on surface water. I insisted in the presentation that this water

[2]Chapter 16 in this book
[3]Chapter 14 in this book.

probably doesn't have very much biodegradable material, so that makes a difference.

The second point I want to make again—this point has already been made by some other people, especially Andy—is that if you have respiration, that doesn't mean there is TOC removal. We're all sitting here, breathing oxygen, but we're not eating anything. (Too bad.) The bacteria are there, but it is not because they breathe that they are removing TOC. That's something that people don't understand.

den BLANKEN: My point is that the oxygen consumption is a very important parameter for measuring the activity of bacteria. When you have an organic substrate—that is to say, glucose or phenol—and it is oxidized, it used oxygen; that's the first step toward oxidation. And I think that's a very fundamental issue.

McGUIRE: Francois said something after our discussion of his paper that I think is important, and that is that he's considering the slow adsorption phenomenon as a working hypothesis. I think that's great. He's not beating the drums and saying that it's the most important mechanism.

I think what Al Stevens said is perfectly true. Maybe we're only quibbling about which is important, biodegradation or slow adsorption. They probably are both happening. I would just like to see some hard data.

PILOT- AND LARGE-SCALE STUDIES

17

Experimental Error Estimates Associated with Pilot Activated-Carbon Investigations of Trace Organic Removals

MICHAEL J. McGUIRE, THEODORE S. TANAKA, and MARSHALL K. DAVIS

The Metropolitan Water District of Southern California, Water Quality Branch, La Verne, CA 91750

The removal of trihalomethane precursor substances as estimated by total organic carbon using granular activated carbon was investigated in a pilot plant study. The study was designed with replicate GAC columns and replicate analytical procedures so that statistical statements could be made on postulated differences between column performances. After 55 days, the TOC removal mechanism appeared to be biological degradation, although the slow adsorption mechanism could not be ruled out. Grab and composite sampling should be evaluated for each granular activated-carbon study to avoid conclusions based on nonrepresentative data. Trihalomethane analytical variability dominated the variability of the overall experiment.

T HE METROPOLITAN WATER DISTRICT of Southern California (Metropolitan) is a public and municipal corporation of the State of California, which provides supplemental water as a wholesaler through 27 member agencies (cities and water districts) to nearly 12 million people in a 13,000-km^2 (4,900-sq. mi.) service area on the coastal plain of Southern California. Approximately one-half of this supplemental water—860 $\times$ 10^6 m^3 (700,000 acre-ft)—is imported through 242 miles of aqueduct from the Colorado River. Initial deliveries of Colorado River water (CRW) began in 1941. In addition, Metropolitan has contracted to receive more than 2400 $\times$ 10^6 m^3 (2 million acre-ft) annually of Northern California water through the 444-mile-long State Water Project. First deliveries of State project water (SPW) began in 1972.

0065-2393/83/0202-0383$07.00/0
© 1983 American Chemical Society

On November 29, 1979, the U.S. Environmental Protection Agency promulgated the trihalomethane (THM) regulation. The regulation established monitoring requirements and a maximum contaminant level (MCL) of 0.10 mg/L for total trihalomethanes (total THM). Figure 1 shows a schematic of the feasibility analysis process that Metropolitan has been engaged in over the past few years in anticipation of the promulgation of the THM regulation. The schematic is very similar to a feasibility analysis schematic presented previously for control of synthetic organic chemicals (1). While the feasibility analysis process appears to be linear, several steps are being pursued in a parallel fashion.

Table I summarizes the extent of the trihalomethane problem in Metropolitan's system (2). Quarterly averages over the past 2.5 years have been very close to the MCL for total THM. What Table I does not show is that many member agencies take water from Metropolitan's system, hold it for several days, and possibly rechlorinate. Data from these systems are just now becoming available and, as expected, some member agency systems contain THM levels higher than the MCL. To meet its responsibility as a wholesale water purveyor, Metropolitan has embarked on a course of action to provide water to its member agencies that will allow them to comply with the MCL without additional treatment.

Changing the source of supply and using existing treatment processes to completely control the THM problem have been investigated and determined infeasible for Metropolitan. Other more exotic treatment techniques have been evaluated on a bench- and pilot-scale basis. In general, there are three ways to control trihalomethanes in drinking water: remove THM after formation, remove precursors before THM formation, and use a disinfectant other than chlorine to prevent THM formation.

This chapter focuses on research at Metropolitan dealing with the second option. The first option is not being seriously considered since the kinetics of THM formation in Metropolitan's case are relatively slow and removal of THM at the treatment plant is infeasible. Alternative disinfectants, particularly chloramines, are also being evaluated. The purpose of this chapter is to present the results of a study on the removel of THM precursor substances as estimated by total organic carbon using granular activated carbon (GAC). Data not included in this chapter have demonstrated a relationship between removal of THM precursors and removal of total organic carbon (TOC). Removals of individual THM components will also be discussed as examples of difficult to adsorb, low molecular weight halogenated methanes. The study was designed with replicate GAC columns and replicate analytical procedures so that statistical statements could be made on postulated differences between column performances.

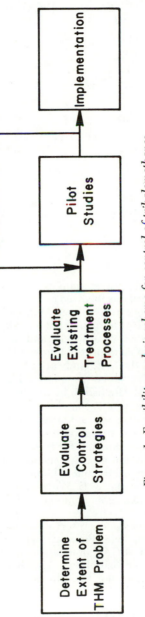

Figure 1. Feasibility analysis scheme for control of trihalomethanes.

Table I. Total Trihalomethane Data Summary

Quarter	Quarterly Average ($\mu g/L$)	Running Annual Average ($\mu g/L$)	Percent of Samples Over 100 $\mu g/L$
Jan. 1979	94	—	47
Apr. 1979	103	—	42
July 1979	114	—	64
Nov. 1979	97	102	47
Feb. 1980	79	98	8
May 1980	90	95	29
Aug. 1980	94	90	41
Nov. 1980	102	91	58
Feb. 1981	86	—	11
May 1981	80	—	18

Experimental

TOC was analyzed in the water samples using a Dohrman DC-80 organic carbon analyzer. TOC removal was followed as a rough surrogate of the removal of THM precursors. In previous research at Metropolitan, a method had been developed to measure the THM formation potential by chlorinating a sample to a desired level, holding it over several weeks, and measuring the resulting THM compounds formed (3, 4). In an effort to improve pH control, a phosphate buffer was added to the samples in the GAC study prior to chlorination. Subsequently, it was discovered that the buffer was contaminated with a THM precursor, and all of the THM formation potential data had to be discarded.

Trihalomethane analyses were accomplished using a liquid–liquid extraction procedure followed by injection on a Varian 3700 gas chromatograph (2–4). Quality assurance procedures were rigorously employed in this work. Standards, spikes, replicates, and blanks accounted for 20–30% of the samples analyzed.

Pilot Plant. A 5-gallon/min all glass, stainless steel, and Teflon pilot plant was used to provide conventional treatment before GAC filtration. The pilot plant is located in a pipe gallery at the F. E. Weymouth Filtration Plant. Rapid mix, chemical addition, flocculation, sedimentation, and dual media filtration are the conventional unit processes.

Following a filtered water storage reservoir are six glass columns 15 cm (6 in.) in diameter and 3 m (10 ft) high. Figure 2 illustrates how the columns were arranged during Phase I to test GAC in a parallel operations mode.

Experimental Design and Sampling. A number of pilot-scale GAC treatment studies have been completed (5). The studies are normally designed to investigate one or more of the following general parameters: comparison of adsorbents, effect of pretreatment, temporal variations of organics, hydraulic parameters, column operation modes, and removal of classes of organics.

Without exception, the studies have been conducted by comparing the results of one column versus another without any replication. There is no

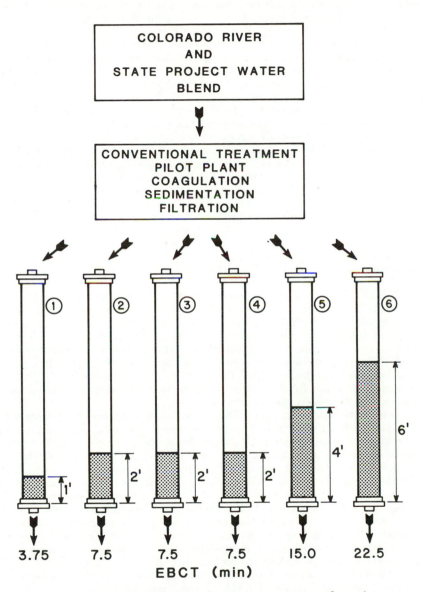

Figure 2. GAC pilot column arrangement in Phase I.

information available in the literature on estimates of the precision of these studies. It has been well established in the designs of most experiments that an estimate of the experimental error is essential to recognize significant differences between treatments (type of GAC, variation in hydraulic parameters, etc.). Until now, investigators relied on their "sense" of what a significant difference is. Of course, this "sense" varies among investigators and is a less than optimum method of decision making. There are a number of case histories in the literature that have produced conflicting conclusions on the effects of various parameters. While some of these conflicts are caused by real differences in the characteristics of the water being tested, it is suspected that some are the result of different "perceptions" of what a significant difference really is.

Another difficulty encountered in evaluating pilot plant data is the usual reliance on grab samples for characterizing the GAC column influent and effluent conditions (e.g., organic compound concentrations). Wide variability in influent/effluent conditions has made it difficult to interpret the actual mass loadings on the GAC columns. Diurnal, seasonal, and yearly variability of compound concentrations and water quality factors are rarely evaluated. Instead, one grab sample is assumed to be representative of an extended period of time (typically 1 week). One reason for the infrequent grab sampling scheme is the high cost and relatively involved analytical requirements for determining organic compounds in water. For the preceding reasons, it was decided to use composite samplers specially constructed for this project.

Figure 2 indicates that three replicate GAC columns with empty bed contact times (EBCTs) of 7.5 min were run in addition to three other columns with EBCTs ranging from 3.75 to 22.5 min. Influent and effluent composite samples were collected using the sampler designed by Westrick and Cummins (6). Grab samples were also collected and compared to the respective composite sample results. The composite samplers were designed to ensure a headspace free sample with no significant losses of volatile trace organic compounds (e.g., trihalomethanes) over the 3.5-day compositing period. The samplers were kept in refrigerators to preserve the sample.

It was anticipated that, over the course of the project period, the quality of the pilot plant influent water would vary considerably. The pilot plant receives the same blend of State Project and Colorado River water that the F. E. Weymouth Filtration Plant receives. Figure 3 shows how the percentage of SPW changed over the project period. The dramatic fluctuations were caused by requirements to shut down either the SPW or CRW system for scheduled preventative maintenance. The purpose of using the composite samplers was to obtain representative samples during the fluctuation periods. While it is true that composite sampling masks the peaks and valleys, it does give an accurate average representation of what a population might be exposed to over a period of time. This is the type of information that toxicologists and epidemiologists require to evaluate a population's exposure to potentially harmful compounds.

Statistics. A variety of statistical techniques were used to evaluate significant differences between the GAC column performances (7, 8). Analysis of variance, unpaired t test, F test, and confidence limits were calculated on influent and GAC column effluent data. In addition, the variance associated with the total experiment was partitioned into the analytical and column variances. From statistical theory we can write:

$$\sigma^2_{total} = \sigma^2_{analytical} + \sigma^2_{column} \qquad (1)$$

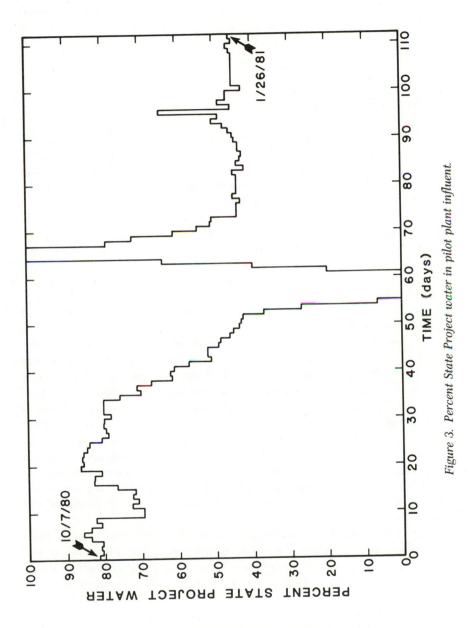

Figure 3. Percent State Project water in pilot plant influent.

This equation is only valid if $\sigma^2_{analytical}$ and σ^2_{column} are statistically independent.

Equation 1 was applied to the breakthrough data from the three 7.5-min EBCT columns to obtain the total variance, σ^2_{total}, and to the analytical precision data to obtain the analytical variance, $\sigma^2_{analytical}$. The variance associated with running the columns, σ^2_{column}, was calculated by subtraction. The purpose of this exercise was to determine which part of the experiment contributed to the majority of the experimental variance.

Results and Discussion

TOC Removals. Variability in the influent TOC data was tested over a 24-h period on December 11 and 12, 1980. Figure 4 shows little variability in the hourly grab samples of the column influent. These data are confirmed in Figure 5 which shows the influent and GAC column effluent TOC data. These data and all subsequent data, unless noted, are derived from composite samples. The influent TOC data are remarkably consistent despite the wide variations in percent blend shown in Figure 3. The TOC content of the two sources of supply was very similar during the Phase I testing period.

The GAC column effluent data in Figure 5 show the classic breakthrough pattern for TOC: (1) immediate low level leakage through the columns, (2) regular, orderly breakthrough during the first phase, followed by (3) a pseudo steady-state.

Figure 6 illustrates the pseudo-steady-state more clearly by the parallel cumulative removal curves. Previous discussions regarding pseudo-steady-state removal have relied on a discussion of biological activity to explain the phenomenon. Recently, an alternate explanation involving the slow adsorption mechanism has been proposed (9). There is direct evidence to support a biological activity mechanism including elevated standard plate counts in GAC column effluents, high bacterial densities measured by standard plate counts on the exhausted carbon surface, and direct observation of bacterial species on the carbon surface.

Figure 7 is a scanning electron photomicrograph of the surface of one granule of activated carbon from Phase I. Visible are cocci, spirillium, and filamentous bacteria.

After the completion of the pilot study, bacteria were washed from the carbon surface, isolated on plate count media, and identified to the genus level using the three-tube technique of Lassen (10). The dominant genera recovered from the carbon surface included *Pseudomonas* and *Moraxella* with some *Micrococcus* and other unidentified bacteria.

The alternate explanation for at least part of the pseudo-steady-state removal of TOC has been described as "slow adsorption" (9). The phenomenon is said to be caused by the large size of the humic acid molecules that are difficult to fit into the much smaller pore system of

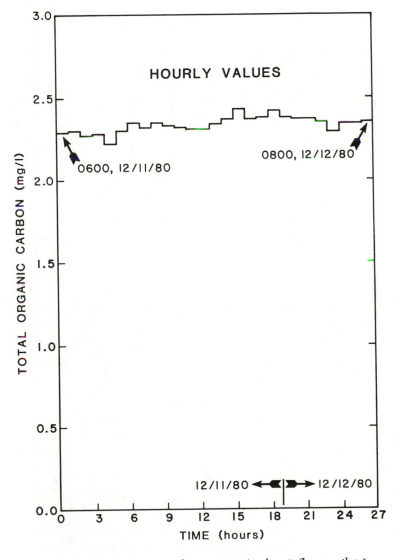

Figure 4. Total organic carbon at GAC/carbon influent; pilot I.

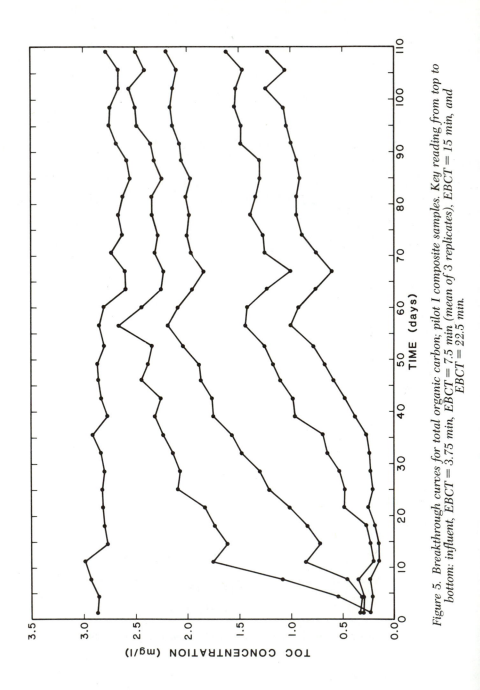

Figure 5. Breakthrough curves for total organic carbon; pilot 1 composite samples. Key reading from top to bottom: influent, EBCT = 3.75 min, EBCT = 7.5 min (mean of 3 replicates), EBCT = 15 min, and EBCT = 22.5 min.

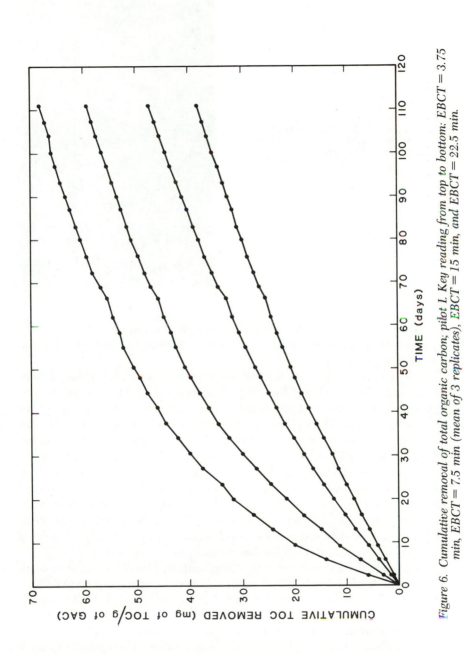

Figure 6.   Cumulative removal of total organic carbon; pilot I. Key reading from top to bottom: EBCT = 3.75 min, EBCT = 7.5 min (mean of 3 replicates), EBCT = 15 min, and EBCT = 22.5 min.

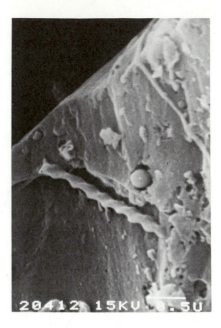

*Figure 7. Scanning electron
photomicrograph of bacteria on
surface of activated-carbon granule
(28,00)XL).*

activated carbon. Thus, the long-term kinetics of humic acid-like molecules are slow and account for at least part of the plateau regions of the
breakthrough curves.

Table II presents a summary of the TOC removal data for the pseudo-
steady-state period (Days 55–110). The amount of TOC removed per
gram of carbon appears to be consistent in the GAC columns except for
the 15-min EBCT. The surface applied rate for all six of the GAC columns
was 2 gallons/min/ft² at the beginning of the test. However, the column
with an EBCT of 15 min lost 25 cm (10 in.) of activated carbon during one
backwash cycle. To keep the EBCT the same, it was necessary to reduce
the surface applied rate to 1.5 gallons/min/ft². This lower surface applied
rate appeared to increase TOC removal on a per gram of carbon basis.
Whether this is due to bacterial activity or "slow adsorption" is not clear.
Other work has shown that compounds removed by a strictly adsorption
mechanism are not affected by variations in linear velocity (11).

Figure 8 illustrates the excellent replication of the TOC breakthrough
data for the three 7.5-min EBCT results. Based on a two-way ANOVA test,
the TOC breakthrough curves for EBCT values of 3.75, 7.5, 15.0, and 22.5
min were judged to be significantly different.

Trihalomethane Removals. Figure 9 illustrates the total THM
results of hourly grab samples of GAC column influent compared to a 24-h
composite sample collected during the same time period. The significant
diurnal variation in total THM values illustrated the need for composite

Table II. TOC Removal Summary

EBCT (min)	Carbon Depth (ft)	Amount TOC Removed Days 55–110 (mg/g)
3.75	1	15.8
7.5	2	16.9
15.0	4	18.1
22.5	6	15.9

sampling. There is no significant difference (based on 95% confidence intervals) between the composite value of 46 μg/L and the mean of the hourly values of 42.6 μg/L.

The more highly variable nature of THM is clearly demonstrated in Figure 10 which shows the influent values and breakthrough curves for chloroform. On a percentage basis, the influent variability of chloroform is significantly greater than that of TOC shown on Figure 5. The chloroform breakthrough curves are also not as "smooth" as the TOC breakthrough curves. The influence of EBCT is demonstrated by the progressively longer time for chloroform breakthrough and exhaustion for each breakthrough curve.

Breakthrough curves for the three other trihalomethanes are shown in Figures 11–13. As bromine substitution increases on the methane molecule, the adsorption of the compound progressively improves. Breakthrough times for the four trihalomethanes for the 3.75-min EBCT were 11, 18, 32, and 49 days, respectively. Traces of bromoform did appear prior to breakthrough; however, this is the usual pattern for trace and highly variable influent organic compound concentrations.

Figure 14 illustrates the increased variability in the 7.5-min EBCT replicate breakthrough curves for chloroform as compared to those for TOC (*see* Figure 8). Also, Figure 14 demonstrates that the effluent chloroform concentrations from all three replicate columns appear to be in excess of the influent concentration on a consistent basis after exhaustion at 39 days. A statistical examination (t test) of the mean breakthrough data for the 7.5-min EBCT column and the influent chloroform concentration established that there was a significant difference at 0.05 α level. This observation is one more illustration of the chromatographic effect discussed elsewhere (*12, 13*). Poorly adsorbed compounds like chloroform are desorbed from the carbon surface by other more strongly adsorbed trihalomethanes or strongly adsorbed fractions of the TOC.

While there was significant variability in the influent chloroform data, a comparison of grab versus composite sample results shown in Figure 15 suggests no significant difference between the two data sets. A t test at the 0.05 α level confirmed the graphical observation. Justifying these results

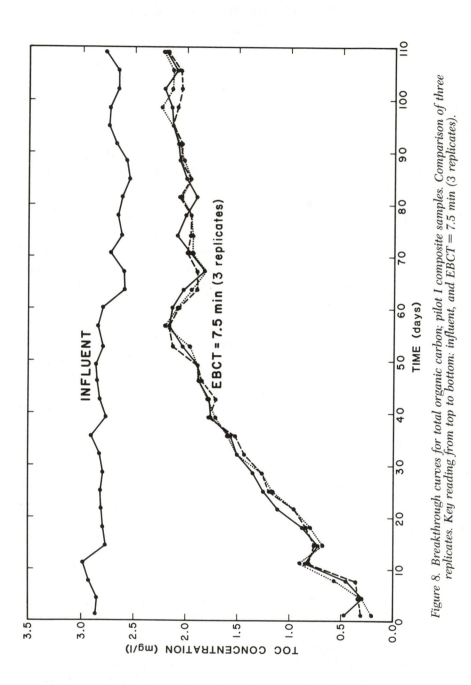

Figure 8. Breakthrough curves for total organic carbon; pilot I composite samples. Comparison of three replicates. Key reading from top to bottom: influent, and EBCT = 7.5 min (3 replicates).

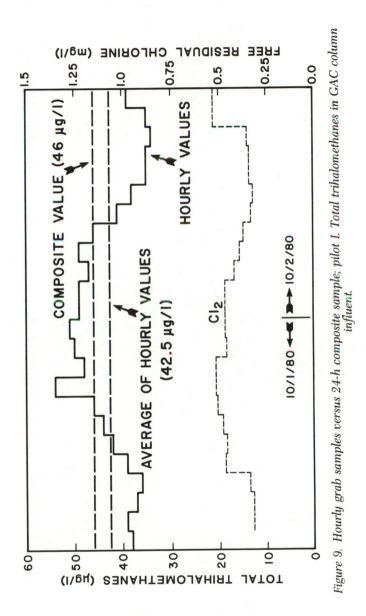

Figure 9. Hourly grab samples versus 24-h composite sample; pilot 1. Total trihalomethanes in GAC column influent.

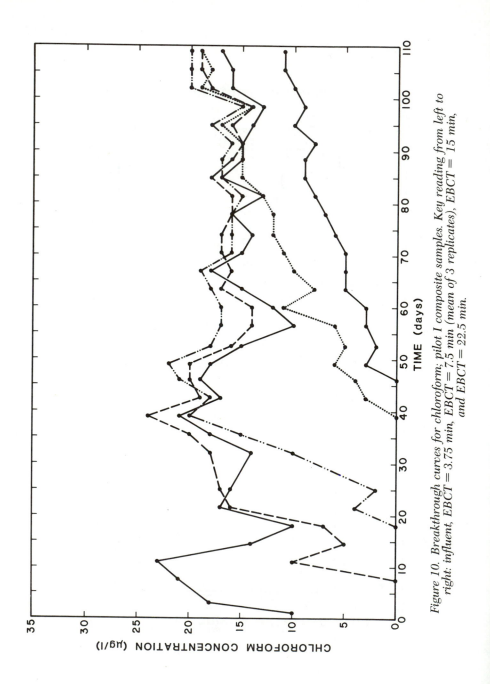

Figure 10. Breakthrough curves for chloroform; pilot I composite samples. Key reading from left to right: influent, EBCT = 3.75 min, EBCT = 7.5 min (mean of 3 replicates), EBCT = 15 min, and EBCT = 22.5 min.

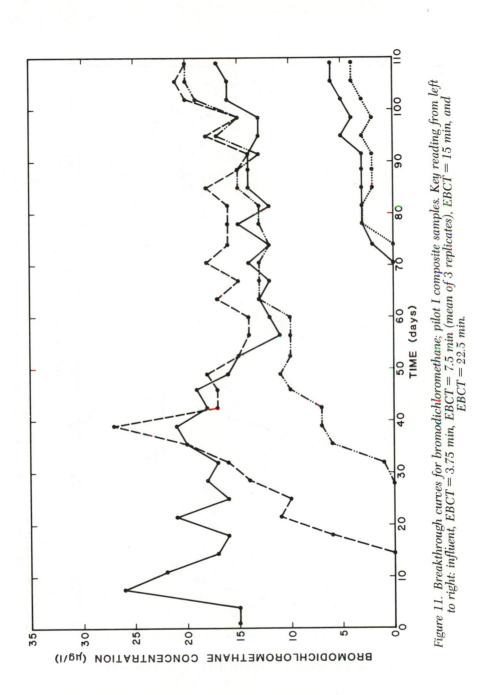

Figure 11. Breakthrough curves for bromodichloromethane; pilot I composite samples. Key reading from left to right: influent, EBCT = 3.75 min, EBCT = 7.5 min (mean of 3 replicates), EBCT = 15 min, and EBCT = 22.5 min.

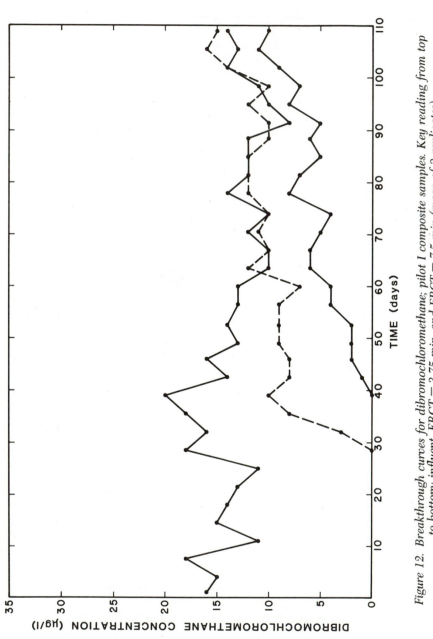

Figure 12. Breakthrough curves for dibromochloromethane; pilot 1 composite samples. Key reading from top to bottom: influent, EBCT = 3.75 min, and EBCT = 7.5 min (mean of 3 replicates).

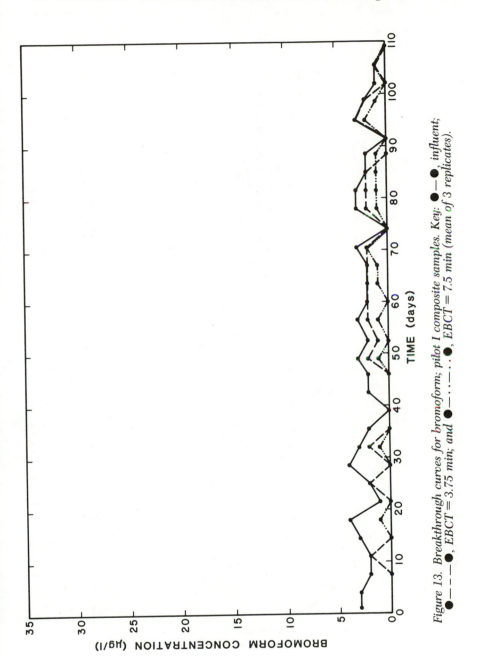

Figure 13. Breakthrough curves for bromoform; pilot 1 composite samples. Key: ●—● influent; ●— — —●, EBCT = 3.75 min; and ●—··—··—●, EBCT = 7.5 min (mean of 3 replicates).

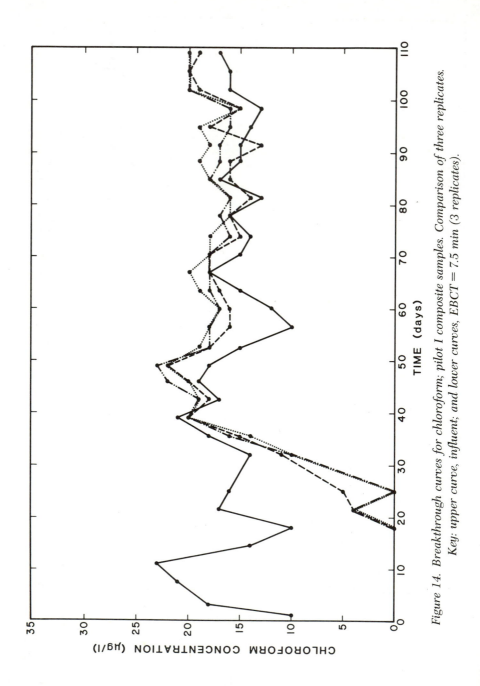

Figure 14. Breakthrough curves for chloroform; pilot 1 composite samples. Comparison of three replicates. Key: upper curve, influent; and lower curves, EBCT = 7.5 min (3 replicates).

with the significant diurnal variation shown in Figure 9 is not straight-forward. It appears that the grab samples were taken at a time during the day that was more or less representative of the 3.5-day composite value. It must be assumed that this is particular to Metropolitan's system and should not be extrapolated to other systems. Every GAC column study should evaluate the need for composite sampling particularly if the removal of highly variable synthetic organic chemicals is a major objective of the project.

Variance Estimates.   Table III lists the variance distribution and coefficients of variation for the five parameters studied in this project. The total variance for the THM compounds appears to be almost exclusively composed of the analytical variance. The anomalous results for $CHClBr_2$ indicate that, for this compound, the analytical variance estimated from the replicate determinations was not a good estimate.

The TOC variance data show that the total variance is composed primarily of the variance associated with running the parallel columns. However, as Figure 8 illustrated, the overall variance is extremely low from a practical point of view. The TOC analytical method is much more precise than the individual THM analyses.

To make comparisons between the total variances for the five parameters, the coefficients of variation were calculated. This statistic was calculated by dividing the square root of the total variance by the overall mean of the breakthrough curve data for the 7.5-min EBCT curves. Table III indicates that the coefficient of variation for the TOC data was at least half of the variation of the THM data. The high bromoform value is the result of dividing by a small overall mean. Figure 13 showed that the bromoform results were generally much lower than the other three THM compounds.

For analyses that are inherently imprecise, such as THM compounds, the total variance was almost exclusively associated with the analysis. Results shown on Table III indicated low relative values for σ^2_{column} which suggests that excellent hydraulic control was maintained throughout the pilot plant. In the design of the pilot plant, a reliable hydraulic design was

Table III. Variance Distribution Summary

Compound	σ^2_{total}	=	$\sigma^2_{analytical}$	+	σ^2_{column}	Coefficient of Variation (%)
$CHCl_3$	1.17	=	1.15	+	0.02	6.6
$CHCl_2Br$	1.00	=	0.91	+	0.09	7.8
$CHClBr_2$	0.31	$\neq$	0.79	+	—	8.7
$CHBr_3$	0.48	=	0.46	+	0.02	80.6
TOC	0.003	=	0.00084	+	0.00216	3.3

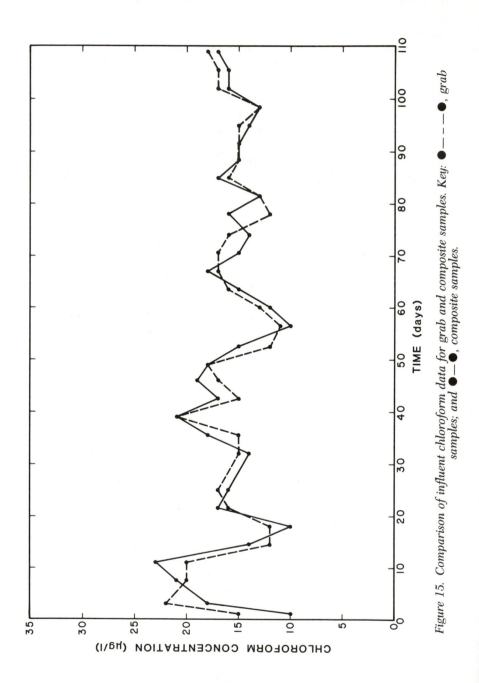

Figure 15. Comparison of influent chloroform data for grab and composite samples. Key: ● – – – ●*, grab samples; and* ●——●*, composite samples.*

recognized as critical to the overall performance of the research studies. It appears that care in hydraulic design has paid off in stable, reproducible results.

Summary and Conclusions

Replication of GAC columns and analytical results is critical if statistically significant results are to be determined.

TOC removal appears to be affected by linear velocity. It is expected from other work that removal of compounds affected primarily by adsorption is not affected by linear velocity. The linear velocity effect will be tested in upcoming experiments.

After 55 days, the TOC removal mechanism appears to be biological degradation, although the "slow" adsorption mechanism cannot be ruled out. More site-specific and mechanistic investigations are needed of the "slow" adsorption phenomenon and its contribution to pseudo steady-state removal of TOC.

The chromatographic effect for chloroform was demonstrated to be statistically significant. The more poorly adsorbed compound is preferentially desorbed by the other more highly adsorbed compounds.

Grab and composite sampling should be evaluated for each GAC treatment study to avoid conclusions based on nonrepresentative data.

The results of the Phase I study showed that the predominant variability of the replicate columns was caused by analytical variability for THM compounds. TOC results were demonstrated to be much more precise than the THM results.

Acknowledgments

The work of Edward G. Means in identifying bacterial genera on the carbon surface and in producing the electron photomicrograph is gratefully appreciated.

Literature Cited

1. McGuire, M. J. In "Activated Carbon Adsorption of Organics from the Aqueous Phase"; McGuire, M. J.; Suffet, I. H., Eds.; Ann Arbor Science: Ann Arbor, Mich. 1980; Vol. 2, 459.
2. Davis, M. K.; McGuire, M. J. "Trihalomethane Monitoring Program Summary Report 1979 and 1980", in-house report, The Metropolitan Water District of Southern California, Feb. 1981.
3. McGuire, M. J.; Shepherd, B. M.; Davis, M. K. "Surface Water Supply Trace Organics Survey—Phase I Maximum Trihalomethane Potential", in-house report, The Metropolitan Water District of Southern California, Feb. 1980.
4. Shepherd, B. M.; Davis, M. K.; McGuire, M. J. "Surface Water Supply Trace

Organics Survey—Phase II Maximum Trihalomethane Potential", in-house report, The Metropolitan Water District of Southern California, Feb. 1981.
5. McGuire, M. J.; Suffet, I. H. "Activated Carbon Adsorption of Organics from the Aqueous Phase"; Ann Arbor Science: Ann Arbor, Mich. 1980; Vol. 2.
6. Westriek, J. J.; Cummins, M. D. *J. Water Pollut. Control Fed.* **1979**, *51*, 2948–51.
7. Sokal, R. R.; Rohlf, F. J. "Biometry"; W. H. Freeman: San Francisco, 1969.
8. Dixon, Wilfrid J.; Massey, Frank J., Jr. "Introduction to Statistical Analysis"; McGraw-Hill: New York, 1969.
9. Benedek, A., "The Slow Adsorption Phenomenon," presented at the 181st National ACS Meeting, Atlanta, Ga., March 29–April 3, 1981.
10. Lassen, J. *Acta. Pathol. Microbiol. Scand. Sect.*, **1975**, *83*, 525.
11. Zogorski, J. S. Ph.D. Thesis, Rutgers Univ. New Brunswick, N.J. 1975.
12. McGuire, M. J.; Suffet, I. H. *J. Am. Water Works Assoc.*, **1978**, *70*, 621.
13. Cairo, P. R.; Radziul, J. V.; Coyle, J. T.; Santo, J. P.; Suffet, I. H.; McGuire, M. J.; in "Activated Carbon Adsorption of Organics from the Aqueous Phase"; McGuire, M. J.; Suffet, I. H., Eds.; Ann Arbor Science: Ann Arbor, Mich. 1980; p. 3.

RECEIVED for review August 3, 1982. ACCEPTED for publication March 3, 1982.

18

Experimental Studies of Distribution Profiles of Organic and Inorganic Substances Adsorbed on Fixed Beds of Granular Activated Carbon

K. ALBEN and E. SHPIRT

New York State Department of Health, Center for Laboratories and Research, Albany, NY 12201

N. PERRINS

New York State Department of Environmental Conservation, Albany, NY 12233

Distribution profiles were obtained for substances on fixed beds of granular activated carbon (GAC) after 26 weeks of operation at the Waterford, N.Y., potable water treatment plant. Aroclors 1016 and 1254, added at 1 μg/L to the influent, were found with maximum surface concentrations (7.6 and 5.4 μg/g GAC, respectively) at the GAC inlet. Chloroform and weak organic acids were found depleted at the GAC inlet, with their maximum surface concentrations (1.5 and 3 mg/g GAC, respectively) at the GAC midsection. Chlorine and calcium, determined by energy dispersive x-ray fluorescence (maximum surface concentrations of 7.3 and 6.9 mg/g GAC, respectively) were preferentially accumulated at the inlet. Calcium detected corresponded to 2% of its input; laboratory experiments suggested the calcium was complexed with organic acids.

GRANULAR ACTIVATED CARBON (GAC) INFLUENT from a moderately contaminated surface water, which has received conventional treatment for human consumption, is generally acknowledged to be a complex mixture of chemicals. Many substances are likely to be present whose relative concentrations and characteristic carbon capacities differ greatly and fluctuate with time. The organics may include humic and fulvic acids and their by-products from chlorination and/or ozonation (trihalomethanes and low molecular weight oxidized fragments), constituents of

0065-2393/83/0202-0407$06.00/0

municipal waste (phenols and fatty acids) and their products of chlorination (chlorophenols), diverse synthetic compounds (volatile solvents and semivolatile acids, bases, and neutrals, such as the priority pollutants), and metabolic products of microorganisms (geosmin, methyl isoborneol, and fatty acids). The inorganics may originate from natural and industrial processes (nutrients, minerals, trace metals, acids, and bases) or from water treatment additives (chemicals for disinfection, coagulation, flocculation, corrosion control, or water softening).

Adsorption isotherms are being measured for a number of these substances. However, pilot column studies conducted under real treatment plant conditions can provide insight into the chemistry of GAC adsorption, particularly when substances are identified whose adsorption is not known a priori.

For this project, extensive analyses were performed on GAC samples from pilot columns at the Waterford, N.Y., potable water treatment plant. Several instrumental methods of analysis were used: thermal desorption and solvent extraction gas chromatography (GC) for volatile and semivolatile organics, high-performance liquid chromatography (HPLC) for organic acids, and energy dispersive x-ray fluorescence (EDXRF) to determine elemental composition.

The samples were thus screened for various substances in different chemical classes, and, in each case, substances adsorbed on the GAC were identified. By analyzing samples collected along the length of the GAC columns, distribution profiles were constructed for these substances. The distribution profiles presented in this paper for chloroform, polychlorinated biphenyls (PCBs, added as a spike to the GAC influent), weak organic acids, elemental chlorine, and calcium summarize the results of several analytical stages of development (1–3). These data are discussed with regard to the origin of the substances adsorbed, their relative level of saturation, and the chemical, physical, and biologic factors that may influence GAC adsorption.

Experimental

Water Supply. Influent to the GAC pilot columns was taken from the finished water supply at the Waterford, N.Y. treatment plant clear well (4). This Hudson River water had been through pre-aeration, coagulation (alum, activated silica, and sodium bicarbonate), flocculation (with powdered activated carbon added), sedimentation, prechlorination, sedimentation, and dual media filtration (anthracite and rapid sand). For the pilot column experiments, this water was continuously spiked with a 50:50 mixture of Aroclors 1016 and 1254, to achieve a total concentration of 1 μg/L. During the final week of pilot column operation, the PCB concentration was increased to 10 μg/L.

GAC Pilot Columns and Sampling Procedures. The design and operation of the pilot columns have been described in detail elsewhere (1–4). Results in this chapter are for samples of Calgon F-400 from four pilot columns (10-cm id; and approximately 120-cm bed depth) connected in series. Influent was pumped at a rate of 76 L/h through the GAC columns (7.5-min empty bed contact time per column and 30-min total contact time). When the pilot columns were shut down after 26 weeks (approximately 9×10^3 bed volumes for all four columns), the columns were divided into 20-cm sections, and GAC samples were collected for analysis.

Particle Size Measurements. Samples of GAC were dried at 100°C, weighed, and sieved, and the fractions were reweighed (*1*).

Organic Chemical Analyses. Initially GAC samples were analyzed for volatile organics by thermal desorption (250°C) GC on a 180-cm packed column of Carbopack C–0.1% SP1000, held at −10°C (5 min) and programmed from 50°C (1 min) to 200°C (30 min) at 8°C/min (*1*). Subsequently, replicate samples were extracted with methylene chloride in glass tubes, which were sealed after the sample end was frozen in liquid nitrogen (*2*). The solvent extracts were analyzed by GC on a 40-m SF96 capillary column, held at −10°C for 5 min and programmed from 5°C to 200°C at 6°C/min. The PCBs were determined by established procedures for solvent extraction, Florisil cleanup, and GC with electron-capture detection (*2,5*). Weak organic acids were extracted in 0.1 *N* NaOH and analyzed by pH gradient HPLC (*2,6*).

Elemental Analyses. GAC samples were dried under vacuum, ground to ≤ 100 μm, mounted on graphite planchettes, and analyzed by scanning electron microscopy–energy dispersive x-ray fluorescence (SEM–EDXRF). A 20-kv electron beam was used for excitation; x-ray fluorescence was typically counted from 0 to 10kv for 325 s. Quantitative EDXRF results presented here are based on synthetic reference samples of Calgon F-400, ground to ≤ 50 μm, and spiked with appropriate inorganic minerals to obtain final concentrations typically ranging from 0.1 to 1% w/w of the elements of interest. These spiked GAC reference samples are considered to represent the matrix of the pilot column GAC samples more closely than the pure inorganic minerals used for quantitation in an earlier paper (*3*). The effect observed with the use of spiked GAC standards was to obtain values for elemental concentrations that are three to four times lower than if results are based on pure inorganic minerals. Even commercially available computer programs written to correct for variations in sample matrices do not correct for differences between GAC samples that are ≥ 95% carbon and inorganic reference matrials that contain virtually no carbon. For these reasons, it was considered important to prepare spiked GAC standards that would bracket the composition of the pilot column samples.

To understand results of SEM–EDXRF analyses of Waterford GAC samples, minicolumns (0.9 cm i.d. by 10 cm long) of virgin Calgon F-400 were exposed to 1.9 L of test solutions containing known concentrations of chlorine and/or calcium, prepared in deionized distilled water and pumped through at a rate of 5 mL/min. Samples taken from the column inlet and outlet were analyzed. In general, the composition of the outlet sample was found by SEM–EDXRF analysis to be essentially that of virgin Calgon F-400. When a positive difference was found for the level of chlorine (or calcium) between a minicolumn inlet sample and its corresponding outlet sample, the result is reported as evidence for the element's uptake by GAC.

Bacteriologic Analyses. GAC samples (0.5 g) were added to sterilized water at pH 7.2. Aliquots were taken, diluted (1/1000), and cultured for 72 h; total bacteria were determined by standard plate counts (*7*).

Results and Discussion

Distribution Profiles of Substances Adsorbed on GAC Pilot Columns. Chloroform, weak organic acids, and the PCBs were the predominant organics found in the sample chromatograms taken thus far. Distribution profiles are given in Figure 1. The profiles show that these substances were all concentrated on the GAC and that chloroform and the weak organic acids had reached the most advanced degree of saturation, if not complete, through the GAC columns. Maximum surface concentrations determined for these substances are given in Table I, together with batch isotherm data reported in the literature (8–10). The maximum PCB surface concentrations found on inlet GAC samples (μg/g) are noted to fall substantially below their batch isotherm capacities (mg/g). By using the PCB adsorption isotherm data, influent concentrations, and cumulative mass inputs, it can be deduced that the GAC should have been saturated to a depth of only 2.4 cm by Aroclor 1016 and 8.4 cm by Aroclor 1254 (2). Sectioning the GAC beds in smaller increments than the 20 cm subdivisions of the Waterford columns would have enabled a more precise determination of the gradient in PCB concentrations at the column 1 inlet, as well as a more accurate determination of maximum PCB–GAC surface concentrations.

Chlorine and calcium were the major elements found at concentrations above their background levels in virgin GAC. Distribution profiles are given for these elements in Figure 2 and for other elements found in GAC samples in Figure 3. Clearly the data in Figure 2 show that chlorine and calcium surface concentrations were enhanced at the inlet to the GAC pilot column, implying that these elements were being removed

Table I. Maximum Surface Concentrations of Chloroform, Weak Organic Acids, and PCBs on GAC Pilot Columns from Waterford, N.Y., Compared to Batch Isotherm Data

Substance	Maximum Surface Concentration (per g GAC)	Batch Isotherm Data (per g GAC)
Chloroform	2.2 mg CHCl$_3$ (1)[a] 1.5 mg CHCl$_3$ (2)	0.2–2 mg CHCl$_3$ for 10–100 μg CHCl$_3$/L (8)
Weak organic acids	3 mg TOC (2)	10–100 mg TOC for 0.1–10 mg TOC/L of humic acids (9)
PCBs	7.6 μg Aroclor 1016 (2)	3.5–15 mg Aroclor 1016 for 1–10 μg Aroclor 1016/L (10)
	5.4 μg Aroclor 1254 (2)	0.75–10 mg Aroclor 1254 for 1–10 μg Aroclor 1254/L (10)

[a]Literature reference.

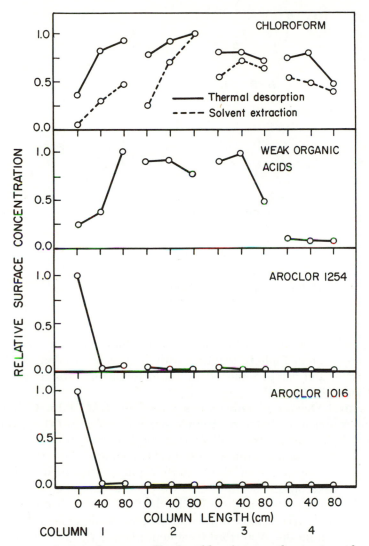

Figure 1. Distribution profiles for chloroform, weak organic acids, and PCBs adsorbed on GAC columns at Waterford, N.Y.

from the influent and concentrated on the GAC. By contrast, the data in Figure 3 show that no consistent trends were evident for Al, Si, S, K, Ti, and Fe: these elements were present on the GAC pilot columns approximately according to their abundance in virgin Calgon F-400. The SEM–EDXRF data obtained for these elements in the Waterford GAC samples were scattered within ± 11–24% of their average value. Maxi-

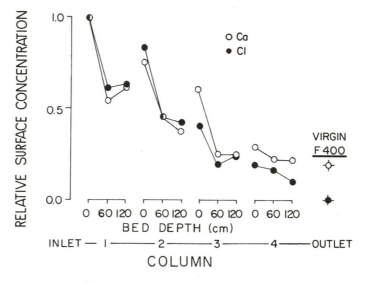

Figure 2. Distribution profiles of chlorine and calcium through GAC
columns at Waterford, N.Y.

mum surface concentrations for the elements found in Waterford GAC
samples are given in Table II, together with their concentrations in virgin
Calgon F-400 and values reported by the manufacturer for commercial
lots of Calgon F-400 (11).

Chemical Origins of Substances Found on GAC Pilot Columns.
The composition of treated water from Waterford, N.Y. (12–14) is given in
Table III. These samples were collected at the point of distribution, which
would correspond to the influent to the GAC columns. From these data,
the presence of substances on the Waterford pilot columns can be better
understood. Chloroform and the PCBs were found on the GAC samples
because they were present in the influent, the PCBs having been added as
a spike (see Table III). Chloroform is the major volatile organic found in
water samples from the Waterford treatment plant (14). The presence of
chloroform is ascribed to humic substances, which account for a large
fraction of the total organic carbon (TOC) of a raw water (15) and which
react with hypochlorous acid to produce chloroform (16). This interpretation is further supported by the TOC content of the water samples (see
Table III).

The weak organic acids also are considered to be related to humic
substances. The high-pressure liquid chromatograms of the organic acid
extracts revealed a low-intensity, early-eluting peak and a high-intensity,
late-eluting peak, which correspond to the successive desorption of
compounds with decreasing acidity (Figure 4). The chromatograms

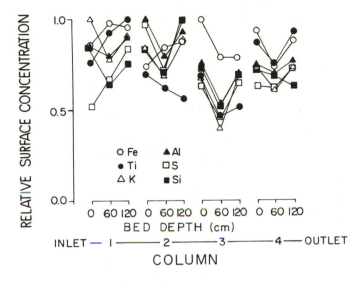

Figure 3. Distribution profiles for silicon, sulfur, aluminum, iron, potassium, and titanium through GAC columns at Waterford, N.Y.

Table II. Elemental Composition (mg/g GAC) of GAC Samples from Waterford, N.Y., and Virgin Calgon F-400

	Waterford GAC		Commercial Calgon F-400 (11)
Element	Pilot columns[a]	Virgin F-400	
Aluminum	—	12.3	—
Calcium	6.9[b]	1.4	1.6
Chlorine	7.3[b]	—[c]	—
Iron	2.2	2.3	5.0
Potassium	0.7	0.8	0.3
Silicon	—	—	15.5
Sulfur	10.9	9.1	8.3
Titanium	0.3	0.6	0.7

[a]Average value.
[b]Maximum value.
[c]None detected, $\leq$ 0.5 mg/g GAC.

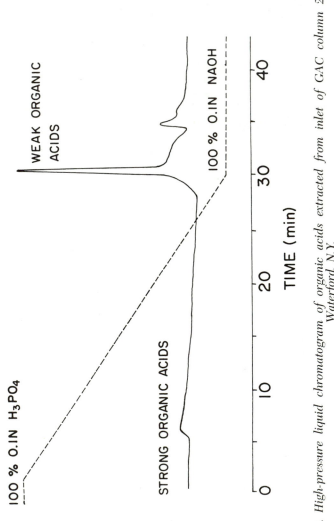

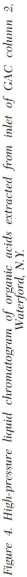

Figure 4. High-pressure liquid chromatogram of organic acids extracted from inlet of GAC column 2, Waterford, N.Y.

Table III. Typical Composition of Treated Water from Waterford, N.Y.

Substance	Concentration (mg/L)	Reference
Organic		
TOC	8.3 (treated)	12
	12.5 (raw)	
$CHCl_3$	0.037	4
	0.048	13
Inorganic		
Residual Chlorine	1.4	12
pH	7.5	
HCO_3^-	69	
SO_4^{2-}	38	
Cl^-	11	
Na	32	
Ca	13	
Mg	2.7	
Al	1.5	
Fe	0.06	
Mn	0.04	

resemble those obtained under similar analytical conditions for humic substances (6).

Several other observations suggest that the organic acids found on the GAC samples are derived from humic substances (2). The organic acids indicated little affinity for a reverse base RP-8 column used to separate phenols, cresols, and chlorophenols. Also, some organic acid extracts coagulated while left to stand in the laboratory. Thus, the fraction of weak organic acids resolved on the XAD-8 column are not the low molecular weight organic acids (≤ 100 amu) that might be expected in water treatment. Finally, the UV spectrum of the colored organic acid extracts is similar to that reported for humic substances (17), with a maximum at 240 nm, followed by a monotonic decrease in absorbance with increasing wavelength.

The presence of chlorine on the Waterford GAC samples (7.3 mg Cl/ g GAC) is attributed to adsorption of both organic and inorganic substances. Elsewhere, similar values for the maximum surface concentration of chlorine on GAC have been reported, as a result of pyrohydrolysis and potentiometric analysis of GAC samples (18, 19). In those studies, both volatile and nonvolatile compounds, some of which are beyond the range of GC analysis, contributed to the total organic chlorine found. Also, it was reported necessary to remove inorganic chlorine by washing the GAC samples with a sodium nitrate solution and taking advantage of the preferential GAC adsorption of nitrate compared to chloride ion (20).

Thus far, we have performed EDXRF analyses of virgin Calgon F-400 exposed to various test solutions (Table IV) and have found that inorganic chlorine is accumulated on GAC from chemisorption of hypochlorous acid. Chloride ion and surface oxides are the major products of reaction between hypochlorous acid and virgin GAC (21), but chloride ion alone is not appreciably adsorbed by GAC (Experiments 1 and 2). However, after exposure to hypochlorous acid, a significant amount of chlorine (approximately 11.7 mg Cl/g GAC) was found and was not removed by washing with distilled water (Experiments 3 and 4).

Table IV. Results of EDXRF Analyses of Virgin Calgon F-400, Exposed to Test Solutions Containing Calcium and Chlorine

Experiment	Influent[a]	Concentration[b] (mg/g GAC)	
		Chlorine	*Calcium*
1	Distilled water	0.5	
2	NaCl, 20 mg/L	0.2	
3	HOCl, 20 mg/L No distilled water wash	13.2	
4	a) HOCl, 20 mg/L b) Distilled water	11.7	
5	a) HOCl, 20 mg/L b) NO_3, 5000 mg/L	0	
6	$CaCl_2$, 20 mg/L		0.4
7	a) HOCl, 20 mg/L b) $CaCl_2$, 20 mg/L		0
8	HOCl, 20 mg/L + $CaCl_2$, 20 mg/L		0
9	$CaCO_3$, 20 mg/L, buffered at pH 9.6 with NaOH–$NaHCO_3$		0.8

[a]a and b denote that solutions were added sequentially.
[b] Concentration above background in virgin F-400.

In further experiments, the ability of nitrate ion to displace inorganic chlorine from the GAC also was confirmed. After a nitrate wash, essentially no chlorine was found on the GAC (Experiment 5). The EDXRF measurements were verified by potentiometric determinations of chloride ion during the experiment (Figure 5). The chloride ion concentration in the GAC effluent remained high while hypochlorous acid was pumped through the column, but it decreased sharply when the column was washed with distilled water. The residual chloride was subsequently displaced by nitrate. Integration of the two peaks indicated that about 10% of the chlorine added as hypochlorous acid was retained on the GAC as a result of the reaction. This value was calculated by correcting the area of the large peaks for the background of chloride ion in solutions of commercial hypochlorous acid. Thus, the chemisorption of hypochlorous acid apparently results in a fraction of chloride ion that is retained in the micropores and that is only removed by flooding the pore volume with a high concentration of another anion.

Experiments to account for the uptake of calcium on GAC have not yet identified the form in which calcium is adsorbed. Integration of the calcium found on the Waterford GAC samples (Figure 2) indicates that only about 2% of the total calcium pumped through the columns in 26 weeks was actually deposited on the GAC. In laboratory experiments (Table IV), it was not possible to bind calcium to the GAC surface (Experiment 6), nor was adsorption enhanced by the presence of hypochlorous acid (Experiments 7 and 8). Deposition of calcium carbonate added as a suspension also was ineffective (Experiment 9). Further experiments are under way to determine if calcium is adsorbed as a complex formed with other organic or inorganic substances in treated raw

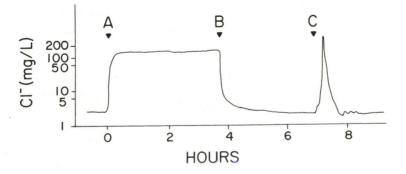

Figure 5. Potentiometric determination of chloride ion concentration in effluent from virgin Calgon F-400 exposed to 2 L each of (A) HOCl, 20 mg/L, (B) distilled water, and (C) NaNO₃, 5000 mg/L.

Table V. Results of EDXRF Analyses of Waterford GAC Column 1 Inlet Samples, Exposed to Conditions for Removal of Chlorine and/or Calcium

| Experiment | Conditions | Percent removal | |
		Chlorine	Calcium
1	Soxhlet extraction for 24 h with hexane	5	0
2	Soxhlet extraction for 24 h with dichloromethane	12	0
3	0.1 N NaOH	41	31
4	0.1 N HNO$_3$	22	91
5	Pyrohydrolysis in helium–steam at 800°C for 30 min	85	0

waters. Recently reported results show that the GAC capacity for fulvic acids is enhanced greatly by the addition of calcium, although mixed effects were observed for the GAC adsorption of low molecular weight acids with calcium present (22).

Complementary experiments were performed to study the ability to remove chlorine and/or calcium from Waterford GAC column 1 inlet samples (Table V). These experiments reiterate the diverse origins of chlorine on the Waterford GAC. From 5 to 12% of the chlorine was removed by solvent–soxhlet extraction procedures appropriate for PCBs, trihalomethanes, and other neutral hydrophobic, semivolatile (chlorinated) organics. It is also interesting that 41% of the chlorine and 31% of the calcium were removed by 0.1 N NaOH. These data appear to suggest organic acids are being extracted that are chlorinated and/or complexed with calcium. However, removal of 22% chlorine and 91% calcium by 0.1 N HNO$_3$ is interpreted to involve separate and probably unrelated adsorption–desorption phenomena. As discussed previously, the nitrate ion is specifically able, by ion exchange, to displace chloride trapped in GAC from chemisorption of hypochlorous acid. Acid conditions in general favor solubilization–extraction of calcium deposits. In the last experiment cited, 85% of the chlorine was removed by pyrohydrolysis at 800°C. This behavior is expected from organic chlorine compounds adsorbed on GAC (16, 17). Thus, the results of these desorption studies agree qualitatively with other data presented in this paper and add yet another pespective to understanding the origins of chlorine and calcium found on Waterford GAC samples.

Relative Levels of Saturation: Effect of Influent Concentrations and GAC Capacities. The distribution profiles for chloroform, the PCBs, and the weak organic acids are considered to reflect basic differences in their influent concentrations (i.e., mass loading) and GAC capacities. Chloroform and the weak organic acids, insofar as the latter are related to the TOC content of the influent, were present at relatively high concentrations (Table III). By contrast, the PCBs (added at 1 μg/L to the influent) were present at quite low concentrations. Also, the maximum surface concentrations (or capacities) reported for chloroform and the weak organic acids are rather low, compared to those of Aroclor 1016 and Aroclor 1254 (Table I). Thus, it is not surprising that chloroform and the weak organic acids advanced well through the pilot columns, whereas the PCBs accumulated at the inlet. In fact, it can be deduced that their distribution profile at the inlet is probably even steeper than the data in Figure 1 indicate (2).

The distributions of chlorine and calcium reveal that, like PCBs, they accumulate at the inlet to the pilot columns. The maximum surface concentrations for chlorine and calcium indicate a significant capacity for the form in which they were adsorbed. As already discussed, the chlorine on the Waterford GAC represents contributions from both chlorinated organics and the reaction of GAC with hypochlorous acid. In laboratory experiments, about 10% of hypochlorous acid in the influent remained chemically adsorbed on the GAC (Figure 5). Chloride ion, formed as the major product of reaction between GAC and hypochlorous acid, was not adsorbed. Similarly, only about 2% of the calcium in the influent was found on the GAC.

Chemical Competition and Displacement, Chemical Reactions, Physical Properties, and Biodegradation. The profiles for chloroform and weak organic acids indicate that complex interactions affect their adsorption. These substances were depleted at the inlet to the GAC system. However, the cause of this behavior may not be the same in each case. Effluent concentrations of chloroform occasionally have been found to exceed influent concentrations (23). In particular, this has been shown to be the case for TTHMs in influent and effluent water samples from column 1 of the Waterford pilot columns, collected during their operation (4). The cumulative adsorption curve calculated for column 1 indicates that TTHMs were continuously displaced between 6 and 26 weeks (2). In contrast, effluent concentrations of TOC have rarely, if ever, exceeded influent concentrations for GAC columns used to treat potable water (23).

From a chemical viewpoint, substances concentrated at the inlet to the GAC system may have caused the displacement of chloroform and weak organic acids. Their adsorption should therefore be studied under controlled conditions that allow them to compete separately with volatile organics with longer GC retention times than chloroform (butanamide, pyridine, toluene, chlorobenzene, chlorofluorobenzene, and ethyl benzene–xylene isomers; *see* Reference 1) and with the PCBs, adsorbed chlorine, and calcium.

It has been hypothesized that surface oxides formed by reaction of hypochlorous acid with GAC alter its capacity for organics (24). However, the effect is only a small decrease in capacity. Distribution profiles for *p*-dichlorobenzene, 1,2-bis(2-chloroethoxy)ethane, and dibutyl phthalate show that they were all concentrated at the inlet of two pilot columns run in parallel, one treating chlorinated influent, the other treating dechlorinated influent. Distribution profiles measured for chloroform on these two columns are both like those reported for the columns at Waterford (13).

Recently it has been shown that hypochlorous acid undergoes on-column reactions with adsorbed humic substances (25). This reaction could account for depletion of the weak organic acids at the inlet to the pilot columns. Any chloroform produced also could be adsorbed downstream. However, this study concluded that most of the hypochlorous acid in the influent reacts with the GAC, rather than with adsorbed organics. Mass balance calculations showed that the presence of GAC significantly limited production of volatile and nonvolatile halogenated organics.

Physical properties of the GAC columns may also influence the extent to which chemical competition and/or chemical reactions occur at the GAC inlet. The mass transfer rate of a substance being adsorbed on GAC varies inversely with particle size (26). The GAC columns at Waterford were stratified from backwashing, as the data in Figure 6 show. The profiles for substances adsorbed on the GAC samples do not reveal a consistent pattern of dependence on particle size. However, experiments are needed to determine if the effect of particle size on competitive adsorption would be strongest at the inlet to a GAC system, where influent concentrations are higher.

Finally, biologic growth is a possible cause of depletion observed at the inlet to the GAC system, particularly for the weak organic acids. Biodegradation has been postulated as a mechanism limiting the total breakthrough of TOC (27). Profiles for total bacteria growing on the Waterford GAC columns are given in Figure 7. In another study, the types of bacteria growing on GAC columns have been partially identified (28). The presence of soil and water saprophytic bacteria was noted, in contrast to the scarcity of coliform bacteria. Further experiments are needed to determine if the growth of these bacteria is related to either humic acids or the weak organic acids found on the GAC samples.

Summary

The complex chemistry of GAC adsorption, under treatment plant conditions, is evident from the analysis of samples from pilot columns at Waterford, N.Y. The analytical techniques used, GC, HPLC, and EDXRF,

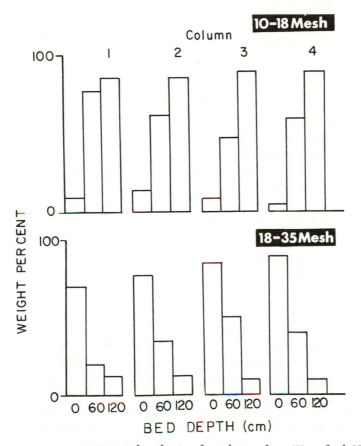

Figure 6. GAC size distribution for columns from Waterford, N.Y.

revealed that GAC adsorbed a number of chemically diverse organic and inorganic compounds.

Distribution profiles for chloroform and weak organic acids showed that they had advanced well through GAC pilot columns, whereas Aroclors 1016 and 1254 were retained at the inlet. Thus far, the relationship of the weak organic acids to humic substances is understood only partially. The relative order of saturation by these organic substances was consistent with available data on their influent concentrations and GAC capacities.

Distribution profiles for chlorine and calcium revealed unexpectedly high surface concentrations at the inlet to the GAC columns. It was concluded that the forms in which chlorine and calcium are adsorbed are very specific and do not reflect their total influent concentrations. For example, both organic and inorganic chlorine compounds were identified and differentiated by their mechanism for adsorption and desorption from

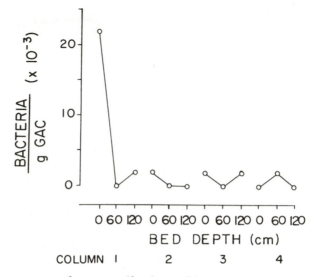

Figure 7. Distribution profiles for total bacteria on GAC columns from Waterford, N.Y.

GAC. The calcium adsorbed represented about 2% of its mass load from the GAC influent. Desorption of calcium from Waterford GAC by dilute sodium hydroxide suggested that it may be complexed with organic acids, but further experiments are needed to identify exactly how the calcium is adsorbed.

Finally, the profiles for chloroform and the weak organic acids revealed the effects of complex interactions on their adsorption. These substances were found depleted at the inlet to the pilot columns. Chemical, physical, and biologic profile data for the columns suggest that competition, chemical reactions, or biodegradation could have affected adsorption. The need for further chemical analyses is evident from the limited ability to resolve mechanisms controlling adsorption.

One hopes that the results and discussion presented in this paper will encourage continued development of instrumental methods of analysis of GAC samples, as well as the use of distribution profiles to understand GAC adsorption. These techniques are proposed to complement the analysis of influent and effluent water samples for the construction of breakthrough curves, so that the information inferred from those analyses can be confirmed directly.

Acknowledgments

Paul Hickey and Min Chen of this Center kindly assisted with the bacteriologic analyses. We are also grateful to Italo Carcich and Jim

Tofflemire of the New York State Department of Environmental Conservation, who supervised the GAC project, and to Thomas Tomayo and Edwin Tifft of O'Brien and Gere for contributing the GAC from the pilot column study. Portions of this paper first appeared in somewhat different form in *Microbeam Analysis-1981* (San Francisco Press, Inc., Box 6800, San Francisco, CA 94101-6800). Portions of this research were supported by New York State Department of Environmental Conservation, through Contract D–141953 to O'Brien and Gere Engineers, Inc.

Literature Cited

1. Alben, K.; Shpirt, E. "Chemistry in Water Reuse"; Ann Arbor Science; Ann Arbor, Mich., 1981; p 264.
2. Alben, K.; Shpirt, E. *Environ. Sci. Technol.*, submitted for publication.
3. Perrins, N.; Alben, K.; Shpirt, E. *Proc. 16th Annual Conference Microbeam Analysis Society*, July 13–17, 1981, Vail, Colo., pp 222–5.
4. O'Brien and Gere Engineers, Inc. "Hudson River Water PCB Treatability Study", N.Y. State Contact No. D-141953 (O'Brien and Gere Engineers, Inc., Syracuse, N.Y., January 1982).
5. U.S. EPA, "Guidelines Establishing Test Procedures for the Analysis of Pollutants; Proposed Regulations", *Fed. Reg.* **Dec. 3, 1979**, *44*, 49464.
6. MacCarthy, P.; Peterson, M.; Malcolm, R.; Thurman, E. *Anal. Chem.* **1979**, *51*, 2041.
7. "Standard Methods for the Examination of Water and Wastewaters", 14th ed., Rand, M.; Greenberg, A.; Taras, M., Eds.; APHA, AWWA, and WPCF: Washington, D.C., 1976, pp 908–12.
8. Chudyk, W.; Snoeyink, V.; Beckmann, D.; Temperly, T. *J. Am. Water Works Assoc.* **1979**, *71*, 529.
9. Boening, P.; Beckmann, D.; Snoeyink, V. *J. Am. Water Works Assoc.* **1980**, *72*, 54.
10. Pirbazari, M.; Weber, W. In "Chemistry in Water Reuse", Cooper, W., Ed.; Ann Arbor Science: Ann Arbor, Mich., 1981; pp 309–40.
11. "Ash and Ash Constituent Analysis for Filtrasorb Carbons (F-400, F-300, F-200, F-100)", August 1978, Calgon Corp.
12. New York State Department of Health, Environmental Health Institute, Routine Water Quality Monitoring Program, Albany, N.Y., 1970–1980, unpublished data.
13. Alben, K., New York State Department of Health, Albany, N.Y., 1980–1982, unpublished data.
14. "National Organics Monitoring Survey", U.S. EPA, Office of Drinking Water, Technical Support Division, Cincinnati, Ohio, 1977.
15. U.S. EPA, "Interim Primary Drinking Water Regulations, Control of Organic Chemical Contaminants in Drinking Water", *Fed. Reg.*, **Feb. 9, 1978**, *43*, 5756.
16. Rook, J., *J. Am. Water Works Assoc.* **1976**, *68*, 168.
17. Schnitzer, M.; Kahn, S. "Humic Substances in the Environment", Marcel Dekker: New York, 1972; pp 55–67.
18. Kuhn, W.; Sontheimer, H. *Vom Wasser* **1973**, *41*, 327.
19. Rook, J. J., "Removal of Chlorinated, Brominated and Iodinated Nonvolatile Compounds by GAC Filtration", presented at the 181st National ACS Meeting, Atlanta, Ga., March 29–April 3, 1981.
20. Maatman, R.; Rubingh, D.; Mellema, B.; Baas, G.; Hoekstra, P. *J. Colloid Interface Sci* **1969**, *31*, 95.
21. Suidan, M.; Snoeyink, V.; Schmitz, R. *J. Environ. Engr. Div. ASCE* **1977**, *EE4* *103*, 677.

22. Randtke, S.; Jepsen, C. *J. Am. Water Works Assoc.*, **1982**, *74*, 84.
23. Sontheimer, H. *J. Am. Water Works Assoc.* **1979**, *71*, 618.
24. Yohe, T.; Suffet, I. *Proc. 97th Am. Water Works Annual Conf.*, June 24, 1979, San Francisco, Calif., Part 1, p 353.
25. McCreary, J.; Snoeyink, V. *Environ. Sci. Technol.* **1981**, *15*, 193.
26. Zogorski, J.; Faust, S. "Carbon Adsorption Handbook", Cheremisinoff, P.; Ellerback, F., Eds.; Ann Arbor Science: Ann Arbor, Mich., 1978; p 753.
27. Tsezos, M.; Benedek, A. *J. Am. Water Works Assoc.* **1979**, *71*, 660.
28. Cairo, P.; McElhaney, J.; Suffet, I. *J. Am. Water Works Assoc.* **1979**, *71*, 660.

RECEIVED for review August 3, 1981. ACCEPTED for publication August 2, 1982.

Pilot Plant Study on the Use of Chlorine Dioxide and Granular Activated Carbon

BEN. W. LYKINS, Jr. and JACK DeMARCO

U. S. Environmental Protection Agency. Organic Control–Field Evaluation Activities, Drinking Water Research Division, Municipal Environmental Research Laboratory, Office of Research and Development, Cincinnati, OH 45268

Chlorine dioxide is shown to be an effective disinfectant and its use allowed a 30–40% reduction in trihalomethane precursors in this pilot plant study in Evansville, Ind. Granular activated carbon removed up to 80% of the remaining precursors at the beginning of a test run when the influent concentration was high (120 µg/L) with no removals at exhaustion after 30 days of use. Performance curves for total organic carbon removal were constructed for virgin and regenerated granular activated carbon; both showed an initial average removal of about 75%, and, after total organic carbon-exhaustion of about 60 days, the rate dropped to 23%. Chlorine dioxide did not produce any organic byproducts other than those noted with chlorine disinfection.

PURSUANT TO PROVISIONS of the Safe Drinking Water Act (Public Law 93-523), extramural research within the U.S. Environmental Protection Agency was designed to satisfy a basic need within the water treatment and supply industry: reduction or prevention of organic compounds in drinking water. As a result of this research objective, an experimental study was initiated in Evansville, Ind. to investigate the use of chlorine dioxide disinfection and posttreatment adsorption by granular activated carbon (GAC). This effort was designed to evaluate a treatment scheme that could drastically reduce or possibly prevent the production of trihalomethanes in the disinfection process in addition to removing other organic compounds by adsorption on GAC. Specific objectives established for this project included:

- Development of a water treatment process using chlorine dioxide as a disinfectant to evaluate the resultant production of trihalomethanes and specific organics as compared to chlorine disinfection.
- Determining if any organic byproducts are formed when using chlorine dioxide as a disinfectant as contrasted to chlorine disinfection.
- Determining the effectiveness of virgin and subsequent re-activated GAC for removal of organic compounds present in the source water, as well as any formed after chlorine dioxide disinfection.

Operation

Full-Scale Treatment Plant Description. One portion of the Evansville Water Treatment Plant was used as a control in this project. The total plant was constructed in a "Y" configuration consisting of two separate treatment schemes with each having a capacity to treat approximately 113.6 million L/day (30 million gallons/day). The South portion of the plant was used as the control, mainly because construction activity was underway at the North plant.

Figure 1 is a flow diagram of the South portion of the full-scale plant. The raw water intakes are located on the Ohio River about 804.5 m (one-half mile) upstream from the city of Evansville. The South plant consists of two primary settling basins, two secondary settling basins, and eight rapid sand filters. Each primary settling basin is 39.6 m (130 ft) in diameter and 5.3 m (17.5 ft) deep with a capacity of approximately 6.8 megaliters (1.8 million gallons). The secondary settling basins are each 27.4 m (90 ft) in diameter with a 4.6 m (15-ft) sidewall depth and a capacity of 2.7 megaliters (0.725 million gallons). Settled water enters the filter building through concrete-lined steel pipes to the sand filters consisting of layers of gravel, sand, and anthracite (mixed media filtration).

Chlorine and alum were added to the raw water with average concentrations of 6 and 28 mg/L, respectively. These concentrations varied, depending on the demand and turbidity. Chlorine dosages were predicated on maintaining a free chlorine residual of 1.5–2.0 mg/L after sand filtration. If make-up chlorine was needed, it was added before the water passed into a common clearwell. Approximately 12 mg/L of lime was added after primary settling for pH control.

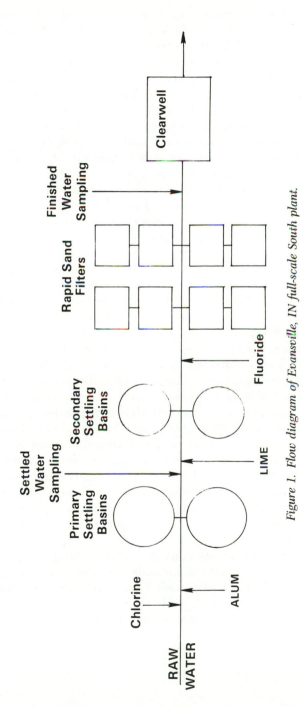

Figure 1. Flow diagram of Evansville, IN full-scale South plant.

Pilot Plant Description.   As a means for studying treatment and operational options at Evansville without disrupting the normal water supply procedure, a 378.5 L/min (100 gallons/min) Neptune Micro-Floc[1] unit was purchased and housed on-site. This unit, or pilot plant, consisted of a rapid mix chamber, a mechanical flocculator, and a settling tube chamber for clarification (Figure 2). Clarified water then passed through a mixed media filter bed to an underdrain system where it was pumped to two parallel pressure GAC contactors or directly to a combination clear-well and backwash tank. Average alum and polymer dosages of 12 and 0.8 mg/L, respectively, were used for turbidity removal. Although not used in the full-size plant, a polymer was necessary because of the short detention time of the water in the pilot plant. For pH control to 8, about 6 mg/L of lime was added to the settling tube effluent. Types of disinfectants and dosages varied depending on the mode of operation. This will be explained in the Pilot Plant Operation section.

Each of two GAC contactors was constructed with a straight column height of 2.4 m (8.0 ft) and an inside diameter of 0.97 m (38 in.). An average carbon bed depth of 2.0 m (6.5 ft) was used to allow for expansion of the GAC during backwash. A hydraulic loading of 3.5 Lps/m^2 (5.1 gallons/min /sq ft) was used to provide a total empty bed contact time of 9.6 min for each contactor. Sample taps consisting of 0.6-cm (1/4-in.) diameter stainless steel pipe were provided at 30-cm (1-ft) increments. A 10-cm (4-in.) diameter opening located just above the Neva Clog under-drain was used for carbon eduction. A similar 10-cm (4-in.) opening was provided at the top of each contactor for carbon replenishment.

For this study, conventional materials normally used in utility construction were also used in the pilot plant. This included carbon steel with an epoxy paint, which was nontoxic and chemical resistant, applied to the inside of the carbon contactors and clearwell to control corrosion.

Pilot Plant Operation.   The project at Evansville was designed to include four distinct sequential phases: training, shakedown, optimization, and operating phase. The training and shakedown phases were necessary to assure efficient operation of the pilot plant and to establish reliable sampling techniques.

The optimization phase consisted of a control study and three different experimental modes of operation. For the control study, the pilot plant was operated in a similar mode to the full-scale plant, namely pre- and postchlorine disinfection, with comparisons lasting for 2 weeks. Comparison of the performance of the three experimental modes of operation to the full-scale plant was 3 weeks duration each and consisted of the modes as shown in Table I. These short-term modes were evaluated

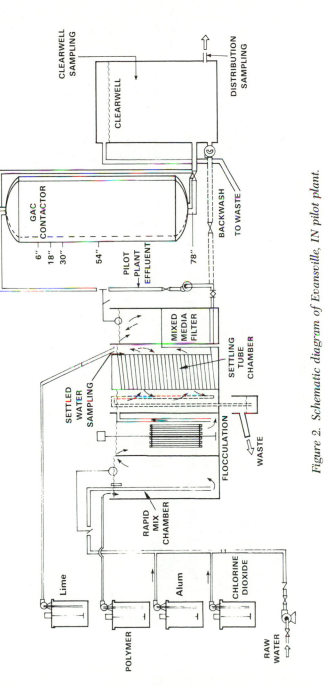

Figure 2. Schematic diagram of Evansville, IN pilot plant.

Table I. Experimental Modes of Operation for Pilot Plant at Evansville, Ind.

Run	Chlorine	Chlorine dioxide	Alum	Polymer	Mix	Settle	Lime	Filter	Carbon	Chlorine	Chlorine dioxide
1	X		X	X	X	X	X	X		X	
2	X		X	X	X	X	X	X	X	X	
3			X	X	X	X	X	X	X		X
4		X	X	X	X	X	X	X	X		X

so that an operational mode could be selected for long-term evaluation during the operating phase.

A total of four runs was completed in the operating phase with each consisting of continuous operation for approximately 3 months. The length of these runs was determined by evaluating the total organic carbon (TOC) and the instantaneous trihalomethanes (InstTHM) for GAC exhaustion. Disinfection in the pilot plant was provided by chlorine dioxide generated from the reaction of sodium hypochlorite (NaOCl) and sodium chlorite ($NaClO_2$) after the pH had been adjusted to 3.0 ± 0.5 with hydrochloric acid. Chlorine dioxide dosages applied to the raw water averaged about 1.5 mg/L to establish a residual of 0.3–0.5 mg/L after mixed media filtration.

For all runs, the performance of the pilot plant through GAC treatment was compared with the performance of the full-scale plant under normal operation. The first run consisted of virgin GAC placed in the two parallel contactors to establish a baseline for subsequent regenerations (reactivations). This parallel comparison provided the opportunity to evaluate the reproducibility of results between the two contactors and to judge minor from major differences when performing evaluations between virgin and regenerated GAC in subsequent comparisons. One contactor (designated T4P) contained 599 kg (1,231 lb) of GAC, while the other one (T5P) contained 567 kg (1,249 lb).

GAC Handling and Regeneration

ICI's Hydrodarco 10 × 30 lignite-based GAC was used. Each contactor, when filled to the designated depth, contained between 545 and 590 kg (1,200 and 1,300 lb) of GAC. The GAC was inducted and educted from the contactors by a water jet that discharged at a rate of 129–140 L/min (34–37 gallons/min). After GAC was introduced into the contactors, it was backwashed for about 2.5 h at 303 L/min (80 gallons/min) to remove the fines.

The GAC to be regenerated was educted into epoxy-coated 208-L (55-gallon) drums containing well screens. The well screens allowed for dewatering of the GAC after eduction and regeneration. On all but one run, the GAC in the drums was shipped to Passaic Valley, N.J. for regeneration.

A 45.4-kg (100-lb) per hour Shirco infrared furnace was used for regeneration of the spent GAC. The GAC was unloaded directly from the 208-L (55-gallon) drums into a small hopper at the base of the carbon-feed screw conveyor. This GAC was dewatered by the conveyor to approximately 50% moisture and fed into the furnace where it dropped onto a woven wire conveyor belt and was leveled into a layer approximately 1.9-cm (3/4-in.) thick. The conveyor belt moved the GAC underneath the infrared heating elements that provided the energy necessary to dry the carbon, drive off adsorbed compounds, and restore the pore structure. Process temperatures were controlled in zones and typically ranged from 1200°F in the drying zone to about 1700°F in the reactivation zone. Residence time of the GAC in the furnace was 20–30 min. Regenerated GAC was quenched and transported back to the drums for shipment to Evansville.

Sampling and Analysis

All sampling was done on a time-sequence basis, so that the same portion of raw water entering both plants was sampled at specified locations. Analysis performed during the extent of the project consisted of quantification of 14 volatile and eight solvent extractable organic compounds as shown in Table II.

The volatile organics were determined by purge and trap procedures using a Tenax trap for adsorption/desorption and detection by electrolytic conductivity detectors. Extractable organics were concentrated for analysis using 15% methylene chloride in hexane after acidification of a 2-L aliquot. After concentration, the hexane was methylated using diazomethane in ethyl ether. The sample was then further concentrated to 5mL and analyzed by electron-capture GC.

In addition, qualitative (detected/nondetected) determinations were performed by MS scans for 32 additional volatiles and 54 extractable organics. Data were also collected on nine inorganic metals (maximum

Table II. Quantifiable Organic Compounds of Interest
Volatile Organic Compounds

Acrolein	Dibromochloromethane
Acrylonitrile	1,2-Dichloroethane
Benzene	1,1-Dichloroethylene
Bromodichloromethane	Ethylbenzene
Bromoform	Tetrachloroethylene
Carbon tetrachloride	1,1,1-Trichloroethane
Chloroform	Trichloroethylene

Extractable Organic Compounds

Bis(2-ethylhexyl) phthalate	1,4-Dichlorobenzene
Butyl benzyl phthalate	Hexachlorobenzene
Di-N-butyl phthalate	Hexachloroethane
1,2-Dichlorobenzene	1,2,4-Trichlorobenzene

contaminant levels), TOC, and other parameters such as turbidity, standard plate count, coliforms, and the disinfectants (chlorine and chlorine dioxide).

Control Study

The performance of the pilot plant without the carbon contactors in operation was compared to the full-scale plant to determine if the full-scale plant could be used as a control for subsequent experimentation with the pilot plant. Chlorine disinfection was used in both plants for a 2-week extensive evaluation. As shown in Table III, the average concentrations of the two plant effluents were comparable.

This close comparison occurred even though the two treatment systems are dissimilar. The main difference between these treatment systems is that the pilot plant employs a much shorter detention time (approximately 37 min) than the full-scale plant (approximately 5 h) with comparable flow differences such that during this study the South portion of the full-scale plant averaged 71.2 million L/day (18.8 million gallons/day) and the pilot plant averaged 303 L/min (80 gallons/min). Because of the contact time difference, an anionic high molecular weight flocculant was used in the pilot plant to accomplish acceptable turbidity removals.

The only major discrepancy between the pilot plant and the full-scale plant effluents was the carbon tetrachloride. The high concentration of carbon tetrachloride in the full-scale plant effluent was suspected of coming from contaminated chlorine gas. This supposition evolved because no carbon tetrachloride was detected in the pilot plant effluent. The two systems were chlorinated differently with calcium hypochlorite being used for chlorine disinfection in the pilot plant and chlorine gas in the full-scale plant.

This problem was eliminated after some excellent detective work by the utility. They were able to trace down the source of contamination and by making the supplier aware that they had the capability and would monitor for carbon tetrachloride, the supplier "cleaned-up" his operation and no further contaminated chlorine gas was received.

Pilot Plant Study

After the optimization phase (discussed previously) was completed, a mode of operation consisting of chlorine dioxide disinfection of the raw water and after GAC treatment was adopted. This mode of operation was used in all four runs in the operating phase.

Virgin GAC Comparisons. The performance of the two parallel

Table III. Performance Comparison of Pilot and Full-Scale Plants
Instantaneous Samples

Parameter	Raw Water	Pilot Plant Effluent	Full-Scale Plant Effluent
Total InstTHM (μg/L)	1.9	33.6	36.9
TOC (mg/L)	2.9	2.1	1.8
Carbon tetrachloride (μg/L)	< 0.1	< 0.1	32.0
1,2-Dichloroethane (μg/L)	< 0.1	< 0.1	< 0.1
Tetrachloroethylene (μg/L)	< 0.1	< 0.1	< 0.1
Turbidity (NTU)	77.0	0.33	0.18
Chlorine (mg/L)	0	1.5	1.9
Median pH	7.5	7.5	7.4
Temperature (°C)	13.3	—	14.6
Coliforms (number/100 mL)	13,000	0	0
Total plate count (number/1 mL)	1,400	2	0

contactors was compared while using identical flows to each contactor that contained approximately the same amount of virgin GAC. The two contactors performed similarly. With an influent TOC concentration of 2–3 mg/L, both contactors were removing about 75% at the beginning of the run. By Day 60, only about 30% of the TOC was being removed, although this removal remained fairly constant, even with a fluctuating influent (Figure 3).

In comparing the performance of the virgin GAC, InstTHMs also were evaluated. For those not familiar with THM terminology, a complete description has been prepared by Stevens and Symons (1). The InstTHM was formed by unreacted chlorine used in the production of chlorine dioxide. Even at the low concentrations shown in Figure 4 (< 5 μg/L) applied to the GAC, equilibrium occurred in about 60 days.

From this observation, it appears that GAC performance for InstTHM cannot be assumed to allow for longer operation before exhaustion if the applied concentrations are low. Because the concentrations are low, however, even at exhaustion, the GAC effluent was below the promulgated standard (0.1 mg/L). This same phenomenon occurred for two of three other runs (Figures 5 and 6). For these two runs, the effluent concentration from the GAC contactors exceeded the applied concen-

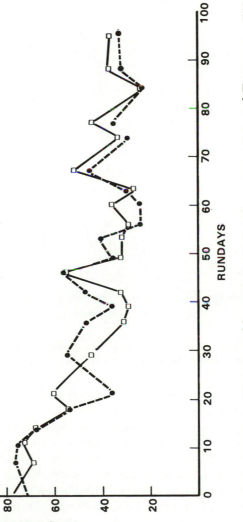

Figure 3. Percentage TOC removal by virgin GAC. Key ●, contactor T4P, and □, contactor T5P.

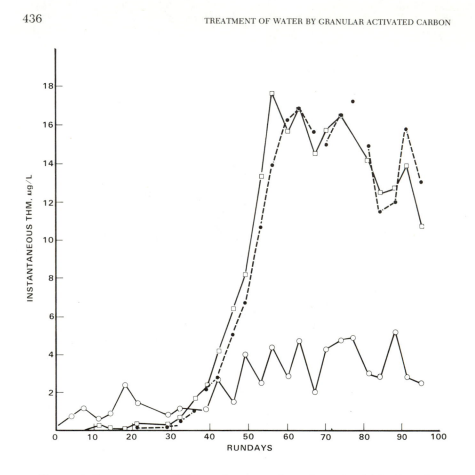

Figure 4. Instantaneous THM performance through virgin GAC. Key: ○, *GAC influent;* ●, *contactor T4P; and* □, *contactor T5P.*

tration in about 15–20 days. Other related studies have shown similar observations at low applied concentrations (2).

GAC exhaustion was determined by the equilibrium of adsorbed TOC and InstTHM with the influent concentration. By evaluating these two parameters, the length of operation of the contactors was determined. As already shown, TOC concentrations were at steady-state in about 60 days and InstTHM concentrations reached equilibrium in about the same time period. By comparison to other GAC studies (2, 3), the concentration of InstTHM at equilibrium was higher than anticipated based on the applied concentration to the GAC. At this time, no explanation can be given for this occurrence.

Although the GAC was exhausted in about 60 days, each run was allowed to continue for approximately 90 days. This extension was needed

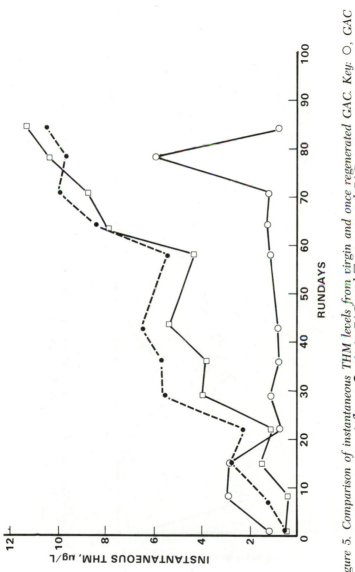

Figure 5. Comparison of instantaneous THM levels from virgin and once regenerated GAC. Key: ○, GAC influent; ●, virgin GAC; and □, regenerated GAC.

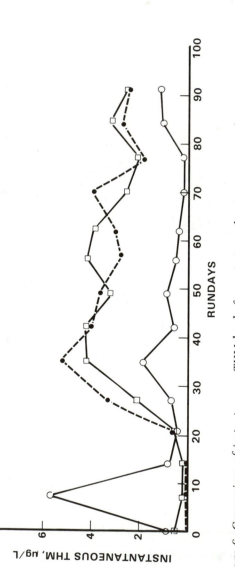

Figure 6. Comparison of instantaneous THM levels from virgin and twice regenerated GAC. Key: ○*, GAC influent;* ●*, virgin GAC; and* □*, regenerated GAC.*

to collect enough data to be sure that the parameter concentrations had actually reached steady-state or equilibrium conditions.

Virgin Compared to Regenerated GAC. The original spent virgin GAC was regenerated subsequently three times. Except for the first regeneration, make-up virgin GAC was required to replenish losses resulting from handling and regeneration. For each run, fresh virgin GAC was added to contactor T4P and its performance was compared to the subsequent regenerated GAC in contactor T5P. Depending on the run, the amount of GAC in these contactors ranged from 499 to 590 kg (from 1,100 to 1,300 lbs).

For all runs when compared to the virgin GAC, the TOC concentration of the subsequent regenerated GAC was practically identical except for the last run where the thrice-regenerated GAC seemed to perform somewhat more effectively (Figures 7–9). This trend was opposite for the terminal THM as shown in Figures 10–12. But in both cases indications were that regeneration had restored the GAC to a state where its performance was similar to the virgin GAC. It should be noted that the terminal THM is essentially the same as the formation potential because of the low instantaneous values.

By evaluating the GAC properties as shown in Table IV for each run, an indication for the slightly improved performance of the regenerated GAC for TOC might be attributed to a change in pore structure. When looking at the iodine number and molasses number, for all runs, a trend has developed whereby the iodine number increased from 617 to 650 mg/g and the molasses number decreased from 357 to 310. This type of trend generally occurs with the subsequent regenerations and does indicate some pore structure change.

Discussion

Samples of raw water, pilot plant effluent before GAC, and full-scale plant effluent were stored for 3 days at ambient temperature and a pH of 8.0 with a free chlorine residual to simulate the residence time at the farthest point in Evansville's distribution system. This gave an opportunity to evaluate a system using pre- and postchlorination as compared to a similar one using pre-chlorine dioxide and post-chlorine disinfection.

The pH in the pilot plant ranged from 6.6 to 8.9. In the full-scale plant it varied from 7.0 to 8.5. Lime was utilized in both plants for pH control. The temperature of the water in both plants varied depending on the season with an average range of 12.6–22.1°C.

As shown in Figure 13, the use of pre-chlorine dioxide disinfection allows about 30–40% more of the THM precursors to be removed by coagulation/settling than with prechlorination, thereby reducing the

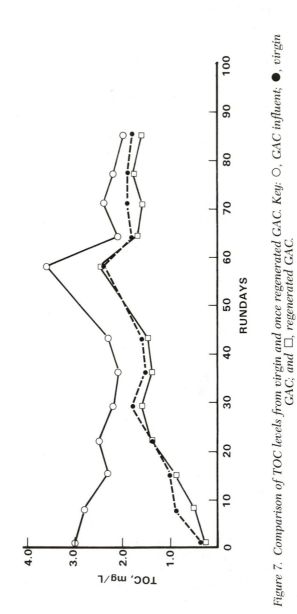

Figure 7. Comparison of TOC levels from virgin and once regenerated GAC. Key: ○, GAC influent; ●, virgin GAC; and □, regenerated GAC.

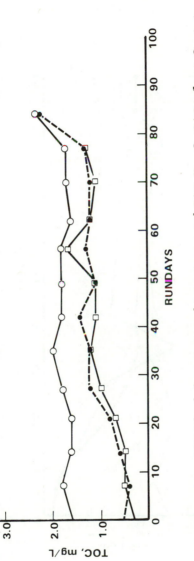

Figure 8. Comparison of TOC levels from virgin and twice regenerated GAC. Key: ○, GAC influent; ●, virgin GAC; and □, regenerated GAC.

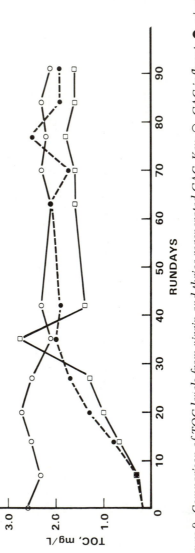

Figure 9. Comparison of TOC levels from virgin and thrice regenerated GAC. Key: O, *GAC influent;* ●, *virgin GAC; and* □, *regenerated GAC.*

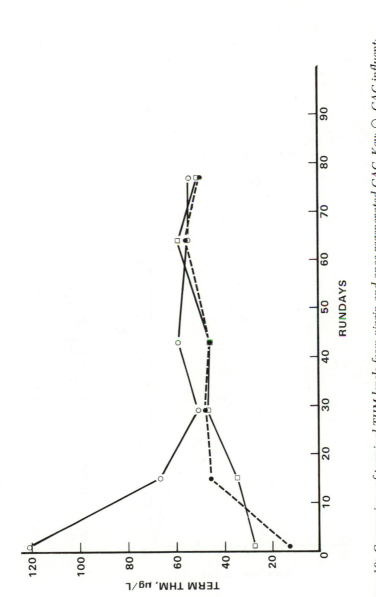

Figure 10. Comparison of terminal THM levels from virgin and once regenerated GAC. Key: ○, GAC influent; ●, virgin GAC; and □, regenerated GAC.

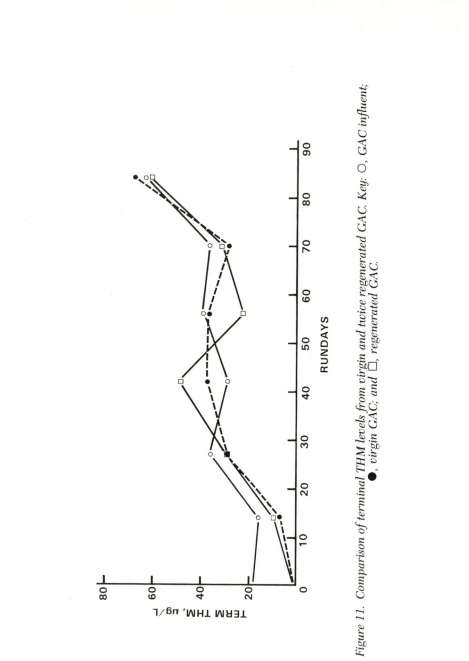

Figure 11. Comparison of terminal THM levels from virgin and twice regenerated GAC. Key: ○, GAC influent; ●, virgin GAC; and □, regenerated GAC.

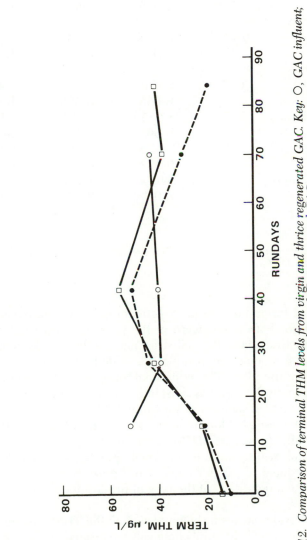

Figure 12. Comparison of terminal THM levels from virgin and thrice regenerated GAC. Key: O, GAC influent; ●, virgin GAC; and □, regenerated GAC.

Table IV. Properties for Virgin and Subsequent Regenerated GAC

GAC Properties	Virgin	Run 1	Run 2	Run 3
Surface area (m²/g)	597	S-UNK	S-636[a]	S-656[a]
		R-675	R-644	R-668
Apparent density (g/cc)	0.427	R-0.554	S-0.434	S-0.443
		R-0.447	R-0.403	R-0.440
Iodine number (mg/g)	617	S-470	S-432	S-467
		R-638	R-596	R-650
Molasses number	357	S-345	S-255	S-250
		R-365	R-324	R-310
Ash content (%)	15.3	S-13.8	S-14.8	S-16.5
		R-13.9	R-15.7	R-17.9
Effective size (mm)	0.80	S-0.74	S-0.70	S-0.70
		R-0.75	R-0.74	R-0.74

Note: S = spent and R = regenerated (before any make-up GAC added).
[a]Sample calcined before analysis.

amount of THM formed during posttreatment chlorination. Because the stored samples for both plants were in the presence of a free chlorine residual for exactly 3 days, the full-scale plant effluent was actually in contact with a free chlorine residual for 5 h longer because of the detention time in the plant. This additional time, however, had no effect on the data shown in Figure 13 as substantiated by the optimization phase control study for stored samples as shown in Table V.

Therefore, the use of chlorine dioxide as a predisinfectant in the pilot plant substantially prevented THM from reaching the concentration present in the full-scale plant under normal operations using chlorine to obtain a free residual of 1.5–2.0 mg/L. No difference was noted in the TOC concentration regardless of the type of predisinfection used (Figure 14).

Chlorine dioxide has been shown to reduce the THM concentration, but what other potentially harmful chemicals are formed by its use? A total of 108 different organic chemicals was evaluated by conventional packed column, GC–MS confirmation and no byproducts attributable to chlorine dioxide were identified. Also, no organics were detected from the use of the epoxy paint.

The fate of chlorine dioxide and one of its inorganic species (chlorite) varied during treatment. In generating chlorine dioxide, a stoichiometric

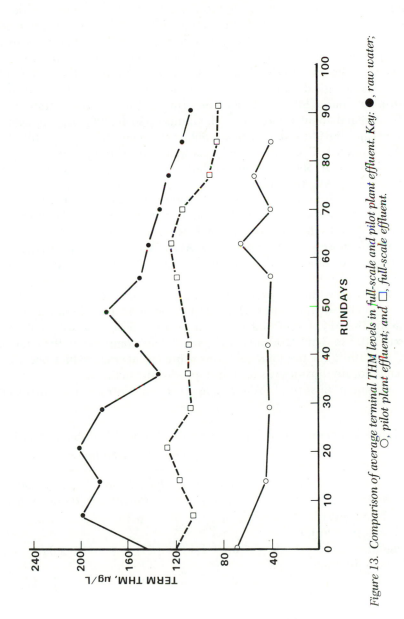

Figure 13. Comparison of average terminal THM levels in full-scale and pilot plant effluent. Key: ●, raw water; ○, pilot plant effluent; and □, full-scale effluent.

reaction was not possible, thereby introducing an average of about 0.8 mg/L chlorite, when an average dosage of 1.5 mg/L chlorine dioxide was applied to the raw water. After filtration, the concentration of chlorine dioxide and chlorite was in the range of 0.3–0.5 and 1.5–2.0 mg/L, respectively, depending on the raw water dosage. The chlorine dioxide was quickly removed by GAC, while the chlorite concentration was reduced to about 0.5 mg/L after a GAC empty bed contact time of 9.6 min.

Bacteriological samples were taken for in-plant analysis of the standard plate count (SPC) and total coliform. Raw water concentrations for SPC averaged about 4,000/mL, with counts up to 40,000/mL. Application of chlorine dioxide reduced this number to about 50/mL after filtration (GAC influent). These average levels increased through the GAC contactors to about 500/mL and 300/mL for the virgin and subsequent regenerated GAC, respectively.

Total coliforms on the other hand averaged about 11,000/100 mL in the raw water and were reduced to about 1/100 mL through filtration with chlorine dioxide disinfection. About 1/100 mL was detected in both the virgin and subsequent regenerated GAC effluents. After post-GAC disinfection with chlorine dioxide, an average SPC of 4/mL and no total coliforms were detected.

Filtration without a predisinfectant had some effect on the bacteriological quality of the water. During this operational mode, a removal of 84% for the SPC (from 290 to 45/mL) and 30% for the total coliforms (from 5,400 to 3,800/100 mL) was noted. This compares to 99% for the SPC and 100% for the total coliforms using the average values presented when chlorine dioxide was used as a predisinfectant.

Although, in several instances, average concentrations for this study

Table V. Performance Comparison of Pilot and Full-Scale Plants
3-Day Stored Samples

Parameter	Raw Water	Pilot-Plant Effluent	Full-Scale Plant Effluent
TermTHM (μg/L)	127.4	76.3	73.3
Carbon tetrachloride (μg/L)	< 0.1	< 0.1	31.1
1,2-Dichloroethane (μg/L)	< 0.1	< 0.1	< 0.1
Tetrachloroethylene (μg/L)	< 0.1	< 0.1	< 0.1
Chlorine (mg/L)	2.8	1.7	2.4
Median pH	8.0	8.0	8.0
Temperature (°C)	13.5	13.6	13.8

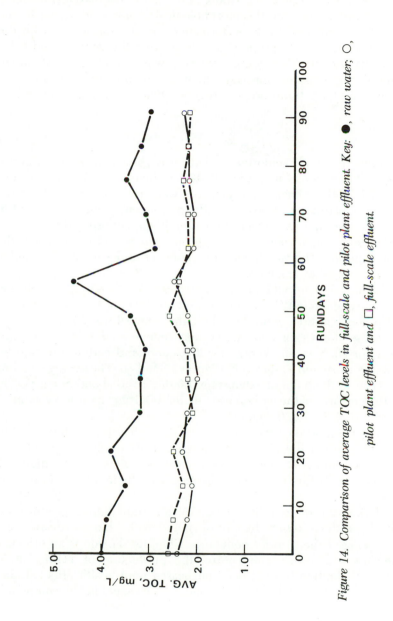

Figure 14. Comparison of average TOC levels in full-scale and pilot plant effluent. Key: ●, raw water; ○, pilot plant effluent and □, full-scale effluent.

have been presented and it appears that the subsequent regenerated GAC performs as well as virgin GAC, a more in depth evaluation is forthcoming. As part of this evaluation, the morphological changes occurring in the GAC will be compared to its performance. On the surface, as with the performance data, GAC properties indicate that the regenerated GAC was essentially restored to virgin state as shown in Table IV. GAC losses incurred during regeneration were about 5% with overall system losses (eduction, handling, transport, and regeneration) averaging about 8%.

Summary

During this study, chlorine dioxide was shown to be an effective disinfectant and its use allowed a 30–40% reduction in THM precursors. Up to 80% of the remaining precursors were removed by GAC at the beginning of a test run when the influent concentration was high (120 µg/L) with no removals at exhaustion after 30 days of use. Other runs did not experience initial removals as high but were exhausted in about the same time period. The variable influent concentrations were attributed to seasonal temperature changes.

With an average TOC influent concentration of 2–3 mg/L, both the virgin and regenerated GAC showed an initial average removal of about 75%. Upon exhaustion, around 60 days, the removal was 23%.

Because the virgin and subsequent regenerated GAC performed similarly for all runs, an average loading curve and carbon use rate for TOC was constructed as shown in Figures 15 and 16. The average loading on the GAC at the point of exhaustion (60 days) was about 20 mg TOC/g GAC. In comparison, the carbon use rate for TOC at exhaustion was about 47.9 mg/L (0.4 lbs/1,000 gallons).

If, however, 30 days were used for GAC exhaustion as indicated by the InstTHM, then the average loading for TOC would be less (13 mg TOC/g GAC) and consequently the carbon use rate would be higher at 95.8 mg/L (0.8 lb/1,000 gallons) indicating less efficient GAC adsorption for TOC.

Chlorine dioxide as a disinfectant in this study did not produce any organic byproducts other than those noted with chlorine disinfection when looking at the priority pollutants. Chlorine dioxide is an effective disinfectant and its use can control the production of THM, but its widespread adoption has not been advocated until health effect studies have been completed. On the other hand, its use should not be completely abandoned before these studies have been completed.

The principal author did an extensive literature survey during this project but most references are not included in the text because only the results of the field study are presented. However, information gathered

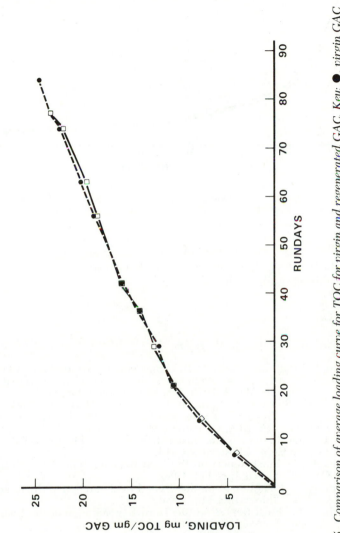

Figure 15. Comparison of average loading curve for TOC for virgin and regenerated GAC. Key: ● *, virgin GAC and* □ *, regenerated GAC.*

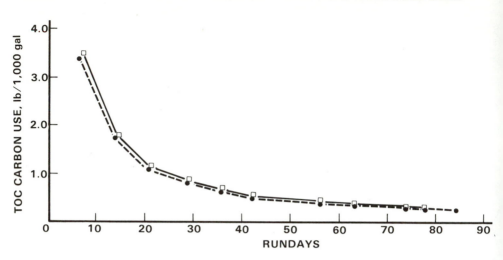

*Figure 16. Comparison of carbon use rate for TOC for virgin and regenerated
GAC. Key:* ●, *virgin GAC; and* □, *regenerated GAC.*

from the literature on the use of chlorine dioxide disinfection was
evaluated during the project and considered in presentation of the results.
This applicable literature is listed (4–26).

Literature Cited

1. Stevens, A. A.; Symons, J. M. *J. Am. Water Works· Assoc.*, **1977**, *69*(10), 546–54.
2. DeMarco, J.; Brodtmann, N. V., Jr. In "Proceedings—Symposium on Practical Application of Adsorption Techniques in Drinking Water Treatment", Reston, Va., April 30–May 2, 1979, U.S. EPA, Washington, D.C.
3. Symons, J. M.; Stevens, A. A.; Love, O. T., Jr.; DeMarco, J.; Clark, R. M.; Geldreich, E. E. "Treatment Techniques for Controlling Trihalomethanes in Drinking Water", U.S. EPA, Cincinnati, Oh., EPA-600/2-81-156, Sept 1981.
4. LaMotte Chemical Products Co. "Instructions for Use of LaMotte-Palin DPD-FAS Titrimetric Unit", Chestertown, Md.
5. McGinnis, F. K. "Final Report on the First Regeneration of the Evansville, Indiana Carbon", Shirco, Inc., Feb 1980.
6. McGinnis, F. K. "Final Report on the Second Regeneration of the Evansville, Indiana Carbon", Shirco, Inc., May 1980.
7. McGinnis, F. K. "Final Report on the Third Regeneration of the Evansville, Indiana Carbon", Shirco, Inc., Jan 1981.
8. McGinnis, F. K.; Horwitz, G. "Effects of Thermal Regeneration on Activated Carbon Properties: A Critical Review of Traditional Physical and Chemical Tests", presented at ACS National Meeting, Environmental Chemistry Division, Atlanta, Ga., April 1, 1981.
9. Juhola, A. J. *Carbon*, **1975**, *13*, 437–42.
10. Love, O. T., Jr.; Symons, J. M. "Operational Aspects of Granular Activated Carbon Adsorption Treatment", Research Report, Drinking Water Research Division, Municipal Environmental Research Laboratory, U.S. EPA, Cincinnati, Oh. July 1978.

11. Palin, A. T.; Darrall, K. B. *J. Inst. Water Engrs. Scientists* **1979**, *33*, 467.
12. Palin, A. T. *J. Am. Water Works Assoc.* **1969**, *61*(7), 483.
13. Wagner, W.; Hull, C. "Inorganic Titrimetric Analysis (Contemporary Methods)"; pp. 173–177, Marcell Dekker: New York, 1971; pp. 174–7.
14. Ramanauskas, E.; Sapragoniene, M. *Chem. Abstr.* **1972**, *77*, 614.
15. Sapragoniene, M.; Ramanauskas, E. *Chem. Abstr.* **1972**, *76*, 532.
16. Sapragoniene, M.; Ramanauskas, E. *Chem. Abstr.* **1972**, *76*, 571.
17. Myhrstad, J. A.; Samdal, J. E. *J. Am. Water Works Assoc.*, **1968**, *60*(7), 205–8.
18. Miltner, R. J., "Measurement of Chlorine Dioxide and Related Products", Research Report, Drinking Water Research Division, Municipal Environmental Research Laboratory, U.S. EPA, Cincinnati, Oh., Jan. 1977.
19. Wheeler, G. L.; Lott, P. F. *Microchem. J.* **1978**, *23*, 160–4.
20. Granstrom, M. L.; Lee, G. F. *J. Am. Water Works Assoc.*, **1958**, *50*(11), 1453–66.
21. Masschelein, W. J. *Water SA*, **1980**, *6*(No. 3, July), 116–29.
22. Masschelein, W. J., "The Use of Chlorine Dioxide in Drinking Water", presented at the Conference of Oxidation Techniques in Drinking Water Treatment, Karlsruhe, Germany, September 11–13, 1978.
23. Augenstein, H. W. *J. Am. Water Works Assoc.*, **1974**, *66*(12), 716–7.
24. Brett, R. W.; Ridgeway, J. W. *J. Inst. Water Engrs. Scientists*, **1981**, *35*(3), 135–42.
25. Roberts, P. V.; et al., "Chlorine Dioxide for Wastewater Disinfection: A Feasibility Evaluation", Final Report for EPA Grant No. R–805426, Wastewater Research Division, Municipal Environmental Research Laboratory, U.S. EPA, Cincinnati, Oh.
26. Gall, R. J. "Chlorine Dioxide: an Overview of Its Preparation, Properties, and Use", Electrochemical Division, Hooker Chemicals and Plastics Corp., Niagara Falls, N.Y., no date.

RECEIVED for review August 3, 1981. ACCEPTED for publication May 4, 1982.

Comparison of the Removal of Halogenated and Other Organic Compounds by Six Types of Carbon in Pilot Filters

J. J. ROOK

Drinkwaterleiding Rotterdam, Galvanistraat 15, Postbus 6610, 3302 AP Rotterdam, The Netherlands

Six types of granular activated carbon were subjected to comparative performance tests in water treatment. Total organic carbon and total organic halogens were determined, as were chlorine, bromine, and iodine levels. The adsorption capacity of the six carbons for trihalomethane was compared, as was chloroform adsorption. The general conclusion of this study was that activated carbon of different makes or origins provides a reliable water purification step for removal of nonvolatile halogenated organic compounds and of mutagenicity. A massive development of planktonic fauna was observed, but the hygienic implication of this is not known.

THIS STUDY COMPARED the performance of six commercially available types of granular activated carbon (GAC) used as terminal treatment in the Kralingen plant of Rotterdam. In this case study, the removal by GAC of halogenated organic compounds, especially the nonvolatile total organic halogens (TOX), that result from pretreatment with chlorine was the main objective. Since halogenated compounds often are associated with potential health hazards, the mutagenicity of the GAC effluents was investigated using the Ames test. The raw water for the plant is abstracted from the Meuse River and stored in open reservoirs with a residence time of 3 months.

During storage, the quality is improved by self-purification, but TOX values as high as 10–14 $\mu g/L$ have been measured. As a means of preservation against biological growth in the pipeline during the 10–14-h period of transportation to the treatment plant, the raw water is chlorinated during summer seasons; when the temperature exceeds 10°C,

1 mg/L chlorine is added. This dosage is below breakpoint (the chlorine demand may vary from 2 to 4 mg/L) so that, in the water arriving at the plant, the residual chlorine is in the combined form. Typical values for combined residual chlorine are 0.2–0.3 mg/L and for ammonium typical values are 0.05–0.2 mg/L. This practice limits total trihalomethane (THM) formation to 10 μg/L.

During the cold season when the transport chlorination is interrupted, THM concentrations remain lower than 1 μg/L. The treatment train comprises coagulation, sedimentation, ozonation (2–2.8 mg/L), and dual media filtration. Water from this stage was taken as the influent for the carbon filters. Table I gives a survey of some quality parameters of the influent water on the GAC. The high maximum value for chloroform was caused by an incidental overchlorination.

Experimental

The six activated carbons tested were obtained from the manufacturers. The few properties available on the commercial data sheets are given in Table II, in which the six brands are coded 1–6. To obtain more information on their internal structure photomicrographs were made by scanning electron microscopy (Figures 1–6).

Particular features are especially characteristic for anthracite-based carbon (1), petrol-coke-based carbon (4), and peat-based extruded product (6). The anthracite type (1) shows a graphite-like microstructure. The petrol coke based (4) typically shows an abundance of small pores. This carbon was especially developed for adsorption of micropollutants of small molecular size. In the micrograph of the peat-based carbon, which is composed of powdered carbon and a binder, the

Table I. Quality of Influent Water (1979)

Parameter	Minimum	Average	Maximum
TOC (mg/L)	1.9	2.6	3.4
UV ext (cm)	0.017	0.028	0.056
pH	7.65	8.0	8.55
NH_4^+ (mg/L)	<0.01	0.06	0.44
O_2 (mg/L)	8.9	11.2	14.6
$CHCl_3$ (μg/L)	0.5	4.4	22[a]
Total THM (μg/L)	<2	8.1	55[a]
Turbidity (FTU)	<0.02	0.06	0.14
Cl^- (mg/L)	46	58	70
Br^- (μg/L)	85	100	130
I^- (μg/L)	2	3.5	5

[a]Exceptional value, caused by accidental high chlorine doses.

Table II. Types of Carbons Tested

Number[a]	Type	Effective Size (mm)	Uniformity	Iodine Adsorption (g/kg) Given	Iodine Adsorption (g/kg) Determined	BET N_2 (m^2/g)
1	European anthracite based	0.4–0.8	1.4	900	1000	900
2	U.S. coal based	0.85	<2	950	950	900
3	European coal based	0.55–0.66	1.9	1000	1050	1100
4	European petrol coke	—[b]	—[b]	1100	1050	1100
5	U.S. lignite based	0.55–0.75	1.8	1050	900	1100
6	European peat based, extruded	cylinders 0.8 mm diameter, variable length		1100	1050	1000

[a]1 = Anthrasorb, type cc 1236, Thomas Ness, United Kindgom; 2 = GAC 30, Carborundum, United States; 3 = F 400, Chemviron, Belgium; 4 = Hydraffin LS Supra, Lurgi, Federal Republic of Germany; 5 = Nuchar WV-G. 42 × 1.68, Westvaco, United States; and 6 =ROW 0.8 Supra, Norit, The Netherlands.
[b]Only size range is given as 0.5–2.5 mm.

binding material is clearly visible. Apparently, the structure of the binder is related to carbon 4. It provides additional micropores.

The filter columns used were 4.8 m high with a diameter of 0.25 m. Each was filled with 145 L of wetted GAC, giving a carbon layer of 3 m. Free board for backwash amounted to 0.8 m allowing 25% bed expansion. After filling the columns, the carbons were backwashed for 24 h to remove fine particles. After this, the height of the layer was adjusted by adding appropriate amounts of carbon. Apparent contact time was fixed at 24 min; accordingly, the percolation rate was 2.5 bed volumes/h, or 7.5 m^3/m^2 h. Sampling taps were positioned at depth intervals of 0.5 m. GAC samples for analysis of TOX were taken from the center by a scoop. Samples for the determination of the effluent total organic carbon (TOC) and other parameters, including bacterial growth and development of higher zooplankton, were taken from the effluent line. Zooplankton was determined by straining 1 m^3 water through a 30-μm plankton net and counting under the microscope.

Bacterial growth was measured by making 22°C plate counts after 3 and 7 days cultivation.

Determination of THM was by purge and trap followed by gas chromatography. The THM formation potential was measured after 48 h of standing at room temperature. The chlorine dose was such that, after 48 h, a free residual of 0.5 mg/L remained.

Measurement of Adsorbed Total Organic Halogen (AOX). Instead of TOX, the term adsorbed organic halogen (AOX) is preferred because essentially the method analyzes that part of TOX that is adsorbable on carbon. At the beginning of the experiment in 1979, the analytical determination of adsorbed organic halogen by pyrohydrolysis of GAC samples (1) was the only available

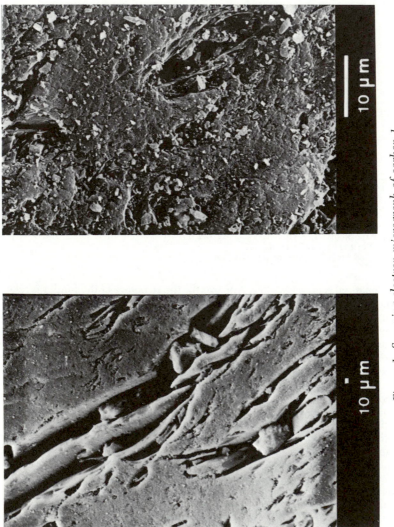

Figure 1. Scanning electron micrograph of carbon 1.

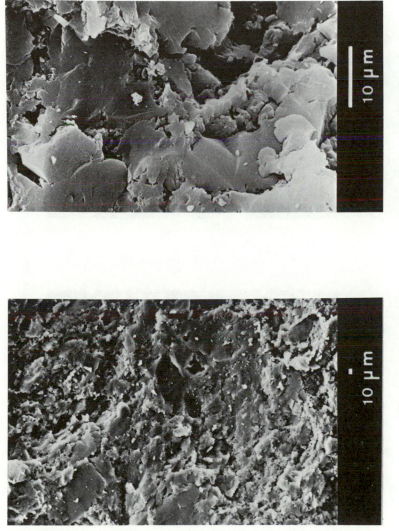

Figure 2. Scanning electron micrograph of carbon 2.

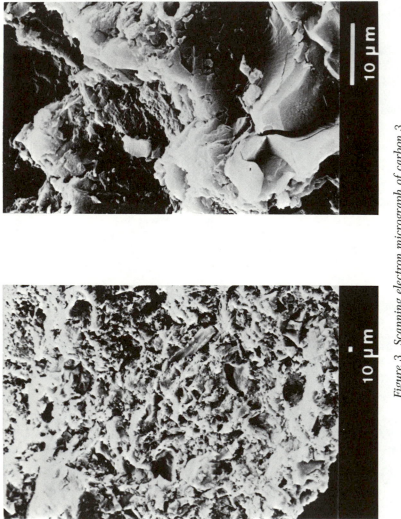

Figure 3. Scanning electron micrograph of carbon 3.

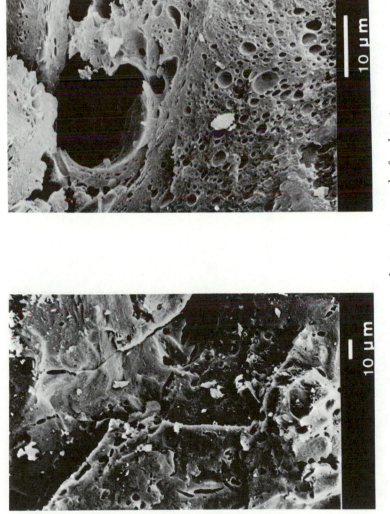

Figure 4.   Scanning electron micrograph of carbon 4.

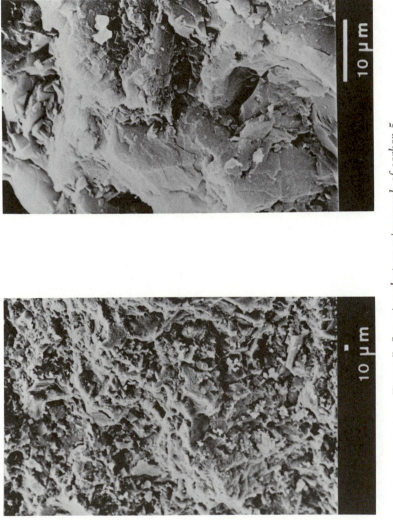

Figure 5. Scanning electron micrograph of carbon 5.

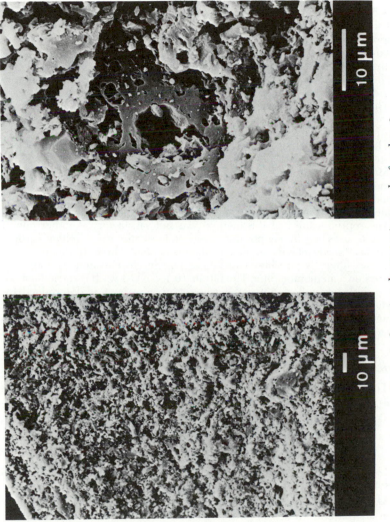

Figure 6. Scanning electron micrograph of carbon 6.

method. A modification was mode, substituting the 24 h or gentle drying with centrifugation of the wet samples. The samples were divided into three portions; 2 g was pyrohydrolyzed giving the total amount of organic and inorganic halides. The halides were measured by microcoulometry in an aliquot of the 100-mL distillate. Further aliquots of the distillate were used for determination of AOBr and AOI by bromide and iodine analysis.

A 5-g portion of the GAC sample was used for the determination of inorganic halide by washing in $0.1N$ nitrate solution. A further 2-g portion was used for determination of the moisture content. The nitrate wash also was effective in removing inorganic bromide.

Bromide was analyzed using a modified method (2) of Fishman and Skougstad (3). The modification consisted of applying a correction for interfering chloride ions that are always in excess. The detection limit for diluted samples was 0.5 $\mu g/L$.

The determination of iodide was based on the principle that iodide ions specifically catalyze the oxidation of arsenite by ceric IV ions which were measured by colorimetry at 420 nm (4). The detection limit was 0.5 $\mu g/L$.

The separate analysis for Cl^-, Br^-, and I^- in the distillate of pyrohydrolyzed carbons allowed determination of AOCl, AOBr, and AOI in the total AOX.

Measurement of AOX in the Aqueous Phase. The analysis for aqueous total organohalides was still under development at the beginning of this work. The method which uses 2 g powdered activated carbon for adsorption from large volumes of water (5–10 L) necessitates a flocculation step by alum as a coagulant (5). The reproducibility of this method was not satisfactory. Another published method (6) uses only 50 mg powdered carbon in smaller water volumes (100–500 mL). It has the advantage that no flocculant is needed. Direct filtration on 0.4-μm membrane filters is possible. The 50-mg powdered activated carbon is easily collected on a membrane filter. The nitrate $(0.1N\ NO_3)$ wash is carried out on the same sample of carbon. Since the amount of adsorbent is determined by an analytical balance before it is added to the sample, the separate determination of moisture is superfluous. By using a microcoulometer, AOX concentrations as low as 0.1 $\mu g/L$ can be detected.

Results and Discussion

Total Organic Carbon. As has been observed often, the residual percentages of TOC measured in the effluents increased steadily with percolated volume. The curves in Figure 7 show that five carbons tested have an identical removal pattern. An exception is carbon 4, which was manufactured for high adsorption of micropollutants. It is the least active with regard to TOC, which apparently is representative of larger molecules.

After percolation of 15,000 bed volumes, the breakthrough curves flatten to plateau values between 70% and 80%. There are no marked differences in performance for these five carbons if compared on a volume base. However, the low density GAC type 6 shows by far the greatest amount of adsorption per unit of mass of adsorbent. It is the only carbon with an apparent density as low as 350 g/L in comparison to the other types with densities between 430 and 470 g/L.

UV Extinction. Measurement of UV extinction at 254 nm is often

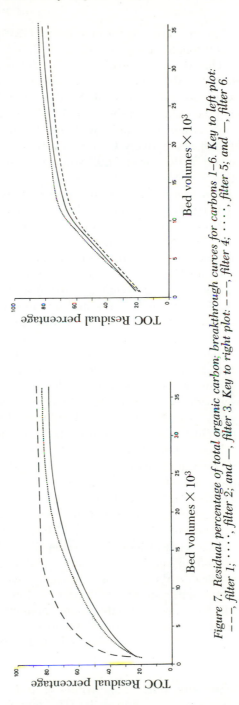

Figure 7. Residual percentage of total organic carbon; breakthrough curves for carbons 1–6. Key to left plot:
— —, filter 1; · · · ·, filter 2; and —, filter 3. Key to right plot: – – –, filter 4; · · · ·, filter 5; and —, filter 6.

advocated as a practical means for monitoring GAC performance. In this study, the breakthrough curves of carbons 1, 2, 3, 5, and 6 did not differ substantially. The curve for carbon 4, however, showed an earlier breakthrough of UV than for TOC analysis. After 10,000 bed volumes, carbon 5 showed the best removal of UV-absorbing compounds with 36% residual level in the effluent; carbon 4 had a 65% residual level which was the worst value. Carbons 1, 2, and 3 were performing with 33%, 36%, and 35% breakthrough, respectively. Carbon 6 with 38% breakthrough for UV extinction ranked at the lower end, which is inconsistent with its high TOC adsorption.

The UV extinction breakthrough curves increased more evenly than for TOC. After 20,000 bed volumes, they reached about 55–65%; after 30,000 bed volumes, they reached values between 70% (carbon 5) and 85% (carbon 4). Carbon 6 remained with 82% near to carbon 4.

THM Formation Potential. This parameter was represented by small differences in UV extinction. The precursor removal was lowest with carbon 4, with 60% breakthrough at 10,000 bed volumes. The other carbons showed breakthrough in the middle range of 40–45%. Carbon 5 was slightly better with 38%. The breakthrough curves rose for five carbons as in the preceding case to values of about 60% breakthrough after 20,000–30,000 bed volumes. Carbon 4 remained at the lower performance of about 75% breakthrough, and carbon 5 had the better performance of lower than 60%. Apparently, the adsorption of precursors matched better with the UV extinction than with TOC analysis.

Chloroform. The behavior of chloroform adsorption, followed by desorption after chlorination was interrupted is illustrated in Figure 8. In the second chlorination season, a limited renewed absorption was observed. The total amounts of chloroform adsorbed in milligrams per kilogram carbon are shown in Figure 9. The lowest adsorption of chloroform occurred on the anthracite-based carbon (1), which adsorbed 11 mg/kg at maximum after 8,000 bed volumes. Carbon 5 had its maximum adsorption of 37 mg/kg after 15,000 bed volumes. The other adsorption maxima were in increasing order: 58 mg/kg for carbon 2, 68 mg/kg for carbon 3, 92 mg/kg for carbon 6, and 93 mg/kg for carbon 4.

The peat-based carbon (6) and the petrol-coke-based carbon (4) were significantly better suited for the small chloroform molecule. This observation is in accordance with the greater amount of micropores shown by their internal structures (Figures 4 and 6). The greater absorptive capacities of carbons 4 and 6 are also reflected in the rise of the curves after the second chlorination had started at 25,000 bed volume (Figure 8).

AOCl, AOBr, and AOI. The profiles of AOCl adsorption on the six carbons tested are given in Figure 10. The curves were determined at

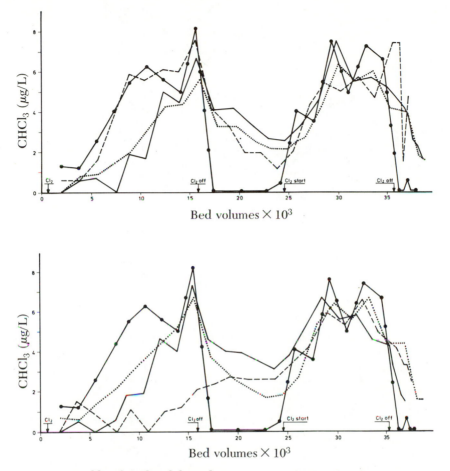

Figure 8. Chloroform breakthrough curves; concentrations in micrograms per liter. Key for top plot: —●—, influent; – – – –, column 1; · · · ·, column 2; and — —, column 3. Key to bottom plot: —●—, influent; – – – –, column 4; · · · ·, column 5; and — —, column 6.

regular intervals, as follows: curve 1 after 4500 bed volumes (b.v.), curve 2 after 9500 b.v., curve 3 after 14,500 b.v., curve 4 after 19,000 b.v., curve 5 after 24,500 b.v., and curve 6 after 30,000 b.v.

Curves 1, 2, and 3 were determined during prechlorination; curves 4 and 5 are without prechlorination. Finally, curve 6 was for the second chlorination period.

Although all six carbons showed different patterns, the increase in AOCl uptake for curves 1–4 is obvious. The fact that curve 4 at 19,500 b.v. still showed a significant increase in AOCl adsorption, although chlorina-

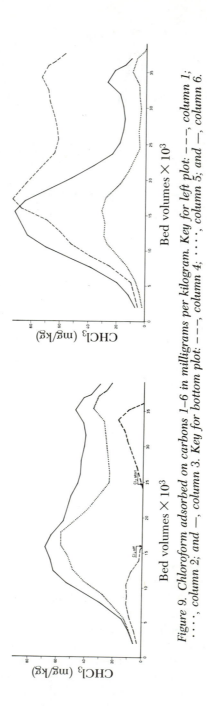

Figure 9. Chloroform adsorbed on carbons 1–6 in milligrams per kilogram. Key for left plot: – – –, column 1; · · · ·, column 2; and —, column 3. Key for bottom plot: – – –, column 4; · · · ·, column 5; and —, column 6.

tion had been interrupted after about 16,000 b.v., is puzzling. An explanation may be that the base load of the raw water still remained on the order of 12–15 ppb. Another explanation may be that the adsorption isotherm may be steep, so that the partition between water and solid lies preferentially on the solid side. However, some elution is observed in curve 5, another 5,000 b.v. later. Immediately after curve 5 had been determined, the second chlorination period started at 24,600 b.v. In all six carbons, the profiles taken at 30,000 b.v. (curve 6) clearly reflected renewed adsorption.

The most active carbons for AOCl adsorption were carbons 2, 4, and 6. The flattened profiles in carbons 3 and 5 may have been caused by more frequent backwash. In order not to disturb the formation of a steep profile, we limited backwash frequency drastically by allowing a higher than usual head loss. Columns 1, 2, and 6 needed only one backwash during the test period. Columns 3, 4, and 5 needed backwash at regular intervals because of clogging. They have been backwashed four or five times during the test period. It is not known if the backwash procedure, which consisted of maintaining 20–25% bed expansion during 1 h, without air scour, allowed for a good stratification of the granules. It may be surmised that the profiles in the good adsorbing carbon 4 might have been steeper with less frequent backwash.

A noteworthy observation is that after 30,000 b.v. had been percolated, the AOCl concentrations measured in the effluents remained very low in all cases and even below the detection limit for carbons 2, 5, and 6.

The AOBr profiles given in Figure 11 match well with the AOCl profiles. The contribution of AOBr to AOX was higher than expected, i.e., 30% of AOCl by weight. However, if, more correctly, the molecular mass is compared, the ratio of organically bound Br atoms is 13% of AOCl.

The amounts of iodine adsorbed are too low to draw profiles. Table III gives the amounts of adsorbed AOCl, AOBr, and AOI and the maximum $CHCl_3$ adsorption. The carbons are given in order of decreasing adsorption. It appears that carbons 2, 4, and 6 rank high, whereas the anthracite carbon 1 ranks low.

Comparison of Carbons. The results given in Table IV show that the coal and lignite-based products are comparable in performance. Only carbons 1 and 4 behave somewhat differently on the parameters chosen. Petrol-coke-based carbon 4 combines high adsorption of organohalogens with low adsorption of medium to large size organic compounds such as precursors and humic substances. Its microstructure (Figure 4) shows abundant small pores. This observation may indicate that the AOX measured on GAC preferentially is of small molecular size. That assumption would also explain the opposed character of the anthracite-based

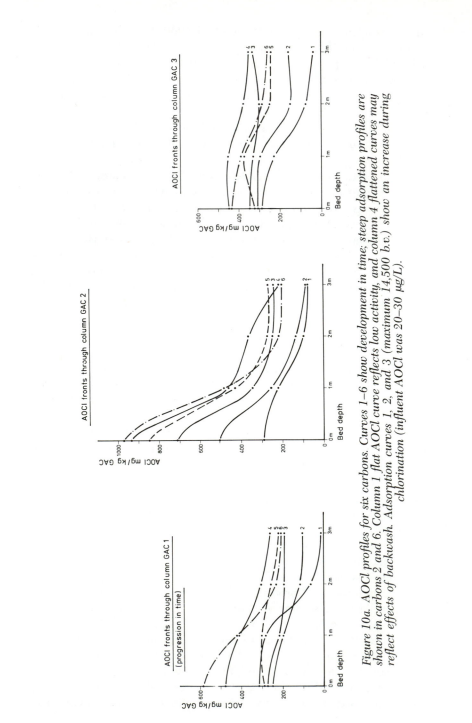

Figure 10a. AOCl profiles for six carbons. Curves 1–6 show development in time; steep adsorption profiles are shown in carbons 2 and 6. Column 1 flat AOCl curve reflects low activity, and column 4 flattened curves may reflect effects of backwash. Adsorption curves 1, 2, and 3 (maximum 14,500 b.v.) show an increase during chlorination (influent AOCl was 20–30 µg/L).

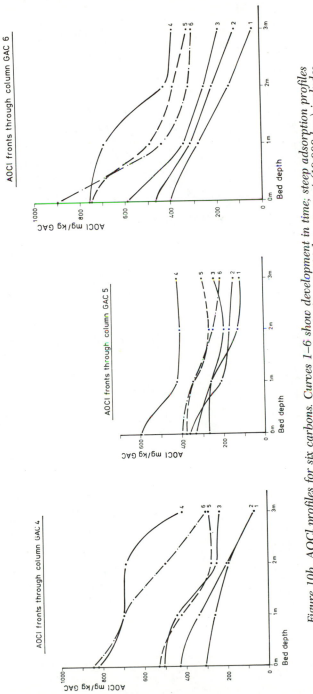

Figure 10b. AOCl profiles for six carbons. Curves 1–6 show development in time; steep adsorption profiles are seen in carbons 2 and 6. Chlorination was interrupted after 16,000 b.v.; curve 4 (19,000 b.v.) includes 3,000 b.v. without chlorine (AOCl was 12–14 μg/L = base load). Curve 5 (24,000 b.v.) shows desorption or decomposition. Chlorination started again at 24,000 b.v.; curve 6 (30,000 b.v.) shows renewed adsorption.

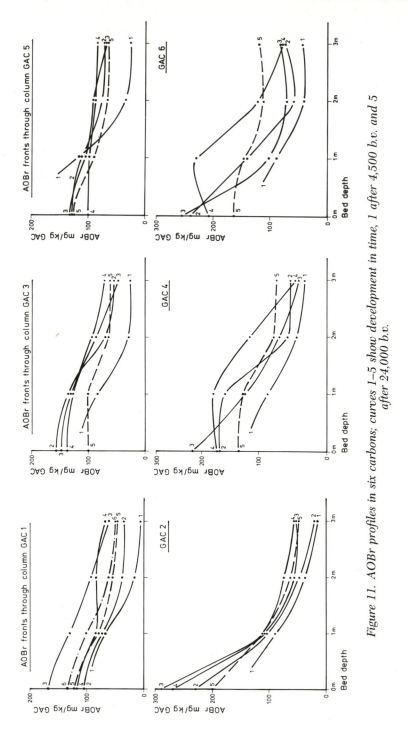

Figure 11. AOBr profiles in six carbons; curves 1–5 show development in time, 1 after 4,500 b.v. and 5 after 24,000 b.v.

Table III. Adsorption of Organohalogens in Milligrams per Kilogram Carbon After 10,000 Bed Volumes

Carbon Number	CHCl$_3$	Carbon Number	AOCl	Carbon Number	AOBr	Carbon Number	AOI
4	93	4	360	4	126	6	23
6	92	2	350	6	126	5	21.5
3	68	6	326	2	105	2	20.5
2	58	3	320	3	102	1	17.5
5	37	5	233	5	98	3	16
1	11	1	208	1	64	4	14

Table IV. Adsorption of Different Organics on Six Carbons (Ranked According to Effect)

Compound	Low		Middle				High	
CHCl$_3$	1	5	2	3			6	4
AOCl	1	5	3	6			2	4
AOBr	1	5	3	2			6	4
TOC	4		5	2	3	1	6	
UV ext.	4	6	2	3	1		5	
THMFP	4		3	6	1	2	5	

carbon 1, which shows fewer micropores, and at the same time ranked lowest in adsorption of large molecules (Table IV). Carbon 6 combined a very strong sorption of small molecules with good adsorption of larger molecules. This behavior of the composite carbon 6 (Figure 6) may be ascribed to the fact that the binding material appears to contribute substantially to the amount of micropores.

The micrographs of the remaining carbons 2, 3, and 5 show identical microstructures, in accordance with their nearly identical adsorption character. Generally, we may conclude that the quality of many activated carbons manufactured for water treatment for general purpose adsorption appears similar. One exception was the German petrol-coke product 4 which showed a preference for small molecules. However, manufacture of that product has ended. A disadvantage of this product was that it gave off a great amount of fines, thus hampering good percolation.

Behavior of Brominated and Iodinated Organics During Treatment. An attempt was made to calculate a mass balance for inorganic bromide and iodide with organically bound halides. This step is not feasible for chloride because its high concentration (50 mg/L) does not allow finding differences in the parts per billion range. Average changes in inorganic bromide and iodide during the season in which chlorination was applied are given in Table V.

Table V. Average Changes in Inorganic Bromide and Iodide During Chlorination Season

Halide	Prechlorination		Ozonation		Total Decrease
	Before	After	Before	After	
Bromide, μg/L	106.5	100	100	100	6.5
Iodide, μg/L	7	5	5	4	3

In the present conditions, bromide ions were oxidized by chlorine but not by 2.5 mg/L ozone. This finding was in contrast to earlier pilot experiments using river water containing 500 μg/L Br^- (2).

The iodide ion was converted by chlorination as well as by ozonation. The deficits are probably converted into organically bound halides, which would correspond with formation of 6.5 μg/L volatile plus nonvolatile organic bromine and 3 μg/L organic iodine.

From the maximum amounts of AOCl, AOBr, and AOI adsorbed on 1 kg carbon after percolation of 10,000 b.v. (Table III) the concentrations that have been removed from the influent water can be calculated: AOCl, 17 μg/L; AOBr, 5.9 μg/L; and AOI, 1 μg/L. The sum of the calculated values would indicate that AOX in the influent water was at least 24 μg/L. In comparison with the values calculated from the oxidative pretreatment, the adsorption of AOBr was good, whereas that of organic iodine seemed modest. This finding need not be interpreted as a weak adsorption. Besides physical adsorption, chemical reactions may take place.

There are indications that decomposition of organic halogen compounds occurs in the carbon columns. During the period that the influent water was chlorinated, the bromide concentrations in the influent and effluent were equal in the first months. Over 4–6 months, a slight increase in bromide was observed in some effluents.

After interruption of chlorination, the bromide ion contents in the effluents of the carbon filters increased by 5–10 μg/L as compared to influents. After 3 weeks, the effluent concentrations had returned to the influent values. The observed increase of inorganic bromide concentrations may be explained by assuming decomposition of the organic brominated compounds on the carbon.

Hydrolytic fission of the organic bromine is a reasonable explanation. This fission may be enhanced by microbial hydrolases. To clarify the role of bacterial metabolism of adsorbed halogenated compounds, solutions of brominated fulvates were prepared in sterile and in inoculated water. The hydrolysis of organically bound bromine in each was determined by measuring the increase in bromide ions.

The inoculated solution clearly showed a steady increase in inorganic bromide during a 10-day period. In contrast, bromide hydrolysis clearly was absent in the sterile solution. The conclusion seems justified that decomposition of bromofulvates is induced by bacterial enzymes.

In the next experiment, the decomposition products of halogenated fulvates were analyzed by GC–MS; the fulvate solution after chlorobromination was split into a sterile and an inoculated portion. After standing 14 days, the GC–MS analysis was repeated. In ether extracts of the freshly chlorobrominated fulvate, the following small molecules were found: THM, bromochloropropene, tetrachloroethylene, chlorinated propanal, chlorobromoacetone, dichloroacetone, monochloroacetate, monobromoacetate, and chlorobenzene. After standing 14 days, most of the compounds with the exception of THM had decomposed in the inoculated solution, in contrast with the sterile solution.

These results all strengthen the reasoning that bacterial metabolism is important in the hydrolysis of organic bromine, and, by analogy, similar biological mechanisms may be expected with chlorinated and iodinated compounds.

Mutagenicity

Mutagenicity was measured after 1 and 2 years of the total run. For each sample, 150 L was concentrated on XAD resin. The eluted extracts were tested in the Dutch State Health Institute (7) with the histidine-deficient mutants of *Salmonella typhimurium T98* and *T100*. The results of both were in the same direction. A sample is judged to give a positive test if its concentrate doubles the number of spontaneous revertant colonies. Table VI gives a survey of the number of induced revertants per liter. Since halogenated compounds are suspected of having mutagenic properties, the AOCl concentrations of the different samples are listed with the Ames test result.

Apparently, the prechlorinated raw water showed a value for mutagenicity of about 20 when AOCl amounted to 20–30 μg/L.

The results of the old production plant I showed that the highest increase in mutagenic activity and in chlorinated compounds was caused by breakpoint chlorination. The subsequent addition of powdered activated carbon (10 mg/L), coagulation, and filtration did not effectively remove the mutagens.

The most interesting data were obtained with the new plant II in which ozonation and activated carbon filters are used. The ozonation did not remove mutagenic activity. It is most striking that, in repeated tests, the effluents of activated carbons of all six types remained negative even after 2 years of use. It is also remarkable that the AOCl concentrations in

Table VI. Mutagenicity and AOCl, Ames Tests, *Salmonella typhimurium* T98

Water Sample	Revertants per Liter	AOCl ($\mu g/L$)
Raw water	—	12–14
after chlorine		
(1 mg/L)	20–30	20–30
Plant I		
After breakpoint		
Chlorine (2–3 mg/L)	70	50–60
After PAC,		
coagulation, and		
filtration	30	35–45
Plant II		
After coagulation,		
ozone, and filtration	15–20	20–28
Filtrate of Six pilot		
GAC filters	negative	0–2.5
Postchlorination	7–12	10–20

the effluents remained near zero, which again suggests an association of AOCl with mutagenicity.

Finally we read from the data that postdisinfection with 0.6 mg/L chlorine leads to reintroduction of significant amounts of mutagens and chlorinated organics.

Consequences of Biology in GAC Filters

The concentration of biodegradable matter on GAC filters leads to abundant bacterial growth. In our six filters, the bacterial counts (22°C) in the effluents all ranged between 500 and 1000/mL. On the carbon granules, bacterial counts of $10^7/g$ were common. As a consequence, zoöplankton graze on the bacterial biomass. The numbers of plankton organisms were counted by straining 1000 L of filter effluent through 30-μm plankton nets. The results are collected in Figure 12 in which we combined related organisms into three main groups: rotifers, crustacea (mostly Canthocamptus and Nauplius larvae), and wormlike organisms (such as Nais and Aelosoma).

It is striking that the development of organisms was practically identical on all six carbons and in numbers ranging up to several thousand per 1000 L.

Organisms in GAC filter effluents per m^3

——— rotifers   — — — wormlike organisms   ········· crustacea

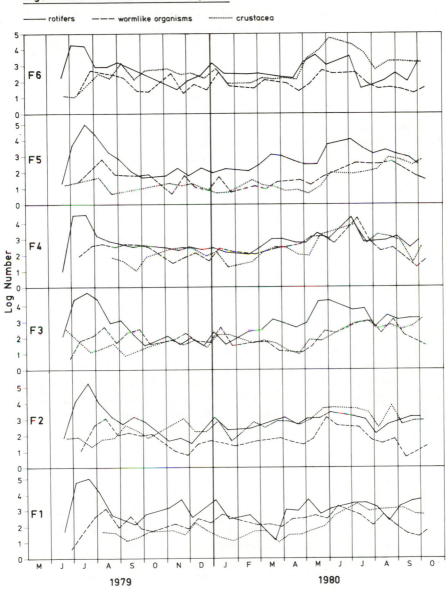

Figure 12. Growth of microfauna on six carbon columns.

In all six GAC columns, the start of zoöplankton growth had an identical time lag. Obviously, a certain amount of bacterial biomass had to develop before plankton growth could start. Another common feature was that rotifers were the first growing higher organisms. This is because this organism can graze on bacterial biomass. As next in the ecosystem, the predators feeding on rotifers developed, especially the crustacean Canthocamptus. This organism as a riparian finds a good foothold on the carbon granules. The only carbon that had lower amounts was the anthracite-based carbon 1.

Free-swimming organisms such as cyclops rarely were found. Seemingly, the environmental conditions of a packed bed are not favorable to cyclops.

A less appealing effect was that, on all carbons, especially the more activated types (e.g., 4), excremental pellets of crustacea were abundantly present. These biological impurities tended to wash out into the carbon-filtered waters, as was found by regular straining of 1000-L samples over plankton nets. The black color of the centrifuged plankton net concentrates first suggested that the sediments consisted of carbon fines. However, microscopic examination showed that they mainly consisted of excremental pellets in which the carbon particles were embedded.

There were indications that clogging of the filters was caused by cell debris and excremental pellets to a great extent. One hour backwash without air scour as applied in the pilot plant was not sufficient to control the abundant zoöplankton growth. Of course, the development of such ecosystems is a consequence of the adsorption of biodegradable matter on the carbon. Biologic fouling of carbon columns must be considered as a hygienic limitation in connection with the long contact times sometimes advocated in biological activated-carbon (BAC) processes.

Summary and Conclusions

Six types of GAC were subjected to comparative performance tests in water treatment. The organic matter was determined by several methods. Total organic carbon (TOC) and UV absorbance were considered to represent mainly the larger molecular fraction of the organic matter.

The total amounts of halogenated organic compounds were determined by ultimate analysis of chlorine, bromine, and iodine. This group was considered to represent the most important pollutants. An association with mutagenicity is proposed.

Trihalomethanes were determined as representatives of small organic molecules. It appears from the adsorption capacities observed (Tables III and IV) that four of the six carbons are comparable in removal of the chosen organic parameters. The anthracite-based carbon, which had the lowest adsorption for nonvolatile organochlorines and trihalomethanes, was one exception. The petrol-coke-based carbon in which high adsorption of trihalomethanes was combined with low adsorption of the larger nonchlorinated molecules (TOC, UV absorbance, and THMFP) was the other exception.

Five carbon filters had in common that the maximum chloroform adsorption from an influent concentration of 4.5 μg/L was reached after percolation of 15,000 bed volumes.

After interruption of chlorination of the influent water, the adsorbed chloroform eluted gradually.

The general conclusion of the experiments is that activated carbon of different makes or origins provides a reliable water purification step for removal of nonvolatile halogenated organic compounds and of mutagenicity.

A small but clearly perceptible increase in bromide concentrations of the carbon filter effluents is explained by hydrolytic fission of adsorbed brominated compounds. Experiments with sterile and nonsterile solutions of chlorobrominated fulvate indicated that bacterial enzymatic activity enhances hydrolysis and further breakdown of this specific substrate. The observations are in support of the presumed biodegradation of halogenated compounds adsorbed in carbon filters.

As an ecological counterpart of the intensive bacterial growth on activated carbon, a massive development of planktonic fauna was observed. The microscopic crustacea produced significant amounts of excremental pellets containing fine carbon particles. The hygienic implication is yet unknown but has to be taken into consideration when the BAC process is used.

Literature Cited

1. Kuhn, W.; Sontheimer, H. *Vom Wasser* **1973**, *41*, 65–80.
2. Rook, J. J. *Vom Wasser* **1975**, *44*, 57–67.
3. Fishman, M. J.; Skougstad, M. S. *Anal. Chem.* **1963**, *24*, 146–9.
4. Sandell, E.B.; Kolthoff, I. M. *Mikrochim. Acta* **1973**, *1*, 9–25.
5. Dressman, R. C.; McFarren, E. F.; Symons, J. M. Evaluation of Determination of TOCl in Water by Adsorption onto Ground Granular Activated Carbon, Pyrohydrolysis and Chloride-ion Measurement, Paper by U.S. EPA at the Water Quality and Technology Conference Kansas City, 1977.
6. Sander, R.; Fuchs, F. Analysevorschrift zur summarischen Bestimmung von organischen Halogenverbindungen in Wässern, *Veröffentl. Lehrstuhl für Wasserchemie,* Karlsruhe 1980, 15, 173–198.
7. Kool, H. J.; Van Kreyl, C. F.; De Greef, E.; Van Kranen, H. J. *Envir. Health. Persp.,* **1982**, in press.

RECEIVED for review August 3, 1981. ACCEPTED for publication April 8, 1982.

A Comparison of Granular Activated Carbon and a Carbonaceous Resin

For Removal of Volatile Halogenated Organics from a Groundwater

P. C. CHROSTOWSKI[1], A. M. DIETRICH[2], and I. H. SUFFET

Drexel University, Environmental Studies Institute, Philadelphia, PA 19104

R. S. CHROBAK

American Water Works Service Company, Inc., Haddon Heights, NJ 08035

This chapter presents results of both laboratory and pilot plant studies conducted on removal of four volatile halogenated organics from a contaminated groundwater. Adsorption capacities are reported along with material on competitive adsorption. The effects of physical process kinetics and mass transfer as shown by breakthrough curves are discussed. The results are interpreted on the basis of contemporary adsorption models and compared to other installations utilizing similar adsorption operations.

G ROUNDWATER SOURCES supply approximately half the United States population with drinking water. In the recent past, most groundwater supplies were thought to be reasonably free from chemical contamination. This attitude has lately begun to change. The President's Council on Environmental Quality (CEQ) has stated that "groundwater contamination from various activities such as the use of pesticides or the improper disposal of toxic wastes is becoming a major problem for drinking water supplies in parts of the country" (*1*). Specific instances of groundwater pollution by organics have been documented for trihalomethane precursors (*2*), volatile halogenated organics (*3–5*) and pesticides (*6*). The extent of this problem has recently been reviewed by the

[1]Current address: Vassar College, Department of Chemistry, Poughkeepsie, NY 12601.

[2]Current address: University of North Carolina, Chapel Hill, NC 27514

0065-2393/83/0202-0481$06.50/0

CEQ (7). The compounds most usually identified are low molecular weight volatile halogenated compounds that are primarily used as industrial solvents and degreasers. Concentrations found ranged from limits of detection upward to the limits of solubility.

In 1978, the Rutgers University Department of Environmental Sciences under contract to the New Jersey Department of Environmental Protection (NJDEP), surveyed groundwater from over 1000 sources in the state. In 400 wells sampled at random, 5% had concentrations of organics that exceeded state limits for potable use (8). One of the contaminated wells surveyed was the Vannatta Street well in the Borough of Washington, Warren County, New Jersey. This nominal 0.8 million gallons/day well is one of three sources serving about 10,000 people in the municipality. The well is operated by the New Jersey Water Company–Washington District, which is a Subsidiary of American Water Works Company. Subsequent analysis of the water revealed the presence of numerous volatile halogenated organics (VHOs) along with a few other organic compounds. The results of these analyses are given in Table I. These findings led to the investigation of methods of removing the organic contaminants.

Unit operations currently in use for removing trace organics are effectively limited to adsorption and air stripping. Because the background organic concentration in this aquifer was very low (TOC < 0.5 ppm), we felt that adsorption would be effective in controlling the organics problem. We decided to evaluate the use of granular activated carbon (GAC) (Filtrasorb F-400, Calgon Corp.) and a synthetic carbonaceous resin (Ambersorb XE-340, Rohm & Haas Company) for effectiveness in removing the organics. GAC has a well-known history of use in this

Table I. Organic Analysis of Vannatta Street Well

Compound	Concentration Range (μg/L)
Tetrachloroethylene (PCE)	30-215[a,b]
Trichloroethylene (TCE)	ND-6[a]
Carbon tetrachloride	ND-34[a,b]
Chloroform	ND-61[a]
Other THMs	0.5[a]
1,1,1,-Trichloroethane (111-TCE)	15-51[b]
1,2-Dichloroethylene	detected[b,c]
Benzene	detected[b,c]
Hexane	detected[c]
Toluene	detected[c]

[a]Brandt Associates.
[b]Drexel University.
[c]Calgon Corporation.

context; XE-340 is reported by the manufacturer to be extremely effective in adsorption of VHOs (9).

The primary purpose of the laboratory part of the study was a direct comparison of the two adsorbents under identical operating conditions. Also, since field conditions rarely approach ideality, we felt that the on-site pilot study would be most fruitful, yielding data relevant to the sensitivity of the adsorbent to hydraulic or chemical perturbation (i.e., backwashing, variations in concentrations, and particulate effects). The characteristics of the adsorbents used are given in Table II. The adsorbents were evaluated for capacity, time to breakthrough, and complicating factors such as the "chromatographic effect" where the effluent concentration from the adsorbent column exceeds the influent concentration in laboratory and pilot scale columns. The tests are limited in scope to one field location and only should be interpreted within the scope of the work completed.

Experimental

Studies were conducted at both the laboratory and the pilot plant level. The scaling factors were 6×10^5:240:1 for full scale:pilot plant:laboratory based on volumetric flow rate. The laboratory studies were conducted at Drexel University; pilot plant studies were conducted by Calgon Corporation in conjunction with American Water Works Service Company at Washington, N.J. The identities and concentrations of compounds studied are shown in Table III. Laboratory and pilot plant studies were designed to complement each other; no attempt was made to duplicate field conditions in the laboratory.

Laboratory Studies.   Water from the Vannatta Street well was received in 55-gallon polyethylene-lined drums. Upon receipt, the pH was measured and sufficient hydrochloric acid was added to lower the pH to less than 3.0. The purpose of this step was to inhibit microbial growth. Approximately 1 pint of 3% hydrogen peroxide was added to each drum to ensure that the system remained aerobic. The water was conducted through Teflon tubing by gravity to two glass reservoirs (*see* Figure 1) with a combined capacity of about 70 L. The reservoirs were spiked with a solution of four compounds of interest dissolved in methanol.

Table II. Characteristics of Adsorbents

Characteristics	XE-340[a]	F-400[b]
Surface area (m^2/g)	400	1050–1200
Pore volume (cm^3/g)	0.34	0.94
Density (lb/ft^3)	37	25
Particle size (mesh)	20–50	12–40
Iodine number	530	1050
Pore size (Å)	31% < 100	—
	69% 100–300	

[a]Rohm & Haas (Ref. 18).
[b]Calgon.

Table III. Volatile Halogenated Organics Studied

Compound	Laboratory	Pilot Plant
111-TCE	95.9 ± 48.4	25.3 ± 3.7
CCl$_4$	121 ± 51.9	3.3 ± 1.5
TCE	38.7 ± 14.8	1.8 ± 0.8
PCE	92.6 ± 35.0	349 ± 63.8
CHCl$_3$	0	0.8 ± 2.0
1,2-Dichloroethylene		
cis	0	trace
trans	0	trace

Note: Values are the arithmetic mean ± standard deviation in micrograms per liter. If below detection limits, they are counted as zero.

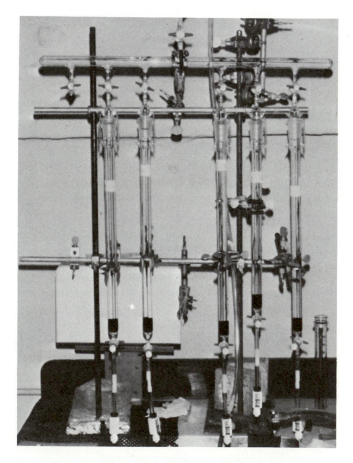

Figure 1. Instrumentation used in the laboratory study.

Table IV. Operational/Hydraulic Data—Laboratory Study

	XE-340			F-400
Condition	*1*	*2*	*3*	*4*
Diameter (cm)	1.9	1.0	1.0	1.0
Height (cm)	10	2.8	2.6	2.9
Q (cm^3/min)	14.3	7.0	3.7	3.6
Volume (cm^3)	29.2	7.3	7.3	7.3
EBCT (min)	2.0^a	1.0		
	14.3442	3.7567	3.7567	3.0555
Bed volumes per day	705	1380	730	710

^{a}Recommended by manufacturer.

The compounds used were 1,1,2,2-tetrachloroethylene (PCE), 1,1,1-trichloroethane (111-TCE), trichloroethylene (TCE), and carbon tetrachloride (CCl$_4$). Methanol concentration did not exceed 1.5 mg/L.

The XE-340 resin was treated by repeated washing with methanol to remove fines followed by two distilled water rinses. Following this treatment, it was boiled for 10-20 min in distilled, organic-free water to remove air from the pores. Methanol was used as the solvent in conjunction with XE-340 because the manufacturer indicated that this solvent was effective for regeneration (*11*). We therefore presumed that methanol would not have any deleterious effects on resin function. Activated carbon was treated by washing repeatedly with distilled, organic-free water to remove fines. Both carbon and resin were packed in the columns to a specified bed depth as an aqueous slurry.

Samples of column effluents and influents were taken daily and stored in head-space free, Teflon-capped vials under refrigeration in the dark. Flow rates through each column were monitored a minimum of once daily and adjusted as necessary. Table IV lists the operating conditions of the columns. Figure 2 shows the variable nature of the influent, which is considered to simulate the variable influent concentrations that presumably would occur under field conditions.

After all the samples had been collected, 5.0 mL of each was extracted with 2.5 mL of hexane in a Teflon-capped vial in the manner of Otson et al. (*12*). Hexane (Burdick and Jackson pesticide quality) was redistilled with lithium aluminum hydride under nitrogen before use. The vials were shaken for 30 min and aliquots of the hexane extract were analyzed via gas chromotography on a Tracor 222 GC. The chromatographic conditions were as follows: a 12-ft × 2-mm i.d. glass column packed with 20% SP-2100/0.1% Carbowax 1500 on 100/200 Supelcoport (Supelco); an electron capture detector with N$_2$ carrier gas; and an oven temperature of 100°C isothermal.

The compounds of interest were resolved without difficulty. For a typical analysis, retention times were 172 s for 111-TCE, 191 s for CCl$_4$, 219 s for TCE, and 401 s for PCE. Quantitative analysis was accomplished by comparison of the peak areas of the sample to those of analytical standards subjected to identical treatment including liquid–liquid extraction. Standards (Chem Service, Chester,

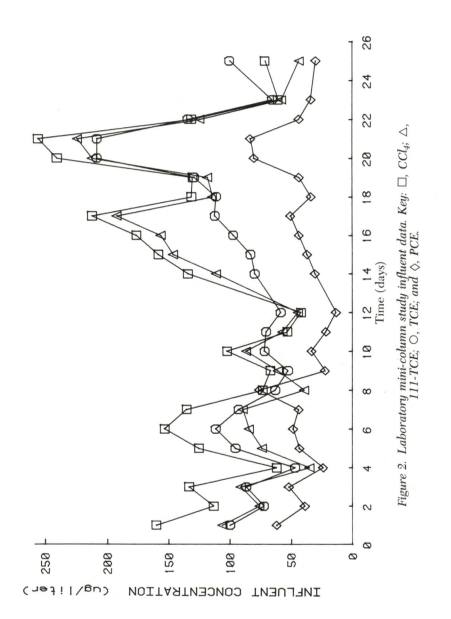

Figure 2. Laboratory mini-column study influent data. Key: □*, CCl₄;* △*, 111-TCE;* ○*, TCE; and* ◇*, PCE.*

Table V. Operational/Hydraulic Data—Pilot Plant Study

	F-400			XE-340
Condition	1	2	3	4
Diameter (cm)	10.2	10.2	10.2	10.2
Height (cm)	91.4	91.4	91.4	43.2
Q (L/min)	0.89	0.89	0.89	1.89
Volume (L)	7.34	14.69[a]	22.06[a]	3.48
EBCT (min)	8.4	16.9[a]	25.5[a]	1.8
Weight (g)	3040	—	—	2093
Bed volumes per day	175	—	—	782

[a]Cumulative values.

Pa.) were made by filling a 10.0-mL volumetric flask nearly to the mark with methanol (Burdick & Jackson, redistilled), weighing, adding the analyte, reweighing, and filling up to the mark.

Precision studies were carried out to test reproducibility of the method. Four replicates of a high concentration standard of TCE gave 74.5 ± 5.3 μg/L (mean $\pm$ standard deviation) or a coefficient of variation of 6.6%. For lower concentration standards, mean $\pm$ standard deviation was 6.1 ± 2.6 μg/L or a coefficient of variation of 44%. Analytical quality control was conducted by evaluating consistence of response factors. The chromatographic response factor (*13*) is defined as the ratio of the chromatographic peak area to the mass of sample represented by the area. This value was calculated for the standards each time a set was run (usually daily or if any chromatographic conditions were altered). These values were then compared with previous values. Additionally, ratios of response factors for the different compounds were compared (e.g., 111-TCE response:PCE response). Those data sets in which response factors were not consistent (agreement within 15%) were discarded and are not reported here.

Several preliminary laboratory column experiments were conducted to determine the optimum operational conditions. The final run, for which data are presented, was conducted over a period of 26 days and involved 1042 L (275 gal) of water.

Pilot Plant Studies. Water from the Vannatta Street well was pumped to a day tank and into the adsorbent columns. The columns consisted of four glass cylinders of 4 in. i.d. $\times$ 5 ft long. The three GAC columns were operated in series and the XE-340 column as a single unit, all in a downflow mode. Operating parameters are shown in Table V. A schematic is shown in Figure 3. The three GAC columns were backwashed to remove fines according to the manufacturer's recommendation. The XE-340 column was also backwashed as needed in accordance with the manufacturer's recommendation[3]. In particular, Rohm and Haas states:

> For maximum adsorption efficiency, the influent solution should be free of suspended particles when the adsorption operation is

[3]Ambersorb Carbonaceous Adsorbents, Technical Bulletin, Rohm and Haas Company, Philadelphia, PA, IE-231/77, August 1977.

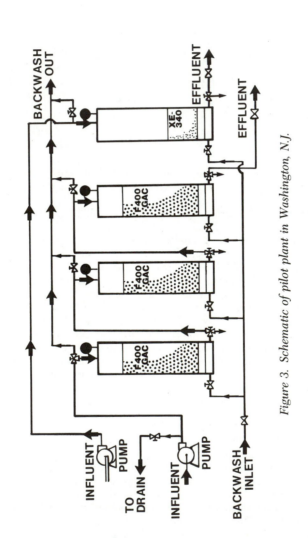

Figure 3. Schematic of pilot plant in Washington, N.J.

carried out in a downflow fixed-bed mode. Therefore, suspended solids must be removed by periodic backwashing of the beds. Otherwise, accumulation of suspended matter on the bed can result in uneven flow, reducing adsorption efficiency.

Grab samples for organics were taken before and after the day tank and after the columns once per week. Flow rate and pressure drop were monitored daily by means of a pressure gage and a calibrated flow meter. Samples were collected in solvent washed and baked glass vials. Analysis was by purge and trap isolation followed by GC-Hall (Calgon Corporation). The conditions were as follows: a 10-ft $\times$ 2-mm i.d. glass column packed with 0.27% Carbowax 1500 on Carbopak C; a Hall detector; He carrier gas; and 50° C for 5 min then to 160° C at 50° C/min.

Confirmation of peak analysis was provided by GC/MS. The influent concentration to the adsorbent columns (effluent from the day tank) is shown in Figure 4.

Data Reduction and Results

Data reduction was performed by the same method for both the laboratory and the pilot plant studies. The raw data for influent and effluent were initially plotted as breakthrough curves.

If the concentration of micropollutant in the effluent (C) is divided by the concentration in the influent (C_0), the data are considered to be normalized with respect to variation in influent concentrations. A plot of this value as a function of the cumulative number of bed volumes is presented (Figure 5). The usual curve fitting techniques may be utilized to détermine the line of best fit described by the breakthrough curve. The morphology of this curve is important and can yield information from which one can draw conclusions in comparison of adsorbents.

The first information extracted from the breakthrough curve was the cumulative number of bed volumes to breakthrough, defined as first appearance of a detectable concentration of the micropollutant in the effluent. These data are present in Table VI. Breakthrough may have occurred earlier and may not have been observed because of scheduling of sampling intervals.

Generally, for water treatment applications, the attainment of a specific effluent concentration is more critical than initial breakthrough. Therefore, specific effluent concentrations and initial breakthrough were both employed to analyze the data. Currently, drinking water maximum contaminant levels (MCLs) for 111-TCE, PCE, CCl$_4$, and TCE have not been established by either EPA or NJDEP. Based on the EPA SNARLS (suggested no adverse response level) (*10*), the working MCLs for 111-TCE, PCE, and TCE were set at 10 µg/L for specific effluent concentrations in this study.

Bed volumes to the attainment of MCLs were determined. In each case the average influent concentration was utilizied to establish a C/C_0

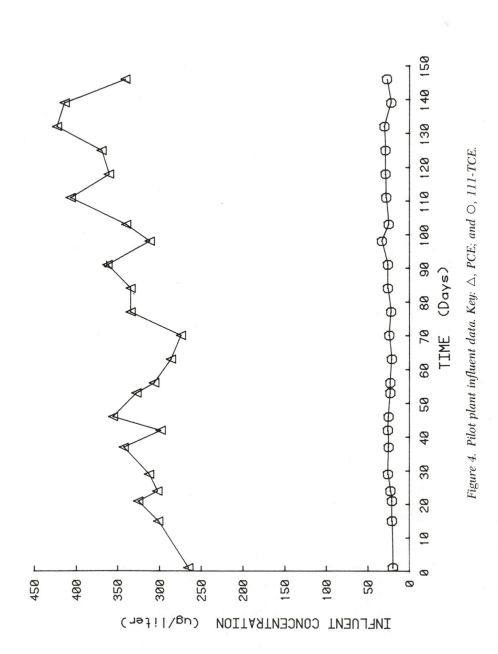

Figure 4. Pilot plant influent data. Key: △, *PCE; and* ○, *111-TCE.*

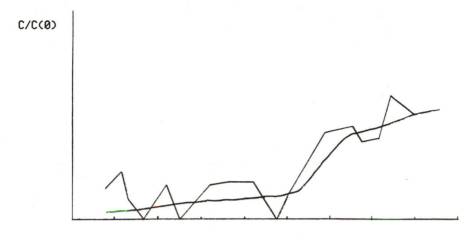

CUMULATIVE NUMBER OF BED VOLUMES

Figure 5. Illustrative breakthrough curve.

value for breakthrough. In several cases, particularly for XE-340 resin where the plot of C/C_0 fluctuated significantly (*see* Figure 6), this approach caused the breakthrough to appear to occur at more than one value of number of bed volumes. In these cases, the higher number of bed volumes was used for comparison. When breakthrough did not occur or occurred before the first sampling time, an approximate value was used (Table VII).

Mass loading values were calculated with reference to the number of bed volumes to MCL breakthrough. Mass loading is defined as:

$$\frac{C_0 N V}{M}$$

Table VI. Number of Bed Volumes to Breakthrough

	Laboratory		Pilot Plant	
Compound	*XE-340*	*F-400*	*XE-340*	*F-400*
111-TCE	<1,400	6,600	12,000	11,000
PCE	1,600	>16,000	12,000	22,000
CCl$_4$	<1,500	3,200	12,000	20,000
TCE	10.600	12,500	—	—
EBCT (min)	2.0	2.0	1.8	8.4

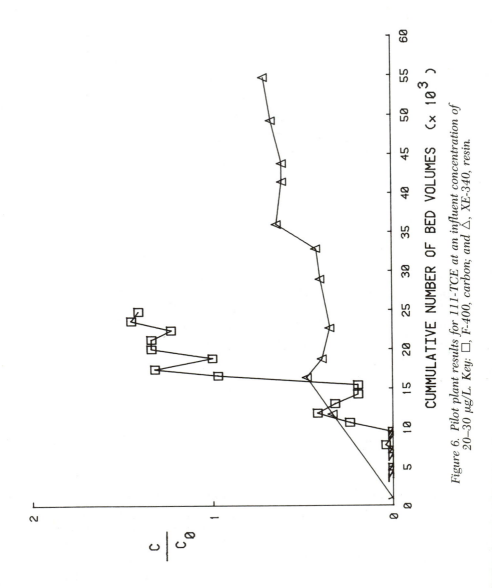

Figure 6. Pilot plant results for 111-TCE at an influent concentration of 20–30 μg/L. Key: □, F-400, carbon; and △, XE-340, resin.

Table VII. Number of the Bed Volumes to MCL Breakthrough

Compound	Laboratory		Pilot Plant	
	XE-340	F-400	XE-340	F-400
111-TCE (10 ppb)	10,000	13,000	33,000	16,000
PCE (10 ppb)	11,000	>15,000	12,000	32,000
TCE (10 ppb)	>16,000	>15,000	—	—
EBCT (min)	2.0	2.0	1.8	8.4

where C_0 is the average influent concentration, N is the number of bed volumes, V is the volume of bed, and M is the mass of adsorbent. This calculation estimates the capacity of the adsorbent for a particular micropollutant. The data are presented in Table VIII.

PCE is the micropollutant of major concern because of its high influent concentration. 111-TCE is also of importance because it was consistantly present in the influent, although it is considered significantly less toxic than PCE. These two compounds have been compared for their removal by XE-340 and F-400 in both the pilot and laboratory studies. Graphic comparisons are presented in Figures 6–8.

With regard to the XE-340 resin, the VHOs eluted at essentially the same time both on the pilot and laboratory scale. On the F-400, however, there was a definite order of elution. For both the pilot plant and laboratory scale, PCE was the last to break through and the last to attain the MCL. PCE was preceeded by CCl_4 and 111-TCE on the pilot plant with the orders reversed on the laboratory scale.

Table VIII. Mass Loading (mg/gm) to MCL Breakthrough

Compound	Laboratory		Pilot Plant	
	XE-340	F-400	XE-340	F-400
111-TCE (10 ppb)	1.87	2.98	1.21	0.98
PCE (10 ppb)	1.98	>3.32	<5.84	29.0
TCE (10 ppb)	>1.21	>1.39	—	—

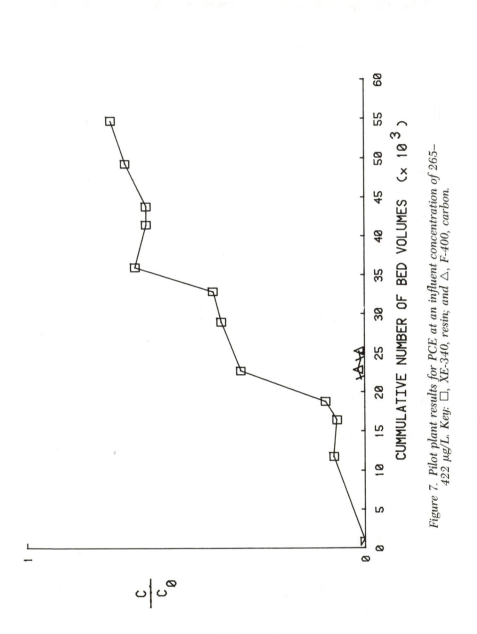

Figure 7. Pilot plant results for PCE at an influent concentration of 265–422 μg/L. Key: □, XE-340, resin; and △, F-400, carbon.

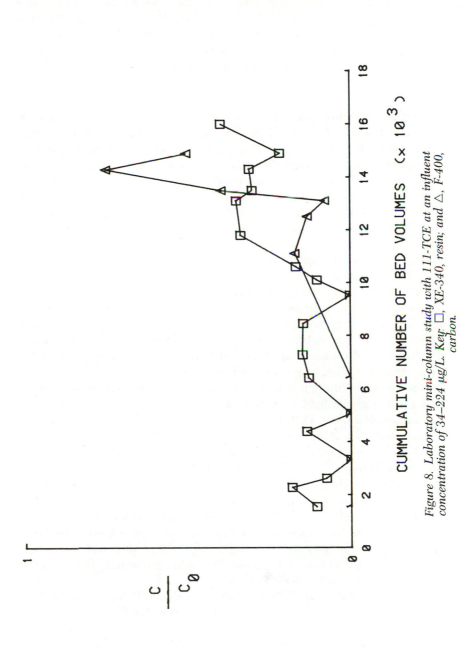

Figure 8. Laboratory mini-column study with 111-TCE at an influent concentration of 34–224 μg/L. Key: □, XE-340, resin; and △, F-400, carbon.

A strong chromatographic effect was noted with regard to PCE and 111-TCE (*see* Figure 6) on F-400 during the pilot scale testing. 111-TCE broke through after 77 days of operation of the pilot plant. By 140 days, the effluent exceeded the influent by factors of 1.2 and 1.8. This effect remained until operation was terminated at 186 days.

Discussion

MCL Breakthrough. When comparing MCL breakthrough of PCE or 111-TCE on GAC and XE-340 for both the laboratory and the pilot plant studies, it is readily seen that the GAC and XE-340 have different characteristic breakthrough curves (*see* Figure 6). The resin, in addition to fluctuating significantly, generally exhibits a steady rise in concentration in the effluent stream that starts at a definite initial concentration. The effluent concentration from the GAC column is less than detectable initially, after which a slight increase and finally a sharp breakpoint rise in effluent concentration is observed. The characteristic curves are similar in structure to those found for phenol on the same adsorbents (*14, 15*).

The advantage of the GAC over the XE-340 is that the GAC breakthrough is a defined phenomenon whereas with XE-340 the fluctuating pattern causes the MCL to be exceeded followed by an apparent period of resin recovery. This activity is possibly related to backwashing of the resin leading to perturbation of mass transfer zones. If this is indeed the case, it appears that GAC is considerably more stable with respect to backwash-induced premature breakthrough; yet the fluctuating breakthrough pattern was observed in the data from another pilot plant study without backwash (*4*).

In the pilot plant, PCE was at a concentration an order of magnitude higher than 111-TCE. Possibly the lower 111-TCE concentration accounts for the increased number of bed volumes to MCL breakthrough on XE-340, whereas the increased concentration of PCE acts as a driving force to displace the 111-TCE from F-400 at a significantly lesser number of bed volume compared to the laboratory scale.

The resin and the carbon seem to be equally effective for removal of TCE at low to moderate concentrations. The bed volumes to MCL breakthrough were not exceeded under any of the conditions tested in the laboratory or pilot studies. CCl_4 adsorption was significantly better on the carbon than the resin in both studies.

Mass Loading. The data presented in Table VIII represent the mass loading of the various micropollutants to the MCL. As with MCL breakthrough, PCE, CCl_4, and TCE exhibit the same loading trends in the laboratory study and the pilot plant; i.e., both studies showed highest capacity for PCE and lowest for CCl_4. Although the trends are the same, the actual loading capacities differ. 111-TCE is again the anomaly, and this is likely again to be caused by competitive and concentration effects.

Carbon generally out-performed resin for adsorption of PCE. The adsorption curves exhibited by the carbon fluctuate less and possess clearer breakpoints. The resin breakthrough is a steady increase in effluent concentration usually without a defined break point. This is due in part to the different characteristics of the resin, such as adsorbent composition, pore size distribution, adsorbent size and shape, and diffusion characteristics (9). It is also possibly related to the hydraulic loading of the adsorbent column.

Rohm & Haas recommended an empty bed contact time (EBCT) of 2 min for their XE-340 resin compared to 7.5 min for Calgon's F-400. The experimental results suggest that this time is too short for XE-340, and given a longer EBCT the resin may perform better. The use of XE-340 in removal of TCE at four locations was reported (16) with EBCTs of from 5 to 9 min, substantially higher than recommended for the present study. A study conducted by Rohm & Haas under conditions similar to the present study used an EBCT of 3.75 min (19). Rohm & Haas claims that their resin acts through absorption as well as adsorption (9). This will affect the performance characteristics of the resin, although further experimental evidence is necessary to actually prove absorption is occurring. The kinetics of interaction of the resin and the micropollutant also are important. Thurman et al. (17) found that, for XAD-8, resin flows greater than 20 bed volumes (BV)/h. caused a decrease in capacity. The capacity decrease was attributed to a nonequilibrium condition at high flow rates. Additionally, Slejko and Neely (18) demonstrated that an increase in flow rate will increase the leakage rate through the resin column.

The above mentioned findings may be applicable here. In both the laboratory and pilot studies the resin flows were operated in excess of 30 bed volume (BV)/h. In each case, leakage of micropollutants through the resin bed occurred almost immediately, while F-400 carbon exhibited no leakage. In studies performed by Neely and Isacoff (9, 19) at Rohm & Haas on Philadelphia tap water and South Brunswick water, XE-340 resin significantly performed better than GAC when operated at 8 and 16 BV/h. Wood and DeMarco (2) found that resin out-performed GAC 3:1 in Florida water for removal of halogenated organic compounds. Their resin column operated at 10 BV/h. Therefore, to optimize the resin's performance it may be necessary to optimize the hydraulic loading with respect to flow rate in addition to empty bed contact time.

Table IX compares the results of several researchers who have used both XE-340 resin and GAC for the removal of aquatic pollutants. Clearly the XE-340 resin is excellent for the removal of chloroform, while GAC functions as a more versatile adsorbent.

Table IX. Literature Comparison of Performance of XE-340 and GAC

Reference	Compounds Removed by XE-340	Compounds Removed by GAC	Parameter
Suffet et al. (20)	CHCl_3	L-sorbo[a]	breakthrough
Snoeyink and Chudyk (21)	CHCl_3	CHCl_3 humic acids	capacity and breakthrough
Isacoff (19)	CHCl_3 CHCl_2Br 111-TCE bis(2-chloroethyl) ether	MIB —	breakthrough and capacity
van Vliet et al. (14)	—	phenol	breakthrough
Wood and DeMarco (2)	four volatile halogenated organics	fulvic acid (THMFP)	breakthrough and capacity
McGuire (15)	methyl ethyl ketone	1-butanol 1,4-dioxane	breakthrough

[a]1,2,4,6-Di-O-isopropylidene-L-sorbose

Wood and DeMarco (2) found that the resin had an adsorptive capacity of approximately three times carbon for removal of individual halogenated organics. These compounds include 1,2-dichloroethene, chloroform, bromodichloromethane, bromoform, 1,1-dichloroethane, and trichloroethylene. However, the resin was not effective for the removal of fulvic acid and its degradation products which are precursors to THM formation. For these compounds GAC possessed a higher capacity.

Suffet et al. (20), Neely and Isacoff (9, 19), and Snoeyink and Chudyk (21) independently found that CHCl_3 was well removed by resin. Larger compounds, such as methylisoborneol, humic acids, and 1,2,4,6-di-O-isopropylidene-L-sorbose were better removed by carbon. In an in-depth study of carbon and resin performance for removal of phenol, van Vliet and Weber (14) found that carbon performed better.

The breakthrough curves show that activated carbon and the resin do not exhibit the same type breakthrough patterns. The resin tends to "leak"; i.e., on initial loading of the column there is an almost immediate appearance of the micropollutant in the effluent. The XE-340 breakthrough pattern tends to fluctuate and rise without achieving the clear classical breakthrough of carbon.

From the standpoint of kinetics, several steps are involved in the adsorption process, including particle phase diffusion, reaction at phase boundaries, pore diffusion, mass transfer from the flowing phase (film diffusion), and axial disperson (22). A visual comparison of the breakthrough curves derived in this study to those presented by Sontheimer (23) indicates that film diffusion may be the rate limiting step with respect to the resin and pore diffusion with respect to the carbon.

Both pore and film diffusion processes depend on the adsorbent and the adsorbate. In addition, film diffusion is strongly dependent on the hydraulic characteristics of the system. The primary property of the adsorbate that relates to mass transfer is diffusivity. The mass transfer coefficients for both types of diffusion are proportional to molecular diffusivity of the adsorbate. In the case of film diffusion, however, this dependence is numerically minimized in comparison to pore diffusion. This condition tends to make the adsorbate more important in the case of pore diffusion (activated carbon) than in film diffusion (resin). This hypothesis is substantiated by the fact that breakthrough appears to be independent of the compound in the resin case; however, there is a definite order of elution in the carbon case. This phenomenon has been noted elsewhere where several compounds with characteristics dissimilar to those used in this study broke through on XE-340 in essentially the same time, but at different times on activated carbon (15). The order of elution from activated carbon in the pilot plant was predicted well by the net adsorption energy concept (24).

The bed volumes to MCL breakthrough, Table VII, indicate that the carbon, F-400, is far superior to XE-340 resin for removal of PCE, which is the principal well-water contaminant in this study. This conclusion is reached in both the laboratory and pilot plant studies. The results for 111-TCE and PCE are ambiguous, with neither study in agreement. 111-TCE is better removed by carbon in the laboratory study, but better removed by resin in the pilot plant. This result may be due to the competitive effects and the differences in concentration of both 111-TCE and PCE in these studies. In the laboratory, the average influent concentrations of PCE and 111-TCE were both approximately 90+ μg/L. This situation possibly equalized the leakage of each compound through the resin as well as delayed the competitive displacement from the carbon.

In addition to fundamental differences between the adsorbents with respect to kinetics, there also appear to be differences with respect to adsorption equilibria. In an evaluation of several different adsorbents for potable water treatment, Weber and van Vliet (25) applied Dubinin adsorption theory to five model compounds, including CCl_4, which may be taken as a paradigm for the compounds in the present study. They found F-400 to have both a greater affinity for CCl_4 than XE-340 (184,280 vs. 178,830 kJ/mol) and a greater maximum capacity (41,940 vs. 11,380 $cm^3/100$ g). Similar results are obtained with application of the Polanyi–Manes theory (2). Thus, it is important to consider the ultimate capacity (equilibrium) and how it is reached (adsorption kinetics) in comparison of adsorbents.

The Chromatographic Effect. A chromatographic effect is evidenced by the presence of an effluent concentration that is greater than the influent concentration for a particular adsorbate at a given time (15) as demonstrated in Figure 6. Displacement and re-equilibrium phenomena can both cause the chromatographic effect. Re-equilibrium may occur in a contactor if an adsorbate influent concentration decreases such that there is a concentration-based driving force that operates to desorb the adsorbate. This action will continue until a new equilibrium is reached. Displacement involves a competition for adsorption sites by two or more compounds, one of which is either better adsorbed on the adsorbent or has a sufficiently large concentration-based driving force to allow it to occupy more sites. In this case, the compound that is less well adsorbed will be displaced.

The chromatographic effect noted in this study appears to operate through the mechanism of displacement. 111-TCE is loaded on the carbon as is PCE. GAC has been shown to have a greater affinity for PCE than 111-TCE by its net adsorption energy (24). At some point, the greater affinity and greater concentration of PCE begin to displace 111-TCE from the bed. Because the concentration of PCE remains high, evidently displacement will persist as long as conditions do not change or until 111-TCE is unloaded from column. If the displaced material were highly toxic or in excess of the MCL, this would present a serious problem to the operator.

A chromatographic effect was not noted on the resin under the same operating conditions. This tends to substantiate the hypothesis stated above that the nature of the compound is less important with respect to the resin; i.e., possible similarities in net adsorption energies for pollutant–resin adsorption may preclude competitive displacement. For a series of five experimental probes, Weber and van Vliet (25) reported pollutant affinities from 184.28 to 1261.30 kJ/mol for F-400 and 178.83 to 568.14 kJ/mol for XE-340; these results support the contention of lesser affinity differences on the resin.

Conclusions

A study of the removal of VHOs from a polluted groundwater reveals that both activated carbon (GAC) and a synthetic carbonaceous resin (XE-340) are capable of treating the water, but to different degrees.

Overall, F-400 appears to have a better capacity for the pollutants than XE-340. Specifically F-400 significantly out-performed the resin for removal of PCE, the primary pollutant of interest, under the experimental conditions reported. An increase of EBCT or lower hydraulic flow rate could allow the resin to perform better.

Fundamental differences in the adsorption mechanisms of each adsorbant are indicated by different shapes of their breakthrough curves and differences in performance data. The order of elution of the compounds on GAC was predicted well by the net adsorption energy concept. On the resin, however, the compounds broke through at essentially the same time. In the pilot plant study a strong chromatographic effect was evidenced on GAC by what appears to be displacement of 111-TCE by PCE. This result was not noted on the resin under comparable conditions. The evidence points to the formulation of a hypothesis that the physical and chemical properties of the adsorbate are considerably more important with activated carbon than with resin.

Acknowledgments

The authors express their appreciation to the following Drexel University personnel: C. Carter for work on the laboratory study, A. Wicklund for reviewing the manuscript, and T. Brunker for graphics. Also, contributions by the following American Water Works Service Company personnel were appreciated: Maryann Matrisin and June Caldwell for typing various drafts of the manuscript, and R. H. Moser and D. L. Kelleher for reviewing the manuscript.

Special acknowledgment goes to Arthur F. Crowley, Manager, and Neil Bamford of the New Jersey Water Company—Washington District for their time and effort involved with sample collection and operation of the pilot plant during the on-site pilot study.

Literature Cited

1. Dickson, *Nature* **1980** 283(5750): 802.
2. Wood, P.; et al. "Removing Potential Organic Carcinogens and Precursors from Drinking Water"; Vol. 1, U.S. EPA 1980 600/2-80-130a.
3. Woodhull, R. S., "Groundwater Contamination in Connecticut"; *Am. Water Works Assoc. Ann. Conf. Proc.*, 1980 p. 291.
4. Ruggiero, D. D.; Feige, W.; Ausubel, R. "Use of Aeration & Resin Treatment of a Groundwater", *Am. Water Works Assoc. Ann. Conf. Proc.*, 1980, 899.
5. Dreisch, F. A.; Munson, T. C. "Contamination of Drinking Water Supplied by Purgeable Halogenated Hydrocarbons", ACS 15th MARM, Washington, D.C., 1980.
6. Guerrera, A. A. "Chemical Contamination of Aquifers on Long Island, New York", *Am. Water Works Assoc. Ann. Conf. Proc.* 1980 p. 83.

7. Council on Environmental Quality, "Contamination of Groundwater by Toxic Organic Chemicals," GPO, 1981.
8. Dallaire, G. *Civil Engineering* **1980** 49(9), 81–86.
9. Neely, J. W. In "Activated Carbon Adsorption of Organics from the Aqueous Phase," McGuire, M. J.; Suffet, I. H., Eds., Ann Arbor Science, Ann Arbor: Mich., 1980, Vol. 2, p 417.
10. SNARL for 1, 1, 1-Trichloroethane (May 9, 1980); SNARL for Trichloroethylene (Nov. 26, 1979); SNARL for Tetrachloroethylene (Feb. 9, 1980) Office of Drinking Water, U.S. EPA, Washington, D.C.
11. Neely, J. W.; Isacoff, E. G. "Regenerability of Ambersorb XE-340", N.J. Section Am. Water Works Assoc. Meeting, Atlantic City, Sept. 1979.
12. Otson, R., et. al., *Environ Sci Tech* **1979** 13(8) 936.
13. McNair, H. M.; Bonelli, E. J. "Basic Gas Chromatography"; Varian, 1968.
14. van Vliet, B. M.; Weber, W. J.; Hozumi, H. *Water Res.* **1980** *14* p. 1719.
15. McGuire, M. J., Ph.D., dissertation, Drexel Univ. Philadelphia, PA, 1977.
16. Symons, J. M.; Carswell, J. K.; DeMarco, J.; Love, O. T. "Removal of Organic Contaminants from Drinking Water Using Techniques Other Than Granular Activated Carbon Alone," MERL, ORD, U.S. EPA, 1979
17. Thurman, E. M.; Malcolm, R. L.; Aiken, G. R. *Anal. Chem.* **1978**, 775, 50167.
18. Slejko, F. L.; Neely, J. W. "A New Stream Regenerable Carbonaceous Adsorbent for Removal of Halogenated Hydrocarbons from Water"; Am. Water Works Assoc. Ohio Conf., Nov. 1976.
19. Isacoff, E. G. *Water Sewage Works*, August, 1979; p. 41.
20. Suffet, I. H.; Radziul, J. V.; Cairo, P. R.; Coyle, J. T. In "Water Chlorination Environmental Impact & Health Effect," Jolley, R. L.: Gorchev, H.; Hamilton, D. H., Eds.; Ann Arbor: Ann Arbor, Mich., 1978 Vol. 2, p. 561.
21. Chudyk, W. A.; Snoeyink, V. L.; Beckman, D.; Temperly, T. *J. Am Water Works Assoc* **1979**, *71*, 529.
22. Perry, R. H.; Chilton, C. H., In "Chemical Engineer's Handbook," 5th Ed., McGraw-Hill, 1973.
23. Sontheimer, H., "Basic Principles of Adsorption Process Techniques", Translation of Reports on Special Problems of Water Technology, Vol. 9, U.S. EPA 1975, 600/9-76-030.
24. McGuire, M. J.; Suffet, I. H., In "Activated Carbon Adsorption of Organics from the Aqueous Phase," McGuire, M. J.; Suffet, I. H., Eds., Ann Arbor: Ann Arbor, Mich. 1980; Vol. 1.
25. Weber, W. J.; van Vliet, B. M. *J. Am. Water Works Assoc.* **1981**, *73*, 420.

RECEIVED for review February 25, 1982. ACCEPTED for publication September 3, 1982.

Dynamic Behavior of Organics in Full-Scale Granular Activated-Carbon Columns

R. SCOTT SUMMERS and PAUL V. ROBERTS

Stanford University, Department of Civil Engineering, Stanford, CA 94305

Three full-scale granular activated-carbon columns were used to characterize the removal of organic substances for 1 year. Long-term partial removal of aromatic organic compounds and organic compounds measured by total organic carbon and total organic halogen occurred. Biological degradation was thought to be a major contributing mechanism. Elution of halogenated one- and two-carbon compounds was observed. There were only minor differences between fresh and regenerated adsorbent with regard to removal of total organic carbon, total organic halogen, and 20 specific organic compounds.

THE PARALLEL PERFORMANCE of three full-scale granular activated-carbon (GAC) columns was characterized with regard to the removal of organic substances over a 1-year period. The parallel arrangement of the GAC system allowed for a direct comparison of fresh, once-regenerated, and exhausted Filtrasorb-300 GAC. Carbon was considered exhausted if it had been in service sufficiently long (1 year, equivalent to 13,000 bed volumes treated) to saturate its adsorption capacity for total organic carbon (TOC). By using a composite sampling system, a more representative characterization of the removal of organic compounds was achieved than was possible with grab samples. The organic compounds were characterized by two collective parameters—TOC and total organic halogen (TOX)—as well as by analysis for 20 specific organic compounds, including 16 on the U.S. EPA's priority pollutant list.

Facilities

Reclamation Facility. The GAC system is part of the Santa Clara Valley Water District's Palo Alto Reclamation Facility. Figure 1 shows a

0065-2393/83/0202-0503$6.25/0

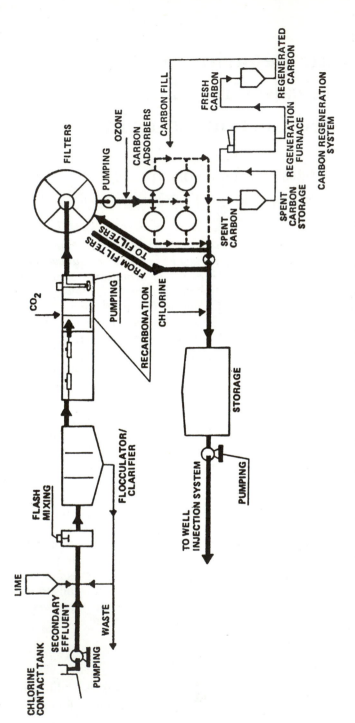

Figure 1. Schematic flow diagram of the Palo Alto Reclamation Facility.

schematic of the basic unit processes in operation during the study. The
Palo Alto Reclamation Facility receives 0.067 m^3/s (1.5 million gallons/
day) of chlorinated activated-sludge effluent from Palo Alto's wastewater
treatment plant. The reclamation facility provided treatment before
granular activated carbon in the following sequence during the present
study: lime clarification at pH 11; recarbonation with CO_2; mixed-media
filtration; and ozonation at a rate of 0.5 mg O_3/mg TOC. This sequence
resulted in a contact time of 2 h for chlorine with the waste stream before
the GAC columns. The effluent from the upflow GAC columns was
filtered and chlorinated before injection into an aquifer as part of a
groundwater recharge system.

The upflow GAC columns are operated at a superficial linear velocity
of 10.8 m/h (4.4 gpm/ft^2) providing an empty-bed contact time of 34 min.
The flow through each of the three columns is 0.022 m^3/s (0.5 million
gallons/day). On-site thermal regeneration of the adsorbent is accomplished with a multiple-hearth furnace.

During the study period, constant conditions were maintained in the
treatment steps prior to activated carbon. No attempt was made to
investigate the effects of changes in pretreatment, e.g., benefits or
disadvantages of preozonation.

Composite Sampler.   In an earlier study, grab sampling of the GAC
influent and effluent resulted in unacceptably high variations in the
measured concentration of specific organic compounds. Such variation
complicates the evaluation of GAC performance. Composite samplers
with floating Teflon covers were fabricated and installed on a common
influent and the three effluent lines of the GAC columns (Figure 2). The
float sample system was based on the design of Westrick and Cummins
(1). The float samplers were designed to minimize the liquid sample–
atmospheric interfacial area, thus reducing the loss of volatile compounds
during the composite period. A line providing continuous flow from the
system sampling point was connected to the refrigerated sampling system.
A sample timer activated the two sample valves, A and B, at a frequency of
30 min over a composite period of 7 days, resulting in a composite of 350
grab samples. A 200-mL aliquot of a saturated solution of sodium thiosulfate was initially added to the sample chamber to minimize the
formation of chlorinated organics in the sample chamber over the compositing period. At the end of the sample period, the float sampler was
thoroughly mixed, and samples for TOX and specific organic compound
analysis were taken through valve C. A separate composite sample
preserved with hydrochloric acid was taken for TOC analysis.

A preliminary study of the stability of samples stored as described
above found that storage at 4°C for 1 week did not appreciably affect the
concentrations of the volatile compounds reported here. The variations in

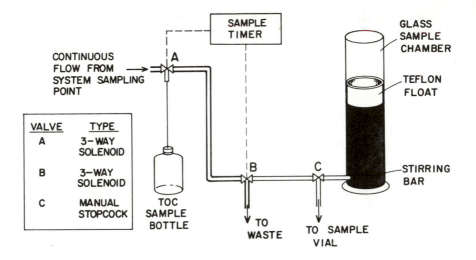

Figure 2. Schematic diagram of the composite sampling system where
$T = 4°C.$

concentration during that period did not exceed the standard deviation of
the gas chromatographic analysis, approximately ± 15%.

The composite samplers were operated continuously with a 7-day
composite period for the first 6 months (7100 bed volumes). Thereafter,
10-day composite samples were taken at 6-week intervals except during
episodes of plant operational problems. Sampling of the exhausted column
for organics other than TOC was discontinued after 4500 bed volumes.
Analysis for TOX in all columns was discontinued after 7100 bed volumes.
Some GAC fines were found in the regenerated column composite sample
during the first month of operation.

Analytical Procedures

The organic constituents were characterized by analysis for 20
specific organic compounds and two collective parameters. Table I lists
the compounds, detection limits, and analytical precisions. The eight
halogenated one- and two-carbon compounds were characterized using
gas chromatography of concentrates prepared by the pentane extraction
procedure described by Henderson et al. (2). The closed-loop stripping
analysis for 11 aromatic compounds as well as for heptaldehyde utilized
the Grob procedure for concentrating trace organics (3). The TOX
analysis measures the purgeable and nonpurgeable fractions separately as
described by Jekel and Roberts (4). Organic carbon was measured using

Table I. Detection Limits and Precision of Analysis

Compounds or Collective Parameters	Detection Limit ($\mu g/L$)	Standard Deviation, % of Mean[a]
Halogenated 1- and 2-carbon compounds:		
chloroform, bromoform	0.05	10–20
1,1,1-trichloroethane, dibromochloromethane, bromodichloromethane, trichloroethylene, and tetrachloroethylene	0.02	10
carbon tetrachloride	0.01	5
Aromatic compounds:		
aromatic hydrocarbons[b]	0.015	90
chlorinated benzenes[c]	0.02	30
Heptaldehyde	0.015	90
Total organic carbon (TOC)	200	5
Total organic halogen (TOX) (measured as Cl)	35	5

[a]Valid for concentrations greater than five times the detection limit.
[b]Ethylbenzene; naphthalene, m- and p-xylene; and 1- and 2-methylnaphthalene.
[c]Chlorobenzene; 1,2-, 1,3-, and 1,4-dichlorobenzene; and 1,2,3- and 1,2,4-trichlorobenzene.

an Oceanography International ampule system adapted to a Dohrmann DC-52 organic carbon analyzer.

Results and Discussion

General Features of Breakthrough Responses. The results of the removal of organics by GAC columns are illustrated by several types of breakthrough curves shown in Figure 3. Evidence of the four phenomena—immediate partial breakthrough, initial breakthrough, steady-state removal, and elution—can be inferred.

Immediate breakthrough refers to a condition where the first sample taken shows the presence of the constituent in the effluent at a concentration substantially above the detection limit. This phenomenon occurred with TOC and TOX, indicating that a portion of the influent TOC and TOX is not amenable to removal by GAC treatment. Cannon and Roberts (5) previously reported that approximately 10% of the TOC in the same wastewater was nonadsorbable, even at high activated-carbon dosages, in equilibrium–isotherm studies.

Initial breakthrough occurs when measurable amounts of the constituent first appear in the effluent; thus, complete removal of the compound occurs prior to this point. In this study, initial breakthrough of organics occurs over a wide range of time, from the first sample taken (immediate breakthrough) to 4 months (4000 bed volumes) into the study.

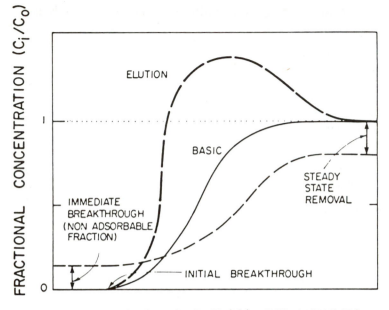

Figure 3. Generalized concept of types of breakthrough responses, illustrating the phenomena of immediate breakthrough, initial breakthrough, steady-state removal, and elution.

Steady-state removal is a condition in which the effluent concentration rises with time and eventually assumes a steady-state value which is less than that of the influent. Evidence of steady-state removal was found for TOC, TOX, and the aromatic compounds. Possible mechanisms for these long-term partial removals are slow adsorption phenomena (6) and biological degradation of organic material by bacteria attached to the GAC in the columns (7). Roberts and Summers (8) in a review of GAC performance found evidence of both immediate breakthrough and steady-state removal of TOC in nearly all cases reviewed.

Elution refers to the situation in which the effluent concentration rises to values greater than that of the influent concentration. Evidence of this phenomenon was found for six halogenated one- and two-carbon compounds. Of the proposed mechanisms for elution, competitive adsorption and re-equilibration to a change in influent concentration are thought to be responsible.

Collective Parameters. The performance of the three GAC columns for the removal of organic compounds measured by TOC and TOX is summarized in Table II. An immediate breakthrough of 1.0–1.5 mg/L TOC upon startup of both the fresh and regenerated columns,

Table II. GAC Performance for Removal of Organics

	TOC	TOX[a]	CHCl_3	Cl_3CCH_3	CCl_4	CHCl_2Br	Cl_2C=CHCl	CHClBr_2	Cl_2C=CCl_2	CHBr_3
Influent concentration[b] (μg/L)	8.7 ± 1.8^c	177 ± 89	10.7 ± 5.0	23.6 ± 27.7	0.10 ± 0.09	4.1 ± 2.9	7.5 ± 8.6	3.1 ± 3.0	19.3 ± 28.8	1.1 ± 1.1
Immediate breakthrough:										
Percent remaining (f)[d]										
fresh	18	24	2	<1	30	0	0	0	<1	0
regenerated	14	27	3	<1	30	0	0	0	<1	0
Initial breakthrough at f = 5[d].										
Bed volumes										
fresh	—	—	300	2000	300	1800	4000	2400	4000	3500
regenerated	—	—	300	1800	300	1600	4000	1600	4000	1600
Exhausted GAC, percent remaining[b]	76 ± 14	79 ± 27	110 ± 20	110 ± 68	88 ± 50	94 ± 26	64 ± 45	74 ± 34	44 ± 68	124 ± 103
Steady state or f = 100[d]:										
Fresh										
percent remaining[b]	67 ± 15	76 ± 30								
bed volumes	5200	4400	2800	3600	4500	3800	5000	7000	5000	7000
Regenerated										
percent remaining[b]	67 ± 11	82 ± 72								
bed volumes	5200	4000	2500	3200	4500	3800	4800	7000	4800	7000

[a] Measured as Cl.
[b] Mean ± standard deviation.
[c] mg/L.
[d] $f = (C_i/C_0) \times 100$.

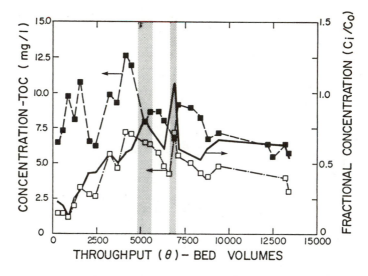

Figure 4. TOC breakthrough response for fresh GAC. The shaded bands represent perturbations of pH and dissolved oxygen consumption. Key: ■, influent; □, effluent; ▦, high pH and low ΔDO; and —, C_i/C_0.

indicates that 15–20% of the influent TOC is not amenable to adsorption by GAC. This immediate breakthrough is illustrated in the TOC breakthrough response for fresh GAC (Figure 4). With increasing service time, the column became saturated with organics and the effluent concentration rose steadily as the volume treated increased to an amount equivalent to 4500 bed volumes (4 months), where, with the exceptions of peaks at 5200 and 6900 bed volumes, it assumed a steady-state value of 65–70% of the influent. This steady-state partial removal of approximately one-third of the influent was maintained for the remaining 9000 bed volumes (8 months) of the study.

Figure 5 shows the relationship between the influent pH and the dissolved oxygen consumption (ΔDO) through the GAC columns. Initially only 1 mg/L of DO was consumed. After 2 weeks of acclimation and growth, the bacteria attached to the GAC began to utilize more oxygen and after 3 weeks, 5 mg/L DO was consumed. The influent pH remained relatively stable during the initial stages, fluctuating between 8 and 6.5 with two exceptions: problems with the CO_2 compressor used in the recarbonation process occurred at 4800 and 6600 bed volumes, resulting in each instance in short-term rises to pH 11. Coincidental with the rise in pH, decreases in the DO consumption and TOC removal in the GAC columns were observed for the duration of the pH perturbation and for the succeeding several hundred bed volumes. This extended reaction to the high influent pH is an indication that biological degradation of organic

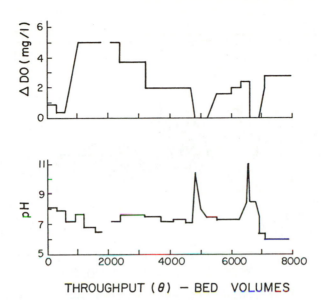

Figure 5. *Relation between dissolved oxygen consumption and pH upsets; data from NASA Water Monitoring System.*

material occurred in the GAC columns under normal pH conditions. After a period of acclimation and regrowth, the bacteria again consumed DO and TOC. Under normal pH conditions, the measured effluent DO concentration range was 0.5-3 mg/L; however, it is possible that some oxygen may have been introduced into the sample stream during flow through the NASA sampling system (9). Thus, it is possible that the DO concentration may have been overestimated, and the availability of DO may have been a limiting factor in the biological utilization of organic material.

The response of all three GAC columns to the pH/ΔDO upsets can be seen in the fractional concentration breakthrough curves for TOC (Figure 6). The periods of lowest TOC removal for all columns coincided with the pH/ΔDO upset periods, with one exception at 2100 bed volumes for the exhausted column. Figure 6 also shows the similarity of the fresh and regenerated carbon with respect to immediate breakthrough of TOC and the general features of the rate of approach to steady state. A statistically similar steady-state condition existed in all three columns after 7100 bed volumes.

The cumulative removal of TOC by adsorption was 34 mg TOC/g GAC and 37 mg TOC/g GAC for the fresh and regenerated adsorbents, respectively. The amount of TOC removed by adsorption is calculated as the difference in the total amount removed in a column and that removed

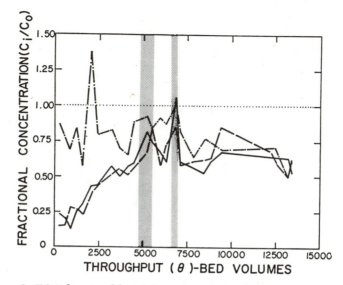

Figure 6. TOC fractional breakthrough curves for fresh (—), regenerated (——), and exhausted GAC (—·—). Conditions: ▨, high pH and low ΔDO.

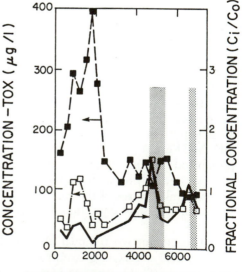

Figure 7. TOX breakthrough response for fresh GAC. Key: ■, influent; □, effluent, —, C_i/C_0; and ▨, high pH and low ΔDO.

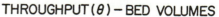

by the exhausted column, presumably by biodegradation. In view of experimental error, the difference between the adsorption capacities of the fresh and regenerated GAC is not significant.

As with TOC, little difference can be seen between the fresh and regenerated GAC columns for the removal of TOX. The breakthrough response for TOX removal in the fresh GAC is shown in Figure 7. Steady-state removals of TOX amounting to 24% and 18% for fresh and regenerated GAC, respectively, were achieved after 4000 bed volumes, while the exhausted column maintained a removal of 21%. The breakthrough curves for TOX also exhibited a response to the pH/ΔDO upsets. The immediate breakthrough of 35–60 μg/L TOX (15%–30% of the influent concentration) in both fresh and regenerated GAC columns indicates that a portion of TOX, similar to that of TOC, is not amenable to GAC treatment.

The cumulative removals of TOX by adsorption were 1.1 mg TOX/g GAC and 1.8 mg TOX/g GAC for fresh and regenerated adsorbent, respectively; the difference is not significant. This rate corresponds to a mass ratio of 0.030 g TOX (as chlorine) adsorbed per g TOC adsorbed, or a mole ratio of approximately 0.010 mole organic halogen adsorbed per mole organic carbon adsorbed.

Halogenated One- and Two-Carbon Compounds. The performance of the GAC columns for the removal of the halogenated one- and two-carbon compounds is summarized in Table II. The three brominated compounds and trichloroethylene were removed completely (detection limit of 0.02 μg/L) by both fresh and regnerated GAC for a substantial period, i.e., 1600–4000 bed volumes. Trichloroethane and tetrachloroethylene exhibited an immediate breakthrough of less than 1% for 2000 and 4000 bed volumes, respectively. Chloroform displayed a slight immediate breakthrough of 2–3%. Carbon tetrachloride exhibited erratic initial breakthrough: 30% after the first 600 bed volumes, then dropping to 10% in both columns. The behavior of carbon tetrachloride was obscured by the low influent concentration of 0.1 μg/L. Trichloroethylene and tetrachloroethylene were both substantially removed by the exhausted GAC column during this study: 13,000–17,500 bed volumes.

Six of the eight halogenated one- and two-carbon compounds in both fresh and regenerated GAC exhibited behavior suggesting elution. Although it is difficult to distinguish competitive adsorption from re-equilibration in a full-scale study where there is uncontrolled variation in the influent concentrations, some tendencies can be discerned.

The breakthrough pattern for chloroform (Figure 8) is an example of elution during the period of relatively steady influent concentration, indicating that competitive adsorption may be the underlying cause. Exhibiting immediate breakthrough to only a slight extent, 2%, the

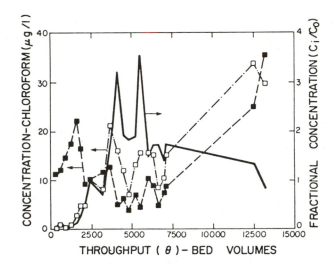

Figure 8. Chloroform breakthrough response for fresh GAC, illustrating elution behavior. Key: ■, influent; □, effluent; and —, C_i/C_0.

chloroform breakthrough curve rose rapidly after 1600 bed volumes to fractional concentration peaks of 3–4. While the initial rapid rise may have been exaggerated by the 30% decrease in the influent concentration, elution continued over a 4000-bed-volume period during which the influent concentration remained stable.

Three halogenated one- and two-carbon compounds (1,1,1-trichloroethane, trichloroethylene, and tetrachloroethylene) eluted at times corresponding to a decrease of 1–2 orders of magnitude in the influent concentration, indicating that re-equilibration of the system to the lower influent concentration may have been responsible. Figure 9 shows the breakthrough pattern for 1,1,1-trichloroethane on regenerated GAC. After showing no immediate breakthrough, the fractional concentration rose rapidly after 2500 bed volumes. Shortly after complete breakthrough, the compound eluted in response to an order-of-magnitude decrease in the influent concentration beginning at approximately 4000 bed volumes. Any reaction to the pH/ΔDO upsets at 4800 and 6600 bed volumes would be masked by the elution phenomenon.

Figure 10 depicts the chloroform mass removal for the fresh and regenerated GAC for the first 7100 bed volumes and for the exhausted GAC for the first 4500 bed volumes of the study. Cumulative mass removed is plotted versus cumulative mass applied, both expressed in units of micrograms of chloroform per gram of GAC. The dashed line represents 100% removal. Two important observations can be made: first, the difference between fresh and regenerated GAC for the removal of chloroform; and second, the apparent net production of chloroform in the exhausted and regenerated GAC.

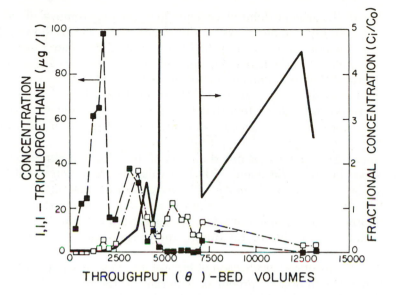

Figure 9. Breakthrough response of 1,1,1-trichlorethane with regenerated GAC, illustrating re-equilibration following decrease in influent concentration. Key: ■, influent; □, effluent; and —, C_i/C_0.

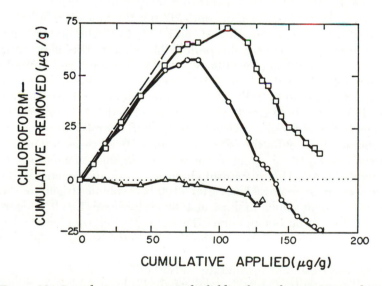

Figure 10. Cumulative mass removal of chloroform, demonstrating elution and net production. Key: □, fresh; ○, regenerated; △, exhausted; and ———, 100% removal.

The fresh carbon exhibited a maximum cumulative removal of chloroform at 106 μg/g applied, while the regenerated GAC showed less removal efficiency after 60 μg/g had been applied and had a maximum removal after 83 μg/g had been applied. The adsorptive capacity for chloroform at the maximum cumulative mass removed was 72 μg/g and 58 μg/g for the fresh and regenerated GAC, respectively. These values agree approximately with those reported for drinking water and with single-solute isotherm data at a similar concentration (10 μg/L) of chloroform (*10*).

Both GAC columns exhibited elution of the adsorbed chloroform. In the period between 2500 and 5000 bed volumes, the regenerated GAC eluted the total amount removed (58 μg/g) during the first 2500 bed volumes. After 5000 bed volumes, there was an apparent net production of chloroform in the regenerated GAC. The exhausted column showed some net production of chloroform during the period between 13,000 and 17,500 bed volumes.

The apparent net production of chloroform may be accounted for by reaction between organic precursor material and the residual chlorine in the GAC influent; the chlorine residual in the influent ranged from 2 to 5 mg/L, as combined chlorine. The phenomenon of net production of chloroform needs to be studied more systematically.

One method of comparison of the fresh, regenerated, and exhausted GAC is based on the cumulative mass removed. Figure 11 is a comparison of the mass removed by activated carbon in several different states (fresh, regenerated, and exhausted), expressed relative to the mass removed by the fresh carbon. The period of comparison is the first 4500 bed volumes for the fresh and regenerated GAC and 13,000–17,500 bed volumes for the exhausted GAC. During this period, the exhausted column continued to remove a significant portion of trichloroethylene and tetrachloroethylene relative to the fresh GAC. Similar results for these two compounds are reported in Table II on a percent removal basis. There were some differences between fresh and regenerated GAC after 4500 bed volumes in the removal of chloroform, 1,1,1-trichloroethane, and carbon tetrachloride but not for the other compounds. In all cases, the fresh GAC removed an equal or greater amount of organics compared to the regenerated GAC. The relationship between fresh and regenerated GAC on the same mass removal basis after 7100 bed volumes (data not shown) led to conclusions similar to those at 4500 bed volumes for all halogenated one- and two-carbon compounds. Two other criteria for comparison of fresh and regenerated GAC are the time to initial breakthrough and the time to adsorption exhaustion (Table II). Only two significant differences between the fresh and regenerated adsorbent can be found using these criteria. The extents of initial breakthrough of dibromochloromethane and bromoform both occur much earlier for the regenerated GAC. Smaller

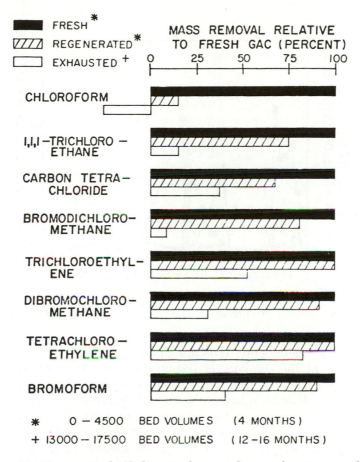

Figure 11. *Mass removal of halogenated one- and two-carbon compounds by regenerated and exhausted GAC, compared to fresh GAC.*

insignificant differences can be found for other compounds using both criteria. As with the mass removal comparison, the fresh GAC exhibits an equal or superior performance for organic removal in all cases.

Aromatic Compounds. The 11 aromatic compounds exhibited a more erratic breakthrough behavior than did the halogenated one- and two-carbon compounds, which hinders a systematic analysis of their breakthrough responses. An example of an aromatic compound breakthrough response is 1,4-dichlorobenzene (1,4-DCB) shown in Figure 12 for the fresh GAC column. As with eight of the 11 aromatic compounds, 1,4-DCB displayed a response to the pH/ΔDO upset. Excluding the periods of pH/ΔDO upsets, over 96% of 1,4-DCB is removed on a long-term average. Table III lists the mean influent concentration, the average per-

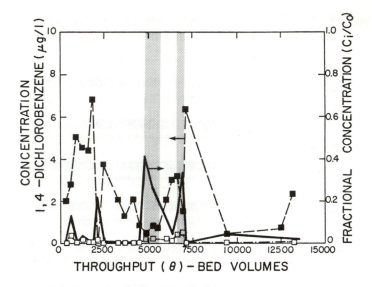

Figure 12. Breakthrough response for 1.4-dichlorobenzene on fresh GAC. Key: ■, *influent;* □, *effluent;* —, C_i/C_0; *and* ▦, *high pH and low* ΔDO.

cent removal including pH/ΔDO upset periods for all three columns, and an indicator of the compounds' response to the pH/ΔDO upset. The chlorinated benzenes and the naphthalene compounds were significantly removed on a long-term basis, whereas ethylbenzene, the xylenes, and heptaldehyde were removed to a lesser extent. Little difference is seen between the removal capacities of fresh and regenerated GAC. In some cases, the exhausted GAC removal capacity showed a significant difference compared to the fresh and regenerated adsorbents, but the exhausted GAC removal capacity is based on a different time period (the first 4 months of the study) than the others, so a direct comparison cannot be made.

Figure 13 shows the comparison of GAC condition on a mass removal basis relative to fresh GAC after the first 4500 bed volumes for eight of the compounds listed in Table III. For the chlorinated benzenes, little difference is seen between any of the three GAC conditions even though the exhausted column had a previous service of 13,000 bed volumes (1 year). Ethylbenzene and heptaldehyde both exhibited different removals for each GAC. A similar comparison after 7100 bed volumes (data not shown) revealed little difference between fresh and regenerated GAC for the removal of any of the compounds including ethylbenzene and heptaldehyde.

The long-term removal trends and the response to the pH/ΔDO upsets indicate that biological degradation may be an underlying removal

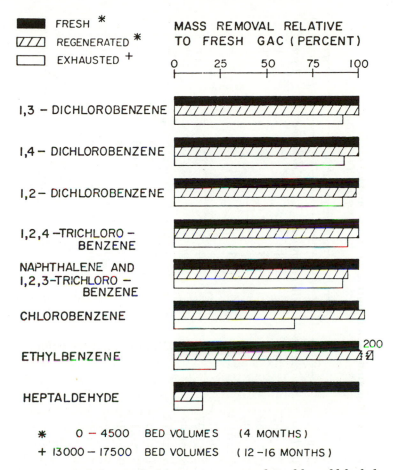

Figure 13. Mass removal of aromatic compounds and heptaldehyde by regenerated and exhausted GAC, compared to fresh GAC.

mechanism. Studies conducted by Bouwer and McCarty (*11*) support this tentative conclusion. A biological degradation study of five of the six chlorinated benzenes reported here (excluding 1,2,3-trichlorobenzene) was conducted in continuous-flow inert-media columns with fixed-film bacteria. At influent concentrations of 10 μg/L, long-term (1 year, or 26,000 bed volumes) removals of 90% or better were achieved with the exception of 1,3-dichlorobenzene. They also report acclimation periods of 10–40 days, which, if applied to the Palo Alto Reclamation Facility GAC system, would imply that adsorption was the controlling chlorinated benzene removal mechanism during the first 300–1200 bed volumes after which biodegradation and adsorption are both important.

Table III. Aromatic Compounds—Concentration and Removal

Compound	Influent[a,b] Conc. (μg/L)	Average Percent Removal			Response to pH/ΔDO Upsets[d]
		Fresh[a]	Regenerated[a]	Exhausted[c]	
1,3-Dichlorobenzene	1.47 ± 1.67	80	85	76	+
1,4-Dichlorobenzene	2.50 ± 1.89	92	89	81	+
1,2-Dichlorobenzene	2.73 ± 2.30	90	91	80	+
1,2,4-Trichlorobenzene	8.33 ± 5.60	90	85	91	−
Naphthalene and 1,2,3-trichlorobenzene	3.22 ± 2.97	90	89	79	−
Chlorobenzene	0.56 ± 0.48	56	62	7	+
Ethylbenzene	0.16 ± 0.12	37	43	8	−
p-Xylene	0.07 ± 0.04	37	37	−29	+
m-Xylene	0.16 ± 0.11	20	11	−146	+
2-Methylnaphthalene	0.10 ± 0.10	62	78	64	+
1-Methylnaphthalene	0.12 ± 0.13	61	83	73	+
Heptaldehyde	0.17 ± 0.33	40	30	48	−

[a] Average over 0–13,000 bed volumes.
[b] Mean ± standard deviation.
[c] Average over 13,000–17,500 bed volumes.
[d] + = response; − = no response.

Collective and Specific Indicators.   Figure 14 shows a comparison of the breakthrough patterns for TOC, TOX, and chloroform for the fresh GAC column. Similar characteristics are observed for TOC and TOX removal. Both exhibited 15–20% immediate breakthrough, adsorption saturation at 4500–5000 bed volumes, steady-state partial removal of 20–35%, and decreased removal in response to the pH/ΔDO upsets. In contrast, chloroform exhibited insignificant immediate breakthrough, a more rapid rise to adsorption saturation followed by substantial elution, and no detectable reaction to pH/ΔDO upsets.

Chloroform is used here as a conservative indicator of the behavior of organic contaminants. The initial breakthrough and adsorption saturation of chloroform occur earlier than the other specific organic compounds and collective parameters. If GAC treatment installations are designed and operated in such a way as to remove chloroform effectively, there is reasonable assurance that most other hazardous specific organic compounds will be removed as well. In other words, chloroform may serve as a conservative indicator in the sense of providing an early warning of impending breakthrough of other specific organics of concern, even though its behavior is by no means representative of organic pollutants in general.

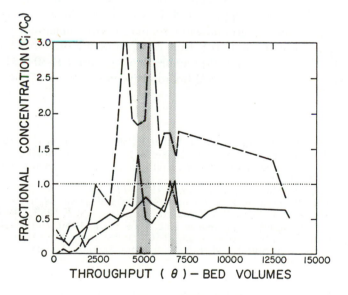

Figure 14. Breakthrough behavior of chloroform (— —), TOX (— · —), and TOC (—) on fresh GAC. Key: ▨, high pH and low ΔDO.

Conclusions

Several phenomena were significant in the removal of organic compounds by granular activated-carbon columns, overlaying the dominant breakthrough pattern of adsorption and saturation. Long-term partial removal of aromatic organic compounds and organic compounds as measured by the collective parameters TOC and TOX occurred. Biological degradation is thought to be a major contributing mechanism in this long-term removal. Elution of halogenated one- and two-carbon compounds was observed. Competitive adsorption and re-equilibration due to a decrease in the influent concentration are thought to be responsible for the elution phenomenon. In some cases, fresh GAC outperformed regenerated GAC, but the differences are small.

Acknowledgments

The authors thank Gary Hopkins for the TOC and TOX analyses and fabrication of sampling system, Martin Reinhard for guidance on organic analytical procedures ,and Richard Harnish for the analysis of the specific organic compounds. The cooperation of the Santa Clara Valley Water District, particularly James Sanchez, and the staff of the National Aeronautics and Space Administration Water Monitoring Station, especially Rick Brooks, is appreciated. This work was carried out as part of a research program on groundwater recharge with reclaimed water, and was supported by the Robert S. Kerr Environmental Research Laboratory of the U.S. Environmental Protection Agency, Grant R-804431, and the State Water Resources Control Board, Agreement No. 8-178-400-0.

Literature Cited

1. Westrick, J. J.; Cummins, M. D. *J. Water Pollution Control Fed.* **1979**, *51*, 12.
2. Henderson, J. E.; Peyton, G. R.; Glaze, W. H. In "Indentification and Analysis of Organic Pollutants in Water"; Keith, L. H., Ed.; Ann Arbor Science: Ann Arbor, Mich. 1976.
3. Grob, K.; Zurcher, F. *J. Chromatog.* **1976**, *117*, 285.
4. Jekel, M. R.; Roberts, P. V. *Environ. Sci. Technol.* **1980**, *14*, 970.
5. Cannon, F. S.; Roberts, P. V. *J. Environ. Eng. Dir.*, ASCE **1982**, *108*(EE4), 766.
6. Peel, R.; Benedek, A. *J. Environ. Eng. Div.*, ASCE, **1980**, *106* (EE4) 797.
7. "Activated Carbon Adsorption of Organics from the Aqueous Phase"; McGuire, M. J.; Suffet, I. H., Eds.; Ann Arbor, Science: Ann Arbor, Mich. 1980; Vol. 2 pp 273–416.
8. Roberts, P. V.; Summers, R. S. *J. Am. Water Works Assoc.* **1982**, *74*, 113.
9. Brooks, R.; Jeffors, E.; Nishioka, K.; Kriege, D. F.; Sanchez, W. I. *Proc. Water Reuse Symposium II*, Washington, D.C., AWWA Research Foundation, August 1981.
10. Symons, J. M. et al., *Treatment Techniques for Controlling Trihalomethanes in Drinking Water*, EPA-600/2-81-156, USEPA, Cincinnati, Oh. 1981.

11. Bouwer, E. J.; McCarty, P. L. *Proc. Environ. Eng. Nat. Conf., Amer. Soc. of Civil Engr.*, Atlanta, Ga, July 1981.

RECEIVED for review August 3, 1981. ACCEPTED for publication April 9, 1982.

23

Experiences in Operating a Full-Scale Granular Activated-Carbon System with On-Site Reactivation

J. DeMARCO[1]

U.S. Environmental Protection Agency, Cincinnati, OH 45268

RICHARD MILLER. DENNIS DAVIS, and CARL COLE

Cincinnati Water Works, Cincinnati, OH 45228

Granular activated-carbon systems were studied in two modes. In the first mode, the systems were acting as a combined filter and adsorber and the filter beds are referred to as a sand replacement system. These carbon beds received coagulated and settled water and the activated carbon acted both as a filter for carryover solids and as an adsorbent material. The second mode employed carbon for adsorption-only by using conventional sand filters prior to the adsorption process and is referred to as a contactor system. Major organic contaminants were determined. The emphasis for data gathering in this study focused on the economic evaluation of the use of the granular activated-carbon systems investigated. The collected data provide an estimate of the actual costs involved in design, construction, and operation of a system that could be full scale at some utilities.

A GROWING NUMBER OF ORGANIC POLLUTANTS in drinking water supplies are being identified as potentially harmful to health, and removing these substances from drinking water appears to be desirable (*1,2*). A granular activated-carbon (GAC) adsorption unit process is especially applicable for use at locations where drinking water sources are subject to a wide variety and varying load of pollutants. The Cincinnati Water Works

[1]Current address: Cincinnati Water Works, Cincinnati, OH 45228

(CWW) site was considered to be representative of a type of system that might require GAC treatment. The water intake is located on the Ohio River below six major tributaries, one of which is the Kanawha River. Numerous large chemical plants that contribute to the vulnerability of the source water to contamination are located in the Kanawha River Valley. As additional evidence of this vulnerability, 39 organic spills have been recorded by the utility laboratory during the past 3 years. Accordingly, the Cincinnati Water Works and the United States Environmental Protection Agency (U.S.EPA) are cooperating in a joint research project to assess the performance of a large scale GAC adsorption system.

The principal research objectives of interest to utility personnel are the determination of the ability of GAC to improve their water quality, the best type of carbon to use, the most effective way to use the carbon, and always the total cost to achieve the improved water quality. No single study at one location can provide a definitive answer that will apply to all locations because results from the use of GAC are highly site specific. However, at any given location, the answer to the question of improved quality and most effective use must involve the application of chemical analyses of the water treated and data interpretation to determine the use of GAC that maximizes water quality improvements while minimizing costs.

The research project includes the study of GAC systems in two modes. In the first mode, the systems are acting as a combined filter and adsorber and the filter beds are referred to as a sand replacement system. These GAC beds receive coagulated and settled water and the activated carbon acts both as a filter for carryover solids from the preceding processes and as an adsorbent material for dissolved organics. The second mode of GAC use under investigation employs carbon for the purpose of adsorption-only by using conventional sand filters prior to the adsorption process. The latter carbon mode is referred to as post-filtration adsorption and the carbon units are referred to as a contactor system.

Major Parameters Used in Carbon Evaluation

The sample handling and analytic methods for all organic parameters used for this paper are described in detail by DeMarco et al. (3). Grab samples were used because they serve the dual purpose of determining the actual ambient influent and effluent concentrations for the organic constituents monitored as well as for estimating the organic loading into and out of GAC beds. Weekly samples were used during the phase of study being reported because the more frequent sampling conducted in earlier phases indicated no significant data loss would occur for the overall objectives sought. From the many organic parameters used in the study, the following were selected for the subsequent discussions of water quality: total organic carbon (TOC), trihalomethane formation potential (THMFP), instantaneous trihalomethanes (InstTHM), simulated distribution trihalomethanes (Sim. Dist. THM), chloroform, and bromodichloromethane. The trihalomethane formation potential was determined by using a terminal trihalomethane (TermTHM) sample that was stored for 7 days at a constant temperature of 29.4°C with a pH buffer of 9.5 and an appropriate InstTHM sample (4). These conditions were selected as the maximum that would be present in the Cincinnati distribution system. InstTHM are the trihalomethanes present in the water at the time of sampling. The Sim. Dist. THM samples were based on TermTHM samples stored for 3 days at ambient distribution system temperature and pH conditions. The Sim. Dist. THM samples are fairly accurate estimates of the THM concentrations that would be present in the distribution system at about 3 days water travel time from the treatment system sampled (3). Considerably more organic analyses were performed, but the concentration and frequency of occurrence of the above mentioned substances provide an adequate example of the evaluation procedures undertaken during this study.

Facilities Description

The study is being conducted at the California Plant of the CWW in Cincinnati, Ohio. The normal water treatment process at the 10.3-m^3/s [235 million gallons/day (MGD)] California plant consists of four sequential unit processes that include long-term presettling followed by coagulation, sedimentation, and sand filtration (Figure 1). The CWW system results in low turbidity water being delivered to the sand filters. Also, the storage provided by the long-term presettling basin is very useful during known spill conditions that frequently occur on river systems. Facilities are available to apply chlorine, aluminum sulfate, and powdered activated carbon to the raw water prior to presettling. The plant also has facilities for feeding chlorine, ferric sulfate, lime, soda ash, and powdered activated carbon between the presettling basins and flocculations basins. Also, chlorine can be added between the sedimentation basin and sand filters as well as between the sand filters and clear wells.

Figure 2 is a schematic of the carbon units under investigation at CWW. Two separate 0.22-m^3/s (5-MGD) sand filters have been converted to sand replacement carbon systems (carbon filters). This conversion was effected by removing the 0.76 m (30 in.) of filter sand and replacing it with 0.76 m (30 in.) of 12 × 40-mesh carbon. A 0.16-m^3/s (4-MDG) carbon contractor (post filtration adsorption system) that treats sand filtered water has also been constructed. These two operating

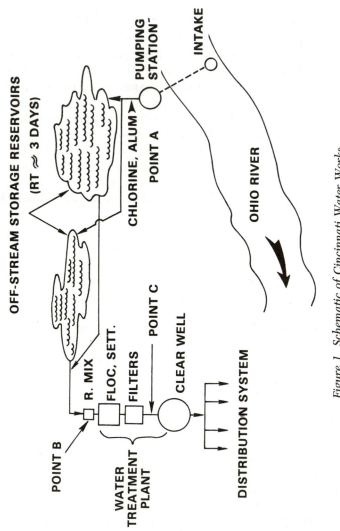

Figure 1. Schematic of Cincinnati Water Works.

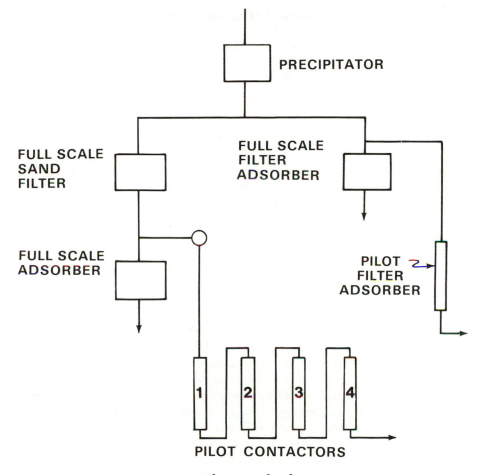

Figure 2. Schematic of carbon units.

modes of carbon application provide a means of comparing the operation of a sand replacement carbon system with that of a post-filtration carbon contactor system. Table I lists pertinent physical data that describe the carbon system now operating. Each of the four $0.4\text{-m}^3/\text{s}$ (1-MGD) carbon contactors has sample ports to allow comparison with the sand replacement system at the same empty bed contact times (EBCT). Pilot carbon columns also are being operated parallel to the full-scale systems to assess their ability to simulate the large systems.

Figure 3 is a schematic diagram of the 227-kg/h (526-lb/h) fluid bed carbon reactivation system being used on-site. To date about 455 metric tons (580 tons) of carbon have been reactivated at the utility. Carbon is moved from the carbon adsorption system to a 29-m^3 (1050-ft^3) spent

Table I. Physical Data for Carbon Systems

Carbon System	Flow	EBCT (min)	Carbon Depth (ft)	Carbon Weight (lb)	Hydraulic Loading (GPM/sq ft)
Post-filtration contactor	1 MGD	16	14.8	42,500	6.94
Sand replacement	5 MGD	7.5	2.5	96,200	2.5
Pilot post-filtration contactor	920 GPD	16	15.1	34.8	7
Pilot sand replacement	170 GPD	7.8	2.6	3.1	2.5

carbon storage tank. Carbon is transported from the contactors to the spent tank by hydraulically pressurizing the units to provide the motive force. Carbon is transported from the sand replacement systems by a manually fed eductor system. The carbon is then dewatered to about 50% moisture and fed to the fluid bed reactivation furnace. The carbon is reactivated to near virgin properties under a controlled combustion process, cooled by water-quenching, and then placed in a 29-m^3 regenerated carbon storage tank for subsequent transfer to the appropriate carbon adsorption system. Figure 4 is a schematic of the furnace system showing the flow process and air test sample locations.

The carbon contactor shells are constructed of 0.95-cm (3/8 in.) steel and were originally painted inside with 20-mil thick coat of Cook Phenicon 980 with a paint performance specification that requires resistance to carbon abrasion and organic leaching. Because of improper application, the organic leaching requirement had to be enforced and proper refinishing was performed on-site for each contactor with the same epoxy-phenolic paint. Contactor influent and effluent piping is steel pipe with a concrete lining. The metal for sampling probes in the carbon contactor and all wetted, carbon exposed surfaces in the carbon reactivation system battery limits is 316L stainless steel. The 316L stainless steel components in the reactor battery limits have not presented any problems to date. Subsequent leaching tests for the painted interior of the contactors showed no discernible organic contaminants. The contactors have been in operation for about 18 months and parts of the interior surfaces of the contactors appear to require repainting. Additional contactor problems encountered include:

1. Inadequate water to carbon ratio. This was resolved by CWW personnel by adding a water injection point at the base of each contactor.
2. Plugging of the fine mesh screens on the back wash effluent manifold and sampling ports. The problem was resolved by removing these screens with no adverse effects observed to date.

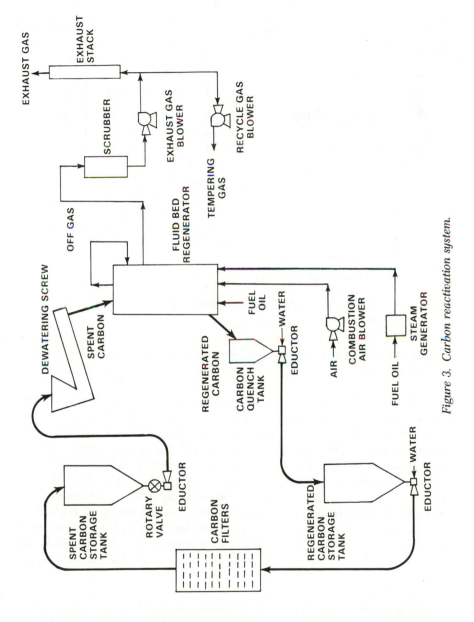

Figure 3. Carbon reactivation system.

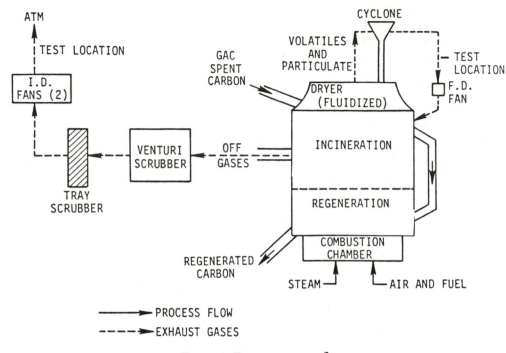

Figure 4. Furnace process flow.

3. Cavitation problems and loss of operating time caused by the piping configuration.

The limitation of a typical sand replacement carbon system is that only a restricted carbon depth can be used. This resulted in the 0.76-m (30-in.) carbon depth and 7.5-min EBCT at the conventional hydraulic loading of 101.8 L/min/m² (2.5 GPM/ft²). Since the post-filtration carbon adsorption system did not have a restriction, an EBCT of between 10 and 20 min was selected based on EPA research data. The final design resulted in a system with a carbon depth of about 4.6 m (15 ft) and a 16-min EBCT with a hydraulic loading of about 285 L/min/m² (7 GPM/ft²). As mentioned, the four metal contactors have multiple ports so that various contact times can be evaluated. Normally, samples are taken to monitor EBCTs of about 4.5, 7.5, and 16.0 min. These two systems were placed in operation on January 14, 1980 and operated following the criteria selected except when prevented by the reactivation system's availability. The two systems were treated like normal water production facilities to the maximum extent possible.

The research project has attempted to maximize data collection for multiple purposes and the operational criteria for carbon reactivation were: equilibrium between the influent and effluent total THM concentration, and presence of an apparent TOC and/or THMFP steady-state or plateau where the effluent concentration does not equal or exceed the influent concentration but stays approximately parallel to the influent level. The degree of effort expended, for some data collection activities, was greater than what normally would be conducted in a routine water plant operation. Whenever possible, activities that generated atypical costs were excluded from consideration as actual normal costs required to operate the systems.

Results and Discussion

Carbon Adsorption. When beginning to interpret carbon performance, a decision was made on the various types of organic analyses to be used to judge the performance of GAC. When a drinking water concentration of a specific organic substance is required to be controlled within given limits, there exists a definitive use of analytic methods to assess performance. However, Drinking Water Regulations currently contain maximum contaminant levels (MCLs) for only a few organics (5, 6). These are the pesticides endrin, lindrane, methoxychlor, toxaphene, 2,4-D, and 2,4,5-TP, plus total trihalomethanes (chloroform, bromodichloromethane, dibromochloromethane, and bromoform). Obviously all of the general and specific organic analyses currently available could be used in assessing treatment performance. For this study, the organics that were most consistently present and that occurred in the highest concentrations were used to compare the two modes of operation under investigation.

The organics discussed in this paper represent a combination of general and specific compound analyses that attempt to show the extent to which GAC reduces the organic content of drinking water. A more comprehensive organics review was previously provided by DeMarco et al. (3).

COMPARISON OF EQUAL CONTACT TIMES. Figures 5–10 present breakthrough curves that show an overview of the concentrations of each measured parameter that passes into the drinking water after GAC treatment by the sand replacement and contactor systems with an equal EBCT of 7.5 min. As expected, the data indicate that a variety of adsorption performances exist, depending on the organic substance, but that in general the organic content of the drinking water was reduced. These figures provide results that satisfy the major objective of comparing the effluent concentrations between the carbon contactor and sand replacement system when operated with the same empty bed contact times of 7.5 min. Although the flow and hydraulic loadings differed as shown in Table I, the contactors were designed so they could be monitored for various EBCTs, thus the equivalent of the sand replacement EBCT was monitored in the contactors. Figures 5–10 show that very few differences exist in the effluent concentration of the sane replacement and the contactor system. Both systems operated at 192 bed volumes/day. Perhaps a slightly better effluent quality could be claimed for the contactor, based on concentration of organic substance leaving the system. Choosing arbitrary criteria as shown in Table II results in the carbon evaluation data shown. The average influent TOC for both systems was about 1900 μg/L. The average chloroform influent concentration was about 14 μg/L for the sand replacement system and 20 μg/L for the contactor system. The difference in the averages for chloroform is primarily because of the longer contact time with free chlorine residual for the influent to the contactor system. Table II data show that, with equal influent TOC concentrations, the specific criteria of maintaining an effluent concentration ≤ 1000 μg/L resulted in almost the same service time, carbon use rate, cumulative mass loading, and percent removal. Using a percent TOC removal criteria resulted in less uniform comparisons, but no consistent major differences were found that refute the similar adsorptive behavior of the matched EBCT systems. Cumulative mass loading and percent removal data plots also could be used to further prove the similarity of adsorptive peformance for the compared system.

EFFECT OF VARIOUS CONTACT TIMES.   The improved effluent water quality with increasing EBCT is shown by the breakthrough curves of Figures 11–15. Selecting an arbitrary effluent criteria of 5 μg/L for chloroform permits further evaluation of the breakthrough curve data. Such an effluent requirement would result in an operating service time of about 15, 57, and 147 days for EBCTs of 4.5, 7.5, and 16.0 min, respectively, in the contactor system for the time period and conditions shown in Figure 14. Thus, increasing the EBCT by a factor of about 3.6 from 4.5 for 16.0 min increases the operating service time by a factor of almost 10. Also, calculation of carbon use rates for the same criteria shows that they decrease from 572 to 367 to 304 lb of carbon per million gallons of water treated from 4.5 to 7.5 to 16.0 min. The best adsorption carbon use rate for 5 μg/L during the test period apparently would occur at an EBCT longer than 7.5 min and appears to be close to 20 min. However, the total least cost system that meets the design criteria is the one sought.

GENERAL ADSORPTION EVALUATION CRITERIA.   The criteria used to assess carbon operation will have an impact on the resulting decisions and must be considered prior to conducting batch, pilot column, or full-scale evaluations. Table III lists additional physical and chemical analyses

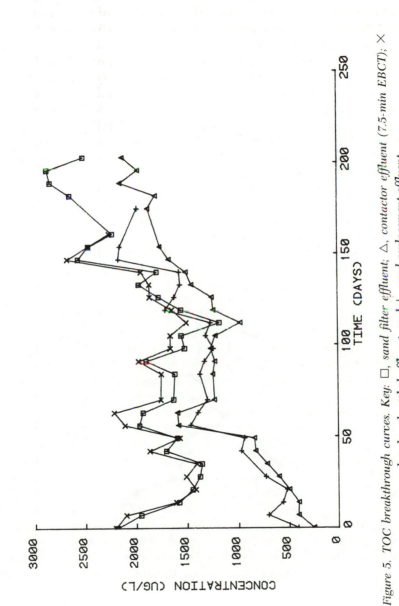

Figure 5. TOC breakthrough curves. Key: □, *sand filter effluent;* △, *contactor effluent (7.5-min EBCT);* ×, *coagulated and settled effluent; and* +, *sand replacement effluent.*

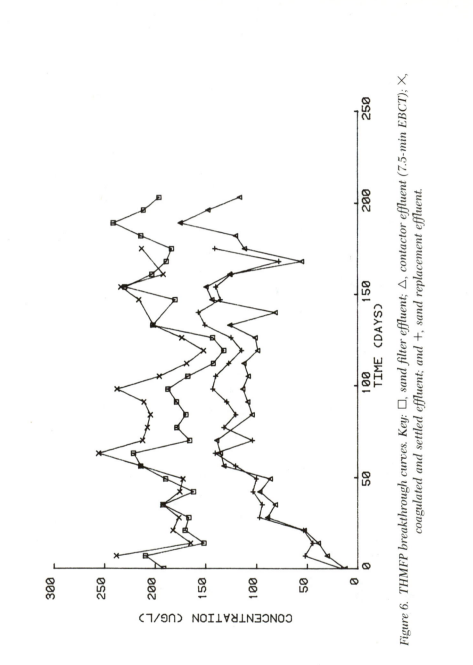

Figure 6. THMFP breakthrough curves. Key: □, *sand filter effluent;* △, *contactor effluent (7.5-min EBCT);* ✕, *coagulated and settled effluent; and* +, *sand replacement effluent.*

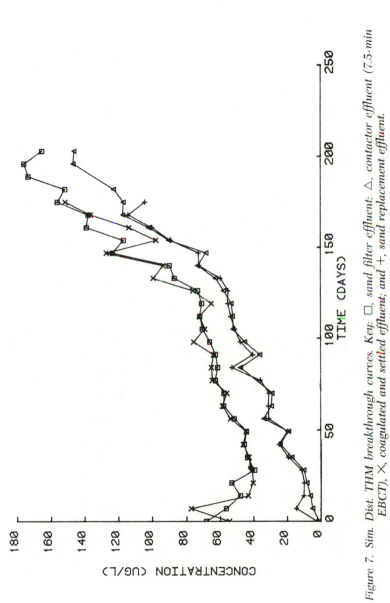

Figure 7. Sim. Dist. THM breakthrough curves. Key: □, sand filter effluent; △, contactor effluent (7.5-min EBCT), ×, coagulated and settled effluent; and +, sand replacement effluent.

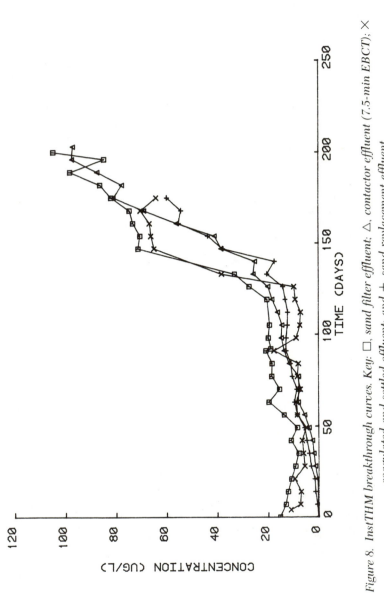

Figure 8. InstTHM breakthrough curves. Key: □, *sand filter effluent;* △, *contactor effluent (7.5-min EBCT);* × *coagulated and settled effluent; and* + *sand replacement effluent.*

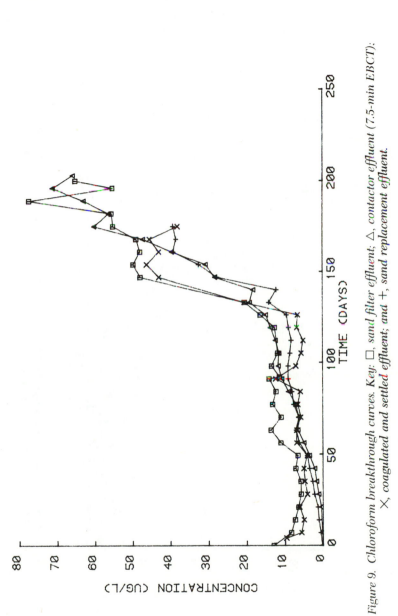

Figure 9. Chloroform breakthrough curves. Key: □, sand filter effluent; △, contactor effluent (7.5-min EBCT); ×, coagulated and settled effluent; and +, sand replacement effluent.

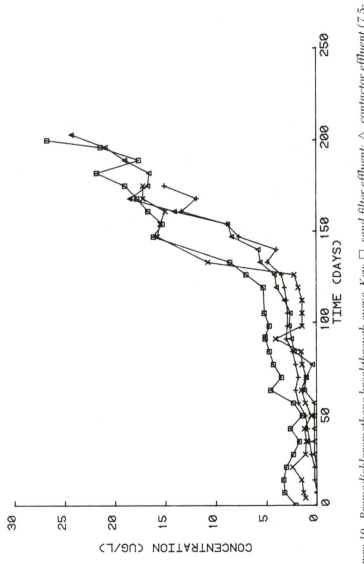

Figure 10. Bromodichloromethane breakthrough curve. Key: □, *sand filter effluent;* △, *contactor effluent (7.5-min EBCT);* ✕, *coagulated and settled effluent; and* +, *sand replacement effluent.*

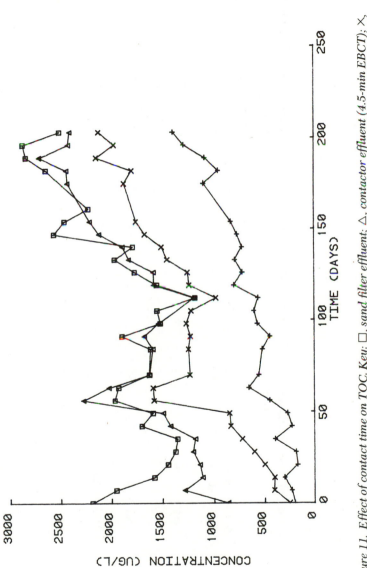

Figure 11. Effect of contact time on TOC. Key: □, sand filter effluent; △, contactor effluent (4.5-min EBCT); ×, contactor effluent (7.5-min EBCT); and +, contactor effluent (16.0-min EBCT).

Table II. Comparison of Sand Replacement (SR) and Contactor (Cont) Carbon Systems Using Various Criteria

Evaluation Criteria	TOC						Chloroform					
	≤1000 μg/L		≤50%		≤75%		≤5μg/L		≤50%		≤75%	
	SR	Cont	SR	Cont	SR	Cont	SR	Cont	SR	Cont	SR	Cont
Service time (days)	51	52	37	43	4	8	53	57	29	50	22	36
Carbon use (lb/10⁶ gallon)	385	405	554	488	4822	2728	361	367	662	420	867	585
Mass load (g/kg)	21.5	21.2	16.9	18.1	2.5	4.9	0.05	0.13	0.037	0.120	0.027	0.096
Percent	40	39	—	—	—	—	10	54	—	—	—	—

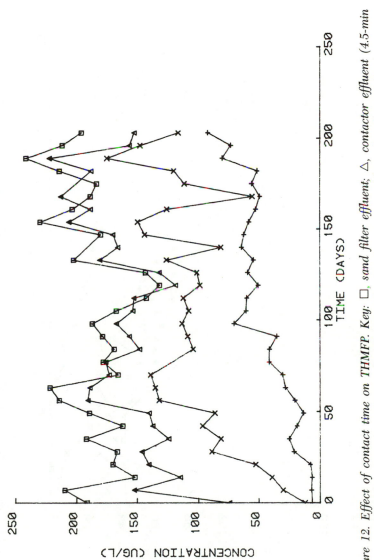

Figure 12. Effect of contact time on THMFP. Key: □, sand filter effluent; △, contactor effluent (4.5-min EBCT); ×, contactor effluent (7.5-min EBCT); and +, contactor effluent (16.0-min EBCT).

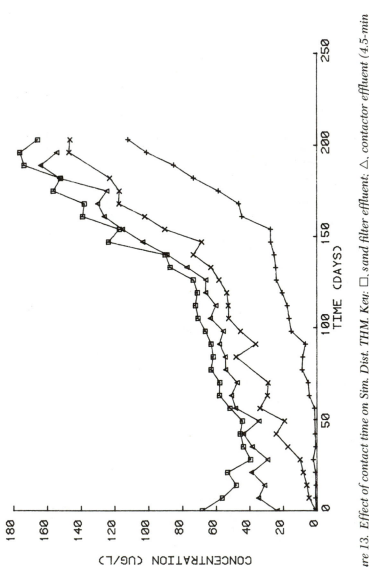

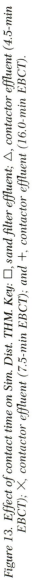

Figure 13. Effect of contact time on Sim. Dist. THM. Key: □, *sand filter effluent;* △, *contactor effluent (4.5-min EBCT);* ×, *contactor effluent (7.5-min EBCT); and* +, *contactor effluent (16.0-min EBCT).*

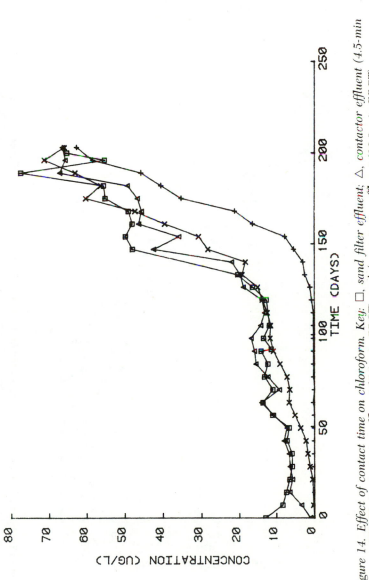

Figure 14. Effect of contact time on chloroform. Key: □, *sand filter effluent;* △, *contactor effluent (4.5-min EBCT);* ×, *contactor effluent (7.5-min EBCT); and* +, *contactor effluent (16.0-min EBCT).*

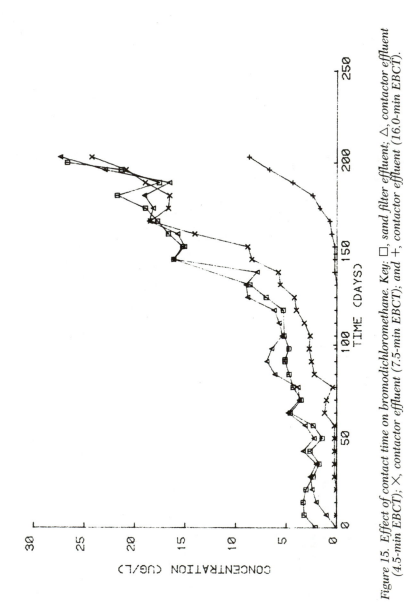

Figure 15. Effect of contact time on bromodichloromethane. Key: □, sand filter effluent; △, contactor effluent (4.5-min EBCT); ×, contactor effluent (7.5-min EBCT); and +, contactor effluent (16.0-min EBCT).

Table III. Carbon Analyses
Surface area (BET)
Iodine number
Modified phenol value
Molasses index
Volatile matter
Ash content
Apparent density
Abrasion number
Sieve analyses

performed on the carbon. Table IV shows the reduction in carbon capacity during the first carbon exposure for both the sand replacement and contactor carbon system using conventional laboratory carbon evaluation techniques. As expected, the surface area, iodine number, and decolorizing index decreased indicating a decrease in adsorptive capacity. An increase was shown, as expected, in the phenol value and volatiles, which also show decreased adsorptive capacity. The apparent density also shows an increase that might be attributed to adsorbed organics as well as inorganics. The increase in ash content indicates the inorganic residues, such as calcium, that were picked up by carbon during the exposure cycle. The ash results show the effect of a combined filtration/adsorption process as compared to a straight adsorption process. The percent of ash on the spent carbon from the sand replacement system was 2.7 times the value for the virgin carbon as compared to the factor of 1.2 shown for the contactor system. Although the ash values of the systems were restored to about virgin values for two cycles, as shown in Table IV, ash build-up will be watched in subsequent cycles.

Table IV. Results of Carbon Analyses

	Sand Replacement			Contactor		
	Virgin	Spent	S/V	Virgin	Spent	S/V
BET surface area (m^2/g)	1102	1052	0.95	1070	1030	0.96
Iodine number (mg/g)	1184	711	0.60	1128	824	0.73
Modified phenol value (ppm)	16.2	38.3	2.4	18	30.2	1.7
Molasses decolorizing index (DI units)	10.5	5.9	0.56	9.6	6.8	0.71
Ash (%)	7.06	19.1	2.7	8.1	9.46	1.2
Volatile matter (%)	2.49	11.0	4.4	3.57	9.82	2.8
Apparent density (g/mL)	0.483	0.600	1.2	0.506	0.558	1.1
Mean particle diameter	1.25	1.34	1.07	1.10	1.11	1.0

Carbon Reactivation.   The 7.5-min EBCT systems generally exceed the previously cited reactivation cycle criteria in about 6–12 weeks depending on influent conditions, and the 0.22-m^3 (5-MGD) sand replacement systems were generally reactivated with that frequency. Each 0.04-m^3/s (1-MGD) contactor has a total EBCT of 16 min; thus, the time between reactivation cycles was longer, generally 3–6 months, depending on influent conditions. The 7.5-min EBCT was constantly monitored for comparison with the sand replacement system. The amount of carbon available for repetitive reactivation consisted of 40,909 kg (90,000 lb) in each of two sand replacement systems and 19,090 kg (42,000 lb) in each of the four contactor systems for a total of about 158,182 kg (348,000 lb).

The initial shakedown and early operation of the fluid bed reactivation followed the pattern described by many investigators. The major problems contributing to down time center around the following areas:

1. Instrumentation required to operate the system automatically. Temperature-related controllers such as thermocouples, recorders, and oxygen analyzers; computer hardware and software; and gas flow controllers.
2. Removing sand from dryer plates and unplugging dryer plate holes because the sand separator does not prevent sand carryover to the furnace.
3. Bearings for blowers and dewatering screw; bent screw shaft.
4. Design of cyclone separator discharge flap valve.

These items, plus many additive minor problems, predictable and unpredictable, have combined to result in a total worst case estimate of 43% operating time (57% down time) during the time period from April 10, 1980 through January 31, 1981. The down time for this period includes all time lost by unavailability of the reactivation system for any reactivator-related reasons. Problems related to the contactors and scheduled down time are not included. Thus, if carbon was scheduled for reactivation and a problem with the contactors prevented reactivation from taking place, then the reactivation system was not credited with down time. The above time period did include problems during shakedown and long time delays that now can be avoided by a proper equipment inventory and experiences learned to date. A more realistic, but still very conservative, estimate of 58% operating time or 42% down time experience is provided during the time period between July 1, 1980 and December 31, 1980. This period was after the initial 3 months of trials. The system operated 24 h/day for a 5-day week from October through December and adjustment for a 7-day week, 24-h/day operation would result in an additional 3% operating time to 60% or a projected down time of 40%. Experiences with similar systems at Hattiesburg, Miss. and Manchester, N.H. seem to forecast continued improvements in operating time.

The primary objective of the reactivation process was to return the spent carbon to nearly its virgin properties with a minimum amount of damage to the basic pore structures. Although all laboratory control tests listed in Table III were performed, the primary carbon quality control tests during the reactivation cycles were apparent density and iodine number. Various operating controls such as temperature, carbon feed rate, and steam feed rate were controlled closely during the reactivation process. Table V provides an example of furnace operating conditions experienced during the carbon feed; use averaged about 482 lb/h based on a backwash and drained density of 0.44 g/cc (27.5 lb/ft^3). Based on the apparent density of 0.48 g/cc (30 lb/ft^3) a carbon feed rate of about 243 kg/h (535 lb/h) was averaged to date. As expected, the duration of run for the 41-metric ton (45-ton) sand replacement carbon system was about twice the 19-metric ton (21-ton) contactor carbon system. The fuel requirements for the operating conditions shown are provided as an example and do not represent an average condition. However, they serve to illustrate the amount of fuel required by the incineration or afterburner function as compared to the total requirements. Fuel requirements were reported by other investigators for a multiple hearth and rotary kiln system used for reactivation of carbon used in wastewater applications at Pomona, Calif. (7, 8). The wastewater applications resulted in heavier organic loadings, but a general comparison of fuel requirement shows that about 19,000 kJ/kg (8000 BTU/lb) was used for the multiple hearth system and about 28,000 kJ/kg (12,000 BUT/lb) for the rotary kiln system as compared to about 14,000 kJ/kg (6000 BTU/lb) shown in Table V. The steam used per pound of carbon and typical operating temperatures are listed.

Table VI shows a comparison of carbon characteristics for the sand replacement carbon through the first two cycles of operation. Sufficient data have not been analyzed yet to draw firm conclusions regarding the effects of multiple reactivations on carbon adsorption capacity. However, the results shown would generally indicate that the carbon was more heavily loaded during the first cycle of operation and that the reactivation process is returning the spent carbon to very near the conditions present in the carbon at the start of each cycle. The analysis of regenerated carbon both before and after addition of make-up carbon (R-1B and R-1A) aids in assessment of the results achieved by each successive adsorption and reactivation cycle. Explaining the changes in iodine number or volatile matter, for example, is easier when the R-1A sample data are included. Without the R-1A iodine number of 1128 mg/g, the spent value in the

Table V. Example of Reactivation System Operating Conditions

Parameter	Value
Duration of run (h)	150–230 sand replacement 90–110 contactor
Furnace loading	
kg carbon/h	219
lb carbon/h	482
Fuel used (BTU/lb carbon)	5973
incineration	3060
reactivation	1475
steam	1438
Steam used (lb/lb carbon)	0.6
Temperatures (°F)	
reactivation zone	1500
incineration zone	1600
dryer	300
dryer loop	250
scrubber stack	varies with water temperature
Total carbon losses	15.9% contactor 18.9% sand replacement

Table VI. Effect of Reactivation on Sand Replacement Carbon

	First Cycle			Second Cycle		
Carbon Characteristics	V	S	R-1B	R1-A	S	R-2B
Surface area (m²/g)	1102	1051	1022	983	1130	1119
Iodine number (mg/g)	1184	711	1011	1128	1093	1111
Modified phenol (ppm)	16.2	38.3	21.7	19.8	32.4	21.1
AWWA CALC (mg/L)	1.84	4.36	2.47	2.25	3.69	2.41
Molasses index (DI units)	10.5	5.9	10.8	13.5	11.2	14.5
Volatile matter (%)	2.49	11.0	3.87	4.79	10.66	5.38
Ash (%)	7.06	19.1	8.31	8.33	10.48	8.44
Apparent density (g/mL)	0.483	0.600	0.482	0.474	0.503	0.471
Mean particle diameter (mm)	1.251	1.336	1.178	1.165	0.932	0.945
Abrasion number (% by mean PD)	67.3	73.6	69.5	73.8	68.4	69.8

second cycle (1093) would appear to be higher than the R-1B value of 1011 from the reactivation in the first cycle. The first adsorption cycle lasted about 6 months, whereas the second cycle lasted only 6 weeks. Therefore, the large decrease from virgin to spent in the first cycle agrees with operating time conditions and likewise the small decrease from 1128 to 1093 mg/g during the second cycle is consistent with operating experience. The same effect is shown by the volatile matter, modified phenol value, and molasses decolorizing data. Also, the effect that virgin make-up carbon has on restoring some of the carbon characteristics may be estimated by comparing the values R-1B and R-1A.

Pore size distribution data will be compared with the general parameters such as iodine number and molasses decolorizing index (DI) to evaluate the relationship of these carbon control techniques to the changes in pore characteristics and performance for removing various specific organics. The increase in molasses DI, for instance, would indicate that there was a slight increase in the surface area in the pores greater than 30 Å. Pore size distribution data for the first cycle of data also showed a somewhat greater surface area between 35 Å and 150 Å. This increase may agree with the molasses DI data. Also, reduced iodine numbers and modified phenol values for cycle 1 could indicate a slight reduction in the surface area available in fine pores. If such a trend were to continue for more than the usual first few cycles, then eventually there would be an increased surface area available to large color body molecules. As mentioned previously, more data are required before a proper evaluation of differences can be performed.

Although the ash content increased by 170% during the first cycle of exposure to drinking water and 26% during the second cycle, the reactivation process returned the values to near virgin quality. The apparent density was the primary practical test used to control the carbon quality during the reactivation process. Frequent determinations were performed during each cycle and results in Table VI are typical of the close approximation of virgin conditions achieved. These results are especially good when the light loadings are taken into consideration. Such light organic loadings would be typical for carbon used at water supply utilities (*9–11*). There appears to be an overall reduction of mean particle diameter from the virgin value to the value shown for the second cycle regenerated carbon (R-2B). The abrasion numbers indicate no softening of particles has taken place thus far.

The measurement of carbon losses is an important consideration in the economic evaluation of the total carbon system installed. In this chapter the total carbon loss is defined as the difference in the spent carbon volume transferred to the spent tank and the regenerated carbon volume returned back to the system. All measurements were performed after backwashing. The sand replacement system volume measurements include corrections for any volume of support sand removed and found by the sand separator system. The contactor system did not have a sand support and, thus, did not require such a correction. The total carbon losses are attributed to both the attrition losses caused by the transport system and carbon loss caused by the oxidation of fines in the transport process and oxidation of the surfaces of the carbon particles by the activating gases. Total carbon losses are shown in Tables V and VII. The range of losses can be seen from the data shown in Table VII. The losses ranged from about 14 to 19% for the sand replacement system. These comparisons, although not without a degree of error, provide an estimate of the total carbon losses experienced in the two systems compared. Such differences will effect the economic comparison that ultimately will be performed. The questions of where the losses occur and how to reduce them become important when trying to minimize operating costs. Attempts were made to measure the carbon losses between the spent and regenerated storage tanks. The confidence in these measurements is not as great as total system loss measurements, and only gross comparisons are planned. Table VIII shows the results achieved to date. Although we may be skeptical about some of the high and low values for each system, we do not have sufficient cause to say that one should disregard them. If a gross comparison is performed for the averages shown, it would appear that about 1% of the carbon is lost outside of the battery limits included by the spent tank to regenerated tank measurements. A special test was conducted whereby the carbon transport loss was measured by moving spent carbon from the contactor to the spent tank and then, with the rotary valve removed, the carbon was educted to

Table VII. Total Carbon Losses

	Contactor			Sand Replacement	
ID	Phase	Percent Loss	ID	Phase	Percent Loss
C	3-0	15.8	21A	1-0	17.1
D	3-0	14.9	15A	3-0	22.6
A	3-0	13.8	21A	3-1	22.6
BB	3-0	18.0	15A	3-1	23.7
C	3-1	15.6	21A	3-2	13.7
D	3-1	18.9	15A	3-2	13.6
A	3-1	14.2			
Average loss		15.9	Average loss		18.9

Note: Total losses calculated by volume measurements in each carbon adsorption bed.

Table VIII. Carbon Losses Across Furnace

	Contactor			Sand Replacement	
ID	Phase	Percent Loss	ID	Phase	Percent Loss
D	3-0	15.9	21A	3-1	31.2
A	3-0	12.0	15A	3-1	16.4
BB	3-0	14.6	21A	3-2	16.6
C	3-1	16.7	15A	3-2	8.7
D	3-1	3.4			
A	3-1	22.8			
Average loss		14.2	Average loss		18.2

Note: Furnace losses calculated by volume measurements in spent and regenerated carbon tanks.

the regeneration storage tank and then back to the contactor. The transport volume losses were 4%. A second loss measurement was conducted under normal conditions with the carbon passing through all of the usual stages which include blow case transport from the contactor vessel to the spent tank, rotary star valve feed to the screw conveyor, fluid bed transport through the normal furnace operating conditions, quench tank immersion, hydraulic eduction to the regenerated tank, and then back to the contactor. The total volume loss experienced was 15.6%; if the 4% transport loss is used as an estimate, then there was about an 11.6% volume loss between the rotary valve at the base of the spent tank and the eductor to the regenerated carbon storage tank.

This special test lends credence to the relative magnitude of the gross estimate of about 1% transport losses outside of the reactivation system battery limits. Also, the total loss for the special test was similar to the average shown for the contactors in Table VII. The estimate of an 11.6% "furnace" loss depends on the validity of the assumed 4% transport loss outside of the reactivation system limits as just described for the purpose of the loss measurements. Although many citations of losses appear, the actual reported data and specifics of measurements are sparse. Directo and coworkers (7, 8) reported total losses of about 7% for their experiences with the multiple hearth and rotary kiln. Nur (12) reported about 10% total losses and 2% transport losses for experiences with the infrared furnace at Pomona, Calif. Inhoffer and McGinnis (13) reported 5% losses for the infrared system at a water supply utility at Passaic, N.J. without any details of how the measurements were performed. Lykins (14) reported total losses most recently of about 14.8% (11% across the reactivation system) at the Manchester, N.H. water supply utility that uses a fluid bed reactivation system. The summation of carbon loss experience described by Culp et al. (15) is generally without specifics of the measurement techniques used or specific data collected. However, the obvious conclusion one should arrive at is that general estimates of losses pose potential for misleading economic evaluations. Thus, the variation in impact of higher than estimated losses should be assessed and decisions made on the range of results presented.

Air Pollution Control Tests. During the normal reactivation cycle, atmospheric emission tests were performed on the fluid bed reactivation furnace. Triplicate tests were conducted downstream from the scrubber control device to determine the concentration and mass emission rates of particulate matter, nitrogen oxides (NO_x), and total gaseous nonmethane hydrocarbons. Also, tests were conducted downstream from the dryer cyclone to quantify particulate and volatile organic concentrations. In addition, the volumetric gas flow rate, temperature, moisture content, and composition (oxygen, carbon dioxide, and carbon monoxide) were measured. Where applicable, sample and analytical procedures used to determine particulates followed those described in EPA Method 5 of the *Federal Register* (16). The duct diameter at the sampling location did not meet the minimum required diameter of 30 cm (12 in.) specified in Method 1 of the *Federal Register* (16). Particulate samples were collected isokinetically at a point of average velocity in the duct. Velocity and temperature data were used to set isokinetic sample rates. The method was approved by the Regional Air Pollution Control Agency prior to testing. Table IX lists the average particulate emission data for the dryer loop and scrubber stack. Concentrations are reported in grains per dry standard cubic foot (gr/dscf) and emission rates in pounds per hour (lb/h). The filterable particulates represent the material collected in the sample probe and on the filter, both of which were heated to about 250°F. The condensible fraction represents that material which passed through the heated filter and was trapped in the impinger section of the sample train at a temperature of 68°F or less. The condensibles are not included as particulates for regulation in Ohio. The results are well within acceptable limits.

Table X lists the results of the organic emissions data collected during the testing. The high moisture content in the dryer loop test required the use of three filtered bag samples instead of the EPA Method 25 sampling tests that were used for the scrubber stack. Analytical procedures were those described in EPA Method 25 of the *Federal Register* (17) for data from both locations.

The nitrogen oxide (NO_x) tests were conducted according to procedures described in EPA Method 7 of the *Federal Register* (18). NO_x concentrations at the scrubber outlet averaged 97 parts per million with a mass emission rate of 0.37 lb/h. Further reseach related to air emission

Table IX. Particulate Emission Data

Type	Dryer Loop	Scrubber Stack
Concentration (gr./dscf)		
filterables[a]	0.5	0.0024
condensibles	0.039	0.0090
organic	0.018	—
inorganic	0.021	—
Mass emission rate (lb./h)		
filterables[a]	1.0	0.01
condensibles	0.08	0.043
organic	0.04	—
inorganic	0.04	—

[a]Only the filterables are considered for particulate emissions in Ohio.

testing may be conducted. All results thus far are well within acceptable limits imposed by Federal, State, and local regulations.

Costs

The general adsorption aspects of carbon have been long known and employed at Cincinnati Water Works by use of powdered activated carbon (PAC). Throughout the years, PAC has been used whenever river conditions warranted the use of an adsorbent. Thus, the study of a continuous operating GAC system and a potentially more effective and constant adsorption treatment barrier was a logical undertaking by water utilities. Since adsorption with carbon was already practiced, the emphasis for data gathering focused on the economic evaluation of the use of GAC systems and the constancy of protection provided for various time periods. The degree of organics removal obviously was equally important and the most comprehensive methods available were used as described elsewhere (3).

Table X. Total Gaseous Nonmethane Hydrocarbons

	Dryer Loop	Scrubber Stack
Concentration (ppm)[a]	3.1	219
Emission rate (lb/h)	0.002[b]	0.29[c]

[a]ppm is parts per million by volume, reported as CH_4.
[b]Emission rate for dryer is in pounds as CH_4 per hour, using measured concentration and average volumetric flow.
[c]Emission rate for scrubber is in pounds as CH_4 per hour, using standard conditions 16 lb/mol, and 385.4 ft^3/lb mol.

Tables XI and XII show some pertinent background data used in the cost calculations discussed later. The useful life cycles shown on Table XI were estimated by CWW personnel based on normal plant practices and operating experiences gained to date. Costs also were calculated using a straight 20-year life as is frequently found in existing literature. Table XII also contains some data used in the cost calculations. The range of power and fuel rates shows the effect of rapid escalation in energy costs. The actual rates and costs incurred were used in arriving at the capital and operating costs reported in subsequent cost tables.

Table XIII contains the costs to design and construct the carbon adsorption and reactivation system already described. No legal, fiscal and administrative, land, or insurance costs were included. However, all other typical design and construction costs, as well as resident engineering costs, are included. The capital costs of preparing two existing sand filters were minimal because no major alterations were performed. The reason that the initial carbon costs were about the same for the sand replacement and contactor system is because of the relative EBCTs previously described. In comparing total costs, one should keep in mind that a typical sand replacement system is being compared to a post-filtration adsorption system that used an EBCT thought to be fairly optimum.

No capital costs for a building in Table XIII or electric costs in Table XIV are shown for the sand replacement system because no additional costs were incurred above what was normally required for a sand filter. Thus, the costs shown reflect only the additional costs required by the use of either GAC system. The sand replacement transport system was manually fed and contributes to carbon losses through excess handling. In a full plant conversion, capital expenditures for a more efficient automatic transport system would appear to be a wise investment.

Table XI. Background Capital Cost Related Data
(Useful Life Estimates for Units)

Sand Replacement System

    Initial carbon inventory 20-year life

Contactor System

Initial carbon inventory	20-year life
Contactors and instrumentation	8-year life (likely conservative)
Building	25-year life

Reactivation System

Furnace, tanks, and controls	15 years
Building	25 years

Note: All capital costs are amortized at 8% over the useful life span indicated.

Table XII. Criteria and Unit Costs

Unit costs

Power	¢/KWH	2.19–2.91
Natural gas	%/100 ft³	31.032–33.809
Water	%/1000 gallons	42.2
Carbon	¢/pound	46–50
Labor	$/hour	
Contactor and furnace		8
General labor		6.50
Maintenance		7–9

Criteria

Furnace throughput	tons/yr	
Total		1200
Sand replacement		816
Contactor		384
Adsorption system throughput	MGY	
Sand replacement		2950
Contactors		1400
Natural gas fuel value	BTU/ft³	1000
Carbon reactivation fuel use	BTU/lb C	6000

The capital costs and annual operating costs of the reactivation system are shown in Tables XIII and XIV along with the costs of the adsorption systems. Also, the capital and operations costs of the reactivation system were prorated between the adsorption systems based on the ratio of the carbon throughput requirements established for each adsorption system to the total throughput. Table XIV data show that the annual operating labor and material costs for the sand replacement system exceed that of the contactor system by more than the 2.5 ratio of the nominal rated capacities of the two systems.

These costs are primarily related to carbon transport and make-up carbon activities. The backwash criterion of 20-psi head loss (developed during pilot column tests) was never exceeded. Therefore, the contactor system pumping costs primarily consist of influent pumping because backwashing was not required during the normal course of contactor operation. The carbon transport costs are primarily the motive water requirements.

Table XV shows a comparison of the annual amortized capital costs based on the useful life cycles shown in Table XI and a straight 20-year life. The majoy difference appears in the contactor system data because of the 8-year life estimated for the entire contactor system. The total direct

Table XIII. Preliminary Capital Cost Estimates

Adsorption

	Sand Replacement (10 MGD)	Contactors (4 MGD)	Reactivation (500 lb/h)
Design and construction	3,400	832,000	740,200
Building cost	—	219,380	219,380
Initial carbon	96,250	84,000	—
Design and construction costs	99,650	1,135,380	959,580
Prorated reactivation cost to carbon systems	652,514	307,065	—
Design and construction total costs for compared systems	752,164	1,442,745	—

Note: Land, legal, fiscal and administrative, and insurance costs were not included.

Table XIV. Preliminary Annual Operating and Maintenance Costs
(Dollars)

	Sand Replacement (10 MGD)	Contactor (4 MGD)	Reactivation (500 lb/h)
Carbon transport	2,530	1,000	23,000
Electric			
Pumping	—	14,700	—
Building	—	3,230	3,230
Furnace	—	—	6,900
Natural gas	—	—	51,400
Operating labor	66,640	16,240	85,400
Operating material	141,400	54,700	°
Maintenance labor	—	—	68,600
Maintenance material	—	—	1,500
	210,570	89,870	240,030
Prorated reactivation costs to carbon systems	163,220	76,810	—
	373,790	166,680	—

Table XV. Comparison of Annual Amortized Capital Costs (Dollars)

	Based on Table XI	*Based on 20-Year Life*
Sand replacement	10,002	10,002
Contactor	169,890	113,961
Reactivation	105,204	96,315
Prorated Reactivation		
Sand replacement	81,340	75,497
Contactor	203,556	144,782

and indirect costs are shown in Table XVI in dollars and Table XVII shows the costs in cents per 1000 gallons. The estimated total operating costs are about 7.5¢/1000 gallons for adsorption only for the sand replacement system and 14.5–18.5¢/1000 gallons for the contactor adsorption system. The criteria of a 6-week and 3-month cycle must be kept in mind when considering the costs shown. Relaxed criteria would result in lower costs with reduced water quality being delivered. The costs including prorated on-site reactivation are about 15¢/1000 gallons for the sand replacement system and 22–26¢/1000 gallons for the contactor system. The carbon reactivation costs were about 14¢/1b based on an annual carbon through-put of about 1200 tons.

The data collected during this study provide an estimate of the actual costs involved in design, construction, and operation of a system that could be full scale at some utilities. The costs in cents per 1000 gallons demonstrated for the small pressure contactor system or sand replacement system should provide an upper limit of costs for using GAC because both economy of scale and variation in limiting criteria will likely result in a reduced total operating cost for systems with a full-scale plant capacity similar to CWW.

The rest of the project will be directed at estimating whether inflation and long-term effects of high operating and low capital costs as compared to low operating and high capital costs will cause any change in the apparent lower total cost observed for the sand replacement system.

Acknowledgments

We appreciate the cooperation and assistance of Donald Houchins and B. J. Fine of the Cincinnati Water Works, in providing data presented from the reseach project. We wish to thank Patricia Pierson for typing of the manuscript.

Table XVI. Total Annual Operating Cost Estimates (Dollars)

	Sand Replacement (10 MGD)	Contactor (4 MGD)	Reactivation	Sand Replacement + Reactivation	Contactor + Reactivation
Direct operating cost/yr	210,570	89,870	240,030	373,790	166,680
Indirect operating cost/yr Based on Table XI data (Based on 20-year life)	10,002 (10,002)	169,890 (113,961)	105,204 (96,315)	81,340 (75,497)	203,556 (144,782)
Total cost	220,572	259,760	345,234	455,130	370,236
(Total cost: based on 20-year life)	(220,572)	(203,831)	(336,345)	(449,287)	(311,462)

Table XVII. Total Annual Operating Cost Estimates (¢/1000 Gallons)

	Sand Replacement (10 MGD)	Contactors (4 MGD)	Sand Replacement + Reactivation	Contactor + Reactivation
Direct operating costs	7.2	6.4	12.7	11.9
Indirect operating costs Based on Table XI useful life data	0.3	12.1	2.7	14.5
Based on 20-year life data		(8.1)	(2.5)	(10.3)
Total operating costs Based on Table XI useful life data	7.5	18.5	15.4	26.4
Based on 20-year life data		(14.5)	(15.2)	(22.2)

Literature Cited

1. "Drinking Water and Health", National Academy of Sciences, Washington, D.C., 1977.
2. "Drinking Water and Health - Recommendations of the National Academy of Sciences", *Federal Register, 42,* 35764 (July 11, 1977).
3. DeMarco, J.; Steven, A.; Hartman, D. J. In *Proceedings of the ACS Division of Environmental Chemistry Symposium on Advances in the Identification and Analysis of Organic Pollutants in Water,* Second Chemical Congress of the North American Continent, Las Vegas, Nevada, August 24–29, 1980, in press.
4. Stevens, A. A.; Symons, J. M. *J. Am. Water Works Assoc.,* **1977**, 69 (10), 546.
5. "National Interim Primary Drinking Water Regulations," *Federal Register,* 40(248), 59566–59588 (December 24, 1975).
6. "National Interim Primary Drinking Water Regulations; Control of Tri-halomethanes in Drinking Water", *Federal Register,* 44(231), 68624–68707 (Nov. 25, 1975).
7. Directo, L. S.; Chen, C. L.; Miele, R. P. "Independent Physical-Chemical Treatment of Raw Sewage", U.S. EPA–600/2–77–137 (August 1977).
8. Chen, C. L.; Directo, L.S. "Carbon Reactivation by Externally-Fired Rotary Kiln Furnace", U.S. EPA–600/2–80–146 (August 1980).
9. Wood, P. R.; DeMarco, J. In "Activated Carbon Adsorption of Organics from the Aqueous Phase," Suffet, I. H.; McGuire, M. J., Eds.; Ann Arbor Science: Ann Arbor, Mich., 1980.
10. Wood, P. R.; DeMarco, J. In "Activated Carbon Adsorption of Organics from the Aqueous Phase," Suffet, I. H.; McQuire, M. J., Eds.; Ann Arbor Science: Ann Arbor, Mich., 1980.
11. Brodtmann, N. V.; DeMarco, J.; Greenberg, D. In "Activated Carbon Adsorption of Organics from the Aqueous Phase," Suffet, I. H.; McGuire, M. J., Eds.; Ann Arbor Science: Ann Arbor, Mich., 1980.
12. Nur, R. Monthly Reports to I. J. Kugelman of MERL, ORD, U.S. EPA, Cincinatti on a Trace Organic Removal Study Using Granular Activated Carbon and On-Site Reactivation by an Infrared Traveling Belt Carbon Regeneration System.
13. Inhoffer, W. R.; McGinnis III, F. K. "Regenerating Granular Activated Carbon at Passaic Valley", Presented at the 1980 Annual Conference and Exposition of the American Water Works Association in Atlanta, Ga., June 1980.

14. Lykins, B. U.S. EPA, ORD Drinking Water Research Division Quarterly Reports.
15. Culp, R. L.; Faisst, J. A.; Smith, C. E. "Final Report on Granular Activated Carbon Installations", EPA Contract No. CI-76-0288 (July 1980).
16. Method 5, *Federal Register, 12*(160), Aug. 18, 1977.
17. Method 25, *Federal Register, 45*(194), Oct. 3, 1980.
18. Method 7, *Federal Register, 43*(160), Aug. 18, 1977.

RECEIVED for review August 3, 1981. ACCEPTED for publication July 13, 1982.

Discussion IV
Pilot- and Large-Scale Studies

Participants

W. Brian Arbuckle, University of Florida
Georges Belfort, Rensselaer Polytechnic Institute
Robert S. Chrobak, American Water Works Service Co.
Jack DeMarco, U.S. Environmental Protection Agency
Francois Fiessinger, Societe Lyonnaise des Eaux et
    de l'Eclairage
Terje Halmo, SINTEF (Scientific and Industrial Research
    Institute at the Norwegian Institute of Technology)
Alan F. Hess, Malcolm Pirnie Consulting Engineers
Benjamin W. Lykins, U.S. Environmental Protection Agency
F. K. McGinnis III, Shirco, Inc.
Michael J. McGuire, The Metropolitan Water District of
    Southern California
Steven J. Medlar, Camp, Dresser and McKee, Inc.
Paul V. Roberts, Stanford University
R. Scott Summers, Stanford University
John S. Zogorski, University of Indiana

DeMARCO: I'd like to begin by asking Mike for comments relative to the composite analyses that you were looking at for the instantaneous trihalomethanes.

Mike, you talked about a refrigerated sample, which I can see in some respects. The concern that some of us have is that when you run an instantaneous trihalomethane you have to quench that sample with a reducing agent. How do you keep a modest amount of reducing agent relative to a changing amount of sample?

McGUIRE: The composite sampler was set up so that a small amount of sodium sulfite was introduced into the bottom of the sampler prior to the operation of the sampler. We knew we were going to be running for approximately three and a half days. We knew that the influent—and this is really true only of the influent to the carbon column, since the effluents

0065-2393/83/0202-0565$06.00/0

are essentially dechlorinated—we knew the influent sample had approximately half a milligram per liter of free chlorine. We were able to control that fairly closely.

So just by calculating the mass of chlorine and knowing how much sulfite was required, we could add a very small volume of high concentration. One interesting thing did happen: We ran out of sulfite in the first couple of runs that we tried, and it wasn't until we started accounting for dissolved oxygen scavenging some of the sulfite that we were able to get the right amount in there. So you do have to take into account some dissolved oxygen reaction getting around the Teflon plug, and also some of the DO in the water that's being pumped into the sampler.

ZOGORSKI: My question really goes to most of the panel members, and, looking at this from a practical viewpoint, there are many areas of concern for the design engineer and for the municipality.

I'd like to discuss an important area: desorption. We see a lot of studies showing desorption occurring, and my question is, can we design carbon filters to resolve this problem? In other words, if the beds are long enough, and so on and so forth, can we get around this desorption problem?

McGUIRE: I think we can. The analytical part plays a big role in controlling the activated carbon columns so that you can avoid desorption, which may cause more problems than you started out with. Mel has reported on one incident of a toluene spill on the Delaware which seemed to displace a number of compounds, and they got a much higher concentration of them in the effluent than they began with the influent[1]. So I think that near-real-time monitoring, or something approaching that, is what you are going to have to do in those complex systems like the Delaware estuary or the Seine River.

Fortunately, at Metropolitan, we really are not faced with those kinds of difficulties as far as the mixed bag of organic compounds is concerned. I think that analysis in the polluted areas is going to hold the key to shutting down carbon columns that have been loaded to the point where they start to desorb excessive quantities of those organic compounds.

CHROBAK: In our particular situation we had the primary contaminant, the tetrachloroethylene, and we had desorption of 1,1,1-trichloroethane. Granted, some people believe that everything should be removed from the water to be totally safe, but when you look realistically at the situation there really isn't much of a hazard with trichloroethane as with tetrachloroethylene.

So you get to a point where you have to make your design based on what you are aiming at. And in this case it was uneconomical to consider

[1] Suffet, I. H.; Wicklund, A.; and Cairo, P. R. In "Water Chlorination Environmental Impact and Health Effects" Vol. 3 Jolley, R. L. et al., Eds., Ann Arbor Science: Ann Arbor, Mich 1980.

removing trichloroethane, yet we were still doing a fair job with tetrachloroethylene. I think those are the tradeoffs you have to work with.

DeMARCO: I think there's one thing that Katherine Alben's information showed, which I believe Tom Yohe addressed, and that is that we see a dynamic situation and we're in various states of depletion in a carbon system. You may or may not be getting a tendency to have this chromatographic effect.

When you design a plant, I believe, you are not going to have a plant design where all the systems go on at Day Zero together. There will be a staggered effect, just as there is when you put any system on-line. There will be some that will be starting up in a virgin state, some that have been partially exhausted, and some completely exhausted.

I think the exact composite mixture is what the designer has to look at in terms of the water that is going out to the consumer. I hope that is one approach—as well as knowledge of when to take the systems off-line— that will actually assist us in terms of providing safe water and not playing around with a loaded gun.

HESS: A question for Mr. Lykins: Did you notice any removal of the chlorine dioxide byproducts in going through the GAC bed?

LYKINS: Yes, we did. Because our generator was not designed for a small system such as we were using, we did see somewhere around one and a half milligrams per liter of chlorite going onto the carbon, and we saw that the effluent coming out of the carbon was about 0.5 milligrams per liter of chlorite. There was no chlorine dioxide coming out of the carbon. And we have no procedure for the chlorates right at the moment.

HESS: I know that EPA has talked in hushed terms about their concern with these byproducts. Will they sometime soon come out with some official position on just what they'd like to see and the cautions that they're going to work with so the water industry will know what ground rules are coming in the future?

LYKINS: Well, the recommendation right at the moment in the THM regulation is 0.5 for the total oxidants. This is a safety factor right at this point. Health-effects studies are proceeding, and maybe in the future we will have some recommendation from the EPA.

ARBUCKLE: I noticed that in some of the studies we've collected a large volume of data on pilot scale, and we spent a large amount of money collecting these data. Also I noticed in this conference that a number of kinetic models were proposed for use in ideal situations (laboratories), and everybody's models worked on their own data.

We are spending a lot of money gathering these data. Has anybody bothered to take the theoretical models and try to apply them to a lot of these real-world data so we can save money in future studies by looking at alternatives?

BELFORT: I think first of all there was some attempt earlier in the conference to put up some models and use them. I would only make the following short comment: Anybody who was here at the model sessions would realize that only now—with the opportunity of using Dr. Chi Tien's model with up to, as he said, approximately a hundred compounds—can we start to look at multicomponent systems. Up till now there really wasn't much use in trying to apply equilibrium models for extremely complicated solutions. I think that if we can verify Dr. Tien's model for many compounds, then for the first time we may have a very useful model to apply to these systems.

It's not that no one wanted to try to apply the models, but I think we've just been too far apart. I think some work is needed in both directions: The theoreticians need to get closer, and perhaps the experimental people could start to supply some data or try to bridge some gaps.

FIESSINGER: I have two minor questions to Jack. First of all, in the data you presented on the carbon losses, you found higher losses in the sand-replacement system. That doesn't agree with what we found, and I think the data you presented for the losses in the sand-replacement system had a higher deviation than the losses for post-filter contactors. You also had fewer data. Do you have an explanation for this?

The second question is, did you make any pore-structure curve studies after reactivation? Did you see any significant difference?

DeMARCO: The differences in the losses that we see in the sand-replacement system and in the contactor system, I would say, are probably at least twofold. One of them was that we had a considerably longer run in terms of moving the carbon from the existing sand-replacement bed into the contactor and into the regenerated building.

Number two: We also had a less-than-optimum method of moving the carbon out of the sand-replacement bed—an eductor system.

As for the second question, we haven't examined the data yet, but we do have the pore-volume distribution curves.

FIESSINGER: If I may say a word on this, we had some changes in the carbon adsorptivity after six or seven reactivations, and we couldn't see any difference in the iodine numbers and in molasses. We made a very extensive analysis of the carbon. (This was done by the Norit Company, which we worked with on this program.) We found a little buildup of iron within the carbon because we were using iron chloride as a coagulant before the carbon filtration. This may have explained a change in the pore structure under reactivation.

DeMARCO: That's one of the reasons why we look at the ash buildup to see what's in it.

SUMMERS: I noticed that in the surface areas between the spent and the regenerated and between the spent and the virgin, there wasn't much difference in the surface areas with the spent carbons. Do you see that same lack of difference in the total pore volume or in the pore-volume distribution—a lack of difference in the spent carbon?

DeMARCO: I'd say it depends on the exhaustion rate. If you look at the two series, the first one went on for about five months of operation, and you saw a larger difference in the apparent density and the iodine number from the start to the finish. After that the reactivation facility got on-line the way it was supposed to, and our cycle, with the criteria that we used, actually led to what we would see at water utilities—a very lightly loaded carbon. Basically I think that's what we would see in contrast to the wastewater facilities, where I believe the loadings would be considerably higher.

McGINNIS: One comment relative to the surface area data: If those samples were done the same way that the Evansville samples were done, the spent surface areas are for a sample on which an effort was made to calcine the organics in place, to carbonize them and get some feel for the effect on surface areas. That's really not a true spent carbon surface area if, indeed, that's the way those experiments were made.

DeMARCO: They were made in the same way, and our report will issue the cautions that Stan Smith gave you when he gave you the data back, because the Evansville project was conducted the same way we conducted the Cincinnati project. You can look at the iodine number, though, instead of the surface area, and then you can avoid that trap.

HALMO: I have a question to Scott concerning his plateauing of the TOC after breakthrough, which he attributes to biological action. It is also a general speculation concerning the discussion at this meeting over slow adsorption and biological action, which no doubt both exist to some extent.

My speculation is based on my own experience at SINTEF with a complex industrial wastewater, namely coke water. In these experiments, in a number of cases, we observe an overshoot of 5 to 15 percent in TOC over influent value. We attribute this to displacement of components by others that have a larger area demand per unit weight—that is, better adsorbed components with a larger area demand. This spurs me to speculate on whether or not the opposite phenomenon may also occur; that is, is it possible that the plateauing in TOC is due to displacement of components from carbon by other components that have a smaller area demand per unit weight of component and are also better adsorbed?

I would assume that this phenomenon may be responsible for a smaller or larger fraction of the plateauing effect, along with slow adsorption and biological action. And since Scott has analyzed a number of single solutes in the influent and effluent, do you feel that this is possible in your case?

SUMMERS: We ran the exhausted column and sampled it for TOC for two years, and, with the exception of its reactions to the pH and microbial population upset, we still saw a partial removal of approximately

25 percent, and it didn't change for a year except for the three points I discussed in my paper[2].

So, yes, your hypothesis may be possible because of the compounds that make up TOC.

HALMO: I think your explanation is pretty plausible that the biological effects would be influenced by the high pH. But so would the adsorption. It's a well-known fact that a number of components are being adsorbed pretty rapidly when you have a high pH.

SUMMERS: That was Andy Benedek's point.

HALMO: I don't think it is any contradiction.

SUMMERS: The contradiction is that after the pH went back down to a normal condition, we still saw some type of high fractional breakthrough. The pH was now the normal condition, so we didn't have a pH that was the same, although it might have been a delayed reaction to the pH; but I would assume it was more likely because the microbes had had a chance to acclimate to normalized pH or because the growth returned to what it had been.

HALMO: I don't object to your hypothesis of biological action. I think it does occur.

SUMMERS: It could be, but with the data that we have everything is masked, and we can't separate the components out.

ROBERTS: In considering what's happened at the symposium, I'd suggest that we might consider our progress from theory through experiment and models to large-scale operation. Under "theory," we've considered matters of equilibrium and rate and models thereof. In "experiment," we've considered bench- and pilot-scale results, and models for their interpretation; and in the area of "large-scale operations," we've considered operations, design, and economics.

And we could look at this in another way as well: I think the theory comes closest to philosophy, and perhaps the large-scale operations, design, and economics come closest to commerce and industry. And there's always a question, covering a patch of ground this wide, as to what one means to the other and how well they are interrelated.

There's also a question as to what the philosophers can tell the artists and craftsmen, and vice versa. Who draws inspiration from whom, and how well? How much good do we derive from the kind of interaction between philosophers and implementers that we've tried to achieve, or I think Mel and Mike have tried to achieve, at this symposium?

MEDLAR: I would like to make a brief comment along the same lines. I think we are all seeking to solve problems, and the problem we have in trying to develop solutions is simply that of bridging the gap between the identification of a problem and the solution. And certainly it's identifying

[2]Chapter 22 in this book.

the theory and understanding the theory—the basic understanding— that's going to lead to solving the problems.

I see it from, let's say, a practicing point of view, of applying what principles we have available to us, in developing systems. The background information that we have, I feel, still needs a tremendous amount of work in theory and research activity.

HESS: I think the situation is no longer one of convenience in terms of time frame, and it's not just "fun to look at" anymore. It's a matter of a real problem where, in many instances, a town is out of a water supply, and the necessity is to do something quickly, whether or not it's the best thing to do. And I think we are going to learn an awful lot over the next couple of years, but meanwhile we'll be having to do something as quickly as possible in an often highly panicky situation, in a political environment, with everything pushing in on the poor utility manager, who has no other options except to try something.

And I think that's the way the water works industry has developed; the profession will always try to do it better, but sometimes we don't have the convenience of time. I think we are on the forefront of technology, and we are going to have to learn by our mistakes in a lot of instances.

ROBERTS: Are you suggesting that this urgency doesn't permit us the time to establish a sound scientific base relevant to a given application? Is that what you're saying?

HESS: I think you have to apply what is already known very quickly. You don't have the luxury of being able to go in and take six months to study it in many situations, especially with these ground waters that are coming up with very high levels of contamination, and the State or the regulatory agency is on the verge of trying to shut them down. People shouldn't be drinking that kind of water. You've got to go with what you have now. Hopefully, something better will come along, and maybe more of an optimization type of thing, but I think for a lot of systems it's a future-shock situation.

McGUIRE: Two and a half years ago, we were just entering the log growth phase of the work on activated carbon because there was a proposed regulation on the street for the inclusion of GAC in water treatment plants. Now, with the existing administration, we see a dramatic slowdown in certain areas of environmental regulation, and that's impacting drinking water regulation also. We've seen dramatic cutbacks just in the past few weeks and months in EPA, and a promise by the current administration to limit regulations, especially those that affect industry.

By question is, is this the end? When the current projects that have been reported on at this symposium are over, the multi-million-dollar, multi-hundred-thousand-dollar projects, is that really going to be the end? Are we going to just pack all this stuff up in a couple of book proceedings and put it on the shelf, or can we expect something to really be done to improve the quality of the drinking water in the United States? Jack, would you like to comment?

DeMARCO: I think Mike has obviously rung what a lot of people perceive as perhaps the death knell of some of the work that has been done. I think that anyone without their head in the sand would know that we in research get a great deal of our impetus, a great deal of our monies, from the pressure situations that exist, either from Congress or from spills and the like.

I'd be foolish to stand here before you and say that without an impetus of this nature, we would go at the log rate at which we've gone in the past. However, in my opinion, we have not answered all of the optimum design questions. We have put forth a large-scale data base that I personally believe will allow us to more efficiently design carbon systems in future plants.

Now, that data has to be thoroughly digested and used not only by us, not only by the the practical engineers, the design engineers who are starting to use it for emergency situations, but also by the theoreticians, the mathematical modelers that we talk about. And to me, somewhere between those poles is our plan for the future. I do not believe that we are finished with our adsorption activities, by any stretch of the imagination. We are embarking on an additional project in Jefferson Parish, Louisiana, which I hope will augment the experiences—and perhaps even go further than the experiences—that we have had full-scale at the Cincinnati Water Works facility.

So, in answer to you, Mike, I don't believe we are willing to pack up our bags and say, well, we have solved all the problems. I don't believe we *have* solved all the problems.

INDEX

Copy Editor/Indexer: L. Luan Corrigan
Production Editor: Anne G. Bigler
Managing Editor: Janet S. Dodd
Jacket designer: Kathleen Schaner

Typesetting by Ampersand, Inc., Rutland, Vt, and
Service Composition Company, Baltimore, Md
Printing and binding by Maple Press, Inc., Manchester, Pa